The Cerrados of Brazil

The Cerrados of Brazil

*Ecology and Natural History
of a Neotropical Savanna*

Editors

Paulo S. Oliveira
Robert J. Marquis

Columbia University Press
New York

Columbia University Press
Publishers Since 1893
New York   Chichester, West Sussex

© 2002 Columbia University Press

Library of Congress Cataloging-in-Publication Data

The cerrados of Brazil : ecology and natural history of a neotropical
savanna / Paulo S. Oliveira and Robert J. Marquis.
p. cm.
Includes bibliographical references.
ISBN 978-0-231-12042-5 (cloth : alk. paper)—ISBN 978-0-231-12043-2
(pbk. : alk. paper)
1. Cerrado ecology—Brazil.   I. Oliveira, Paulo S., 1957–
II. Marquis, Robert J., 1953–

QH117 .C52 2002
577.4'8'0981—dc21   2002022739

Columbia University Press books are printed on permanent
and durable acid-free paper.

Printed in the United States of America

Contents

Preface

THIS IS A BOOK ABOUT THE CERRADO BIOME, A MAJOR BRAZIL-
ian savanna-like ecosystem for which no such summary exists. Biologists
outside Brazil know little about the *cerrados*, despite the fact that the
biome covers approximately 22% of the country's surface area, or 2 million km². Even though much of the attention of conservationists has
focused on rainforests such as the Amazon and Atlantic forests, the cerrados are currently one the most threatened biomes of South America due
to the rapid expansion of agriculture. Nearly 50% of the cerrado region
is currently under direct human use, and about 35% of its total natural
cover has been converted into planted pastures and crops. The average
annual rate of land clearing in the cerrados during 1970–1975 was nearly
twice the estimated deforestation rate of the Amazon forest during
1978–1988. Overall biodiversity for the Cerrado Biome, including all its
physiognomic forms, is estimated at 160,000 species of plants, animals,
and fungi. Endemicity of cerrado higher plants has recently been estimated at 4,400 species, representing 1.5% of the world's total vascular
plant species. Endemic vertebrates range from 3% (birds) to 28% (amphibians) of the species recorded. The cerrados are also unique in that they
serve as corridors for species inhabiting neighboring biomes such as the
Amazonian and Atlantic rainforests. For example, although endemicity is
low among birds, 90% of the species breed in the cerrado region. Given
their geographic extent, it is surprising that the cerrados remain largely
ignored at the international level. Because of the threatened status and
rich biodiversity of this Neotropical savanna, and the lack of familiarity
with cerrado ecosystems at the international level, a volume that compiles
the known natural history, ecology, and biogeography of this biome is
extremely timely.

This is perhaps the first volume in English covering a tropical ecosystem in which the vast majority of the contributors are from the region in
question. The foreign exceptions include scientists that are very familiar
with the cerrados and have long-lasting collaborations with Brazilian
researchers. The volume is broad in scope and raises relevant ecological
questions from a diversity of fields, indicating areas in which additional

research is needed. Such a wide thematic approach should provide the international audience with a broad ecological framework for understanding the cerrado savanna. The editors hope that such a book will make an important contribution for ecology, and for tropical biology in particular, stimulating future research in the cerrados.

The idea of preparing a book summarizing research on cerrado biology arose in 1997 in San José, Costa Rica, during a most exciting meeting of the Association for Tropical Biology. As the book project developed, a number of people helped us shape the scope of the volume, establishing the main research areas to be covered, adjusting chapter contents, and writing the book proposal. At the early stages we have benefited greatly from the encouragement as well as the technical and editorial experience of Susan E. Abrams of the University of Chicago Press and Peter W. Price of Northern Arizona University. Helpful suggestions were also given by Keith S. Brown, William A. Hoffmann, Regina Macedo, Ary T. Oliveira-Filho, and Guy Theraulaz. Humberto Dutra helped with the preparation of the book index, and Glauco Machado and André Freitas helped with the scanning and printing of the figures. Mailing costs were covered in part by the Ecology Graduate Program of the Universidade Estadual de Campinas.

Each chapter was substantially improved by the comments and suggestions of external reviewers. They include Steve Archer, John A. Barone, Kamaljit S. Bawa, John G. Blake, Keith S. Brown, Ray B. Bryant, Phyllis D. Coley, Philip J. DeVries, Peter E. Gibbs, Guillermo Goldstein, Gary S. Hartshorn, W. Ronald Heyer, Peter Kershaw, W. John Kress, Thomas H. Kunz, Diana Lieberman, Arício X. Linhares, Vera Markgraf, Ernesto Medina, Daniel C. Nepstad, Ary T. Oliveira-Filho, James L. Patton, A. Townsend Peterson, Ghillean T. Prance, Peter W. Price, James A. Ratter, José F. Ribeiro, Juan F. Silva, Robert B. Srygley, and Laurie J. Vitt. We appreciate the time they took to give critical reviews.

Finally, we thank Science Editor Holly Hodder and Assistant Editor Jonathan Slutsky, formerly of Columbia University Press, for their initial encouragement and advice on the development of this project. Current Assistant Editor Alessandro Angelini helped at the final stage of the editing process, and Diana Senechal copyedited the entire manuscript. We are especially grateful to Julie S. Denslow and Lucinda A. McDade, reviewers of the book proposal for Columbia University Press, for their careful and constructive suggestions concerning the initial book project.

Paulo S. Oliveira
Robert J. Marquis

The Cerrados of Brazil

1

Introduction: Development of Research in the Cerrados

Paulo S. Oliveira and Robert J. Marquis

THE FIRST DETAILED ACCOUNT OF THE BRAZILIAN CERRADOS was provided by Danish botanist Eugene Warming (1892) in the book *Lagoa Santa*, in which he describes the main features of the cerrado vegetation in the state of Minas Gerais. Since the publication of Warming's book a number of descriptive studies from several cerrado regions in Brazil have been published. The vast majority of this literature is in Portuguese and oriented mostly toward botanical aspects of the cerrado. The studies can be roughly categorized into two major groups: (1) Surveys of woody floras, frequently providing also the general physiognomic characteristics of the vegetation (thorough reviews of this literature are given by Eiten 1972; Goodland and Ferri 1979). (2) Studies on plant ecophysiology focusing particularly on mineral nutrition, fire, and water economy at the plant-soil and plant-atmosphere levels; and on how these factors can account for the characteristic xeromorphic aspect of cerrado woody plants (extensive lists of these studies are given by Labouriau 1966; Ferri 1977; Goodland and Ferri 1979).

The cerrados gained international attention in the early 1970s after the influential works of Goodland (1971), Eiten (1972), and Ratter et al. (1973). These studies established quantitative parameters (i.e., canopy and ground cover, tree density, species richness) to characterize the several physiognomic forms of the cerrado vegetation; provided quantitative and comparative data toward the analyses of shifts in floristic composition along intergrading physiognomic communities (both over geographical and local scales); and enhanced the notion that the *cerrado complex* is the interactive product of climatic, topographic, and edaphic factors.

1

One may say with justice that these works have set the very basic grounds for modern ecological research in the cerrados.

PATTERNS OF RESEARCH PRODUCTIVITY

To understand the development and scope of scientific research in the cerrados, we have analyzed the bibliography in the form of journal articles appearing in the citation databases of the Institute of Scientific Information (ISI). We compiled the list by using *cerrado* and *cerrados* as "Topic Search" terms. Our goal was to detect changes in the quantity of published research papers over time, as well as in the subject matter treated. First we examined the general research productivity from 1966 to 1999, and assigned each study to one of seven major subject areas, as indicated in table 1.1. We treated zoology, botany, and mycology as separate areas to illustrate the allocation of research effort toward studies of animals, plants, and fungi.

In a second phase of the analysis we assigned each article in the *ecology* category to one of six main research areas, in accordance with the thematic scheme employed by McDade and Bawa (1994), as summarized in table 1.2. Studies linked with agriculture, cattle, and wood industry, however, are not placed under the *ecology* category, because their research

Table 1.1 Major Subject Categories Used to Analyze Patterns of Research Productivity in the Brazilian Cerrados

Major Subject Areas	*Fields of Research*
Ecology	General ecology, interspecific ecology, community ecology, physiological ecology, ecosystem ecology, applied ecology, and conservation
Zoology, botany, and mycology	Species descriptions, species lists, systematic biology, anatomy, morphology, physiology, genetics, and chemistry of organisms
Soils	Chemical and physical properties of soils, geology and geomorphology, and soil microbiology
Agriculture, sylviculture, and livestock	Any research linked with the use of cerrado areas for the raising of crops, commercial trees (timber industry), or cattle
Gas emission and landsat mapping	Satellite sensing of fires, smoke, and regional energy budgets, gas emission, climate, and landsat mapping of vegetation

Table 1.2 Main Research Areas Used to Analyze Patterns
of Ecological Research in the Brazilian Cerrados

Areas of Ecological Research	Specific Types of Research
General ecology	Biology of individual species, life history, demography, and behavior
Interspecific ecology	Interactions between species, including pollination, frugivory, herbivory, parasitism, and predation
Community ecology	Structure, dynamics, and organization in space and time of plant and animal communities
Physiological ecology	Physiological adaptations of organisms to the abiotic environment
Ecosystem ecology	Nutrient cycling, energy flow, and physical features of the habitat
Applied ecology and conservation	Conservation of natural resources, biodiversity

Source: Based partially on McDade and Bawa (1994).

approach and goals were generally not related to ecological issues (although the results could have a major ecological impact in the environment; see below).

Although such thematic divisions are widely used in ecology textbooks and professional journals, obviously there are other ways of arranging research papers, as well as other recognizable thematic categories. In fact, as McDade and Bawa (1994) stress, the distinctions between such ecological thematic categories are sometimes arbitrary, and a given paper could probably be assigned to more than one category. In general, however, the assignment of papers to a category was quite easy.

A final note on the accuracy of this bibliographic analysis. The assembled literature is of course incomplete, because it does not include several of the Brazilian publications which are not compiled by the ISI, including local journals, books, and symposium volumes. We believe, however, that such a compilation of articles does provide a general pattern of research productivity in the cerrados.

The results show that research on the cerrados has increased markedly over the last two decades, especially over the past ten years (see fig. 1.1A). Studies linked with the use of cerrado areas for agriculture and pasture accounted for 24% of the papers (see fig. 1.1B). The ever-increasing

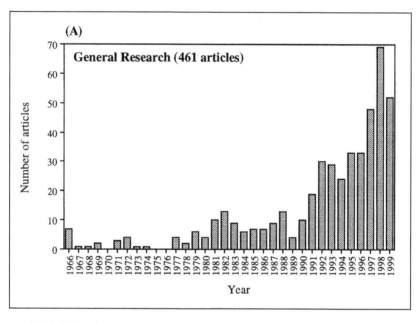

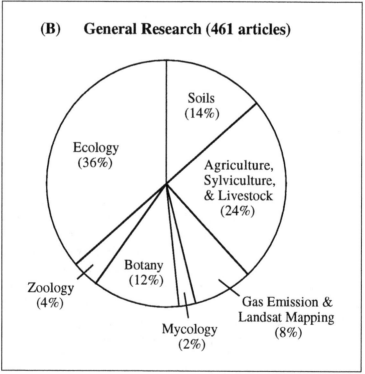

Figure 1.1 Research productivity in the Brazilian cerrados as compiled by the citation databases of the Institute of Scientific Information (ISI), using *cerrado* and *cerrados* as topic search terms. (A) General research over time. (B) Distribution of research articles by major thematic categories.

exploitation of natural cerrado areas for growing crops, trees (*Pinus* and *Eucalyptus*), and cattle, and the clearings caused by these practices, has urged the necessity of satellite measurements of gas emission and vegetation cover within the cerrado region in the late 1990s (fig. 1.1B). Research on soil properties and soil microbiology comprised 14% of the papers compiled.

Studies on ecology, zoology, botany, and mycology comprised 54% of all publications assembled, ranging from less than five papers in 1990 to about 35 papers per year in the late 1990s (see fig. 1.2A). This burst of biological research on the cerrados results from the founding of the first ecologically oriented graduate programs in Brazil in the 1970s. Some of these programs included field courses of 4–5 weeks in natural reserves where students developed field projects, some of which eventually led to theses. Such initiatives have resulted in the remarkable development of natural history and ecological research in a number of Brazilian ecosystems, including the cerrados. Originating mostly from the graduate programs of the public Universities in São Paulo (Southeast Brazil) and Brasília (Central Brazil), numerous student theses were developed in the cerrado savanna. In the state of São Paulo, 203 university theses were produced between 1966 and 1999. In the University of Brasília (UnB), located at the very core of the cerrado distribution, 62 theses were produced between 1997 and 1999. (Data assessed through the library databases of the public Universities of São Paulo, and the University of Brasília; compiled by using *cerrado* and *cerrados* as search terms.)

Ecological research in cerrado has concentrated mostly in the three major fields of community ecology, general ecology, and interspecific ecology (see fig. 1.2B), which are also among the main ecological research areas investigated in Central American tropical forests (McDade and Bawa 1994; Nadkarni 2000). Perhaps for historical reasons, studies on community ecology have been plant-oriented and have focused mainly on vegetation structure and dynamics, including paleoecology. Ecological studies on vertebrates were usually grouped under general ecology and, to a lesser extent, community ecology. They have been mostly oriented toward mammals, birds, and lizards, and generally have dealt with patterns of space use, feeding behavior, guild structure, and biogeography. Invertebrate research, generally incorporated into interspecific ecology, comprises studies on insect-plant interactions, in particular herbivory, pollination, and multitrophic associations. A comparatively small number of studies have reported results on physiological ecology (mostly plants), ecosystem ecology (nutrient cycling, fire ecology), and conservation (biodiversity inventories). Research areas that are clearly poorly represented include animal

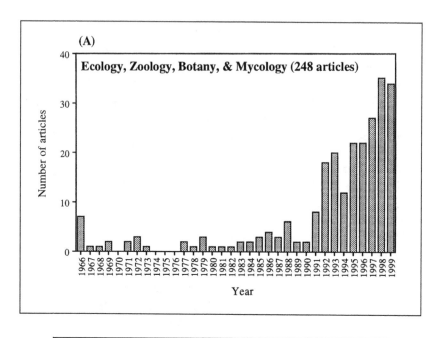

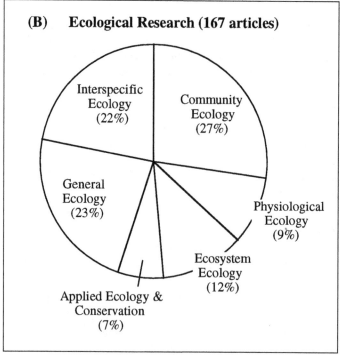

Figure 1.2 Ecology and natural history research in cerrados, as compiled by the citation databases of the Institute of Scientific Information (ISI), using *cerrado* and *cerrados* as topic search terms. (A) Number of articles in ecology, zoology, botany, and mycology over time. (B) Distribution of ecological research by subject matter.

ecophysiology, chemical ecology, invertebrates (except butterflies, and social insects), large mammals, wildlife management, aquatic biology and hydrology, and landscape ecology.

SCOPE AND ORGANIZATION OF THE BOOK

The purpose of this book is to provide a picture of the Cerrado Biome based on broad synthetic treatments by experts from a diversity of research areas. Although the book has chapters whose approach is by necessity mostly descriptive, it also focuses on basic conceptual issues in evolutionary ecology and ecosystem functioning, and points toward future research avenues. Authors were instructed to write for an interdisciplinary audience, giving broad synthetic views within their specialties and making the text palatable enough to attract the interest of nonexperts as well as graduate students. As such, it is intended to provide an in-depth summary of current understanding for researchers versed in the field, as well as an introduction to cerrado biology for the mostly uninitiated international community. The book also provides a synthesis of the extensive cerrado literature in Portuguese, generally not easily accessible by the international audience. Similar volumes exist for African savannas alone (Sinclair and Norton-Griffiths 1979; Sinclair and Arcese 1995), and for Australian and African savannas (Werner 1991), but there is no equivalent for Brazilian cerrados. Most of the literature on neotropical savannas emphasizes the savannas of the northern parts of the South and Central Americas (see Sarmiento 1984), which do not have the extension and the rich biodiversity of the savannas of central Brazil (Dias 1992; Myers et al. 1999). Moreover, most studies on neotropical savannas have focused mainly on vegetation-related processes. A recent attempt toward a more multidisciplinary approach can be found in Solbrig et al. (1995).

This volume treats the historical origins and physical setting, the role of fire, major biotic taxa, insect-plant interactions, and functional processes at different levels of organization (population and community) and scale (local and landscape). The book is organized in five sections, as follows:

Part I provides the historical background and presents the main abiotic properties of the cerrado region. Geology, geomorphology, climatic influence, palynology, fire ecology, and history of human influence are treated in chapters 2–5.

Part II focuses on the plant community and begins with the description of the vegetation physiognomies and the origins of the cerrado biome (chapter 6), followed by the main attributes of the herbaceous layer (chapter 7). Population characteristics of trees in the absence and presence of fire, including spatial patterns and growth and mortality rates, are treated in chapters 8 and 9. The section concludes with the ecophysiological strategies of cerrado woody plants in chapter 10.

Part III gives a general picture of the animal community, focusing on what are probably the five best-known animal taxa of the cerrados. Chapter 11 examines the communities of plant-feeding Lepidoptera (best-known invertebrate group) in conjunction with the complex landscape mosaics in the cerrado region. The diversity, biogeography, and natural history of the four best-known major vertebrate groups (amphibians, reptiles, birds, and mammals) are treated in chapters 12–14.

Part IV covers those species interactions in the cerrado that are currently best documented: namely, insect-plant systems. Chapters 15 and 16 deal with herbivorous insects, and chapter 17 treats the flowering plant pollination systems.

Chapter 18 of Part V closes the book by examining the state of preservation of the cerrado ecosystem, the current threats to its biodiversity, and the appropriate strategies to be adopted based on the identification of priority areas deserving immediate conservation actions.

We would like to comment briefly on a nomenclatural norm to be followed throughout the book. The Portuguese word *cerrado* means "half-closed," "closed," or "dense," and the name is particularly appropriate because this vegetation is neither open nor closed (Eiten 1972). The whole biome is characterized by an extremely variable physiognomy, ranging from open grassland to forest with a discontinuous grass layer. Between these two extremes lies a continuum of savanna formations spanning the entire range of woody plant density, referred to collectively as the *cerrados*. As we shall see in chapter 6, there are several physiognomic "types" of cerrado vegetation that can be recognized along this gradient (Goodland 1971) and that are commonly designated by Portuguese terms. For instance, dry grassland without shrubs or trees is called *campo limpo* ("clean field"); grassland with a scattering of shrubs and small trees is known as *campo sujo* ("dirty field"). Where there are scattered trees and shrubs and a large proportion of grassland, the vegetation is termed *campo cerrado* ("closed field"). The next stage is known as *cerrado* (*sensu stricto*) and consists of a vegetation dominated by 3–8-m-tall trees and shrubs with more than 30% crown cover but with still a fair amount of herbaceous vegetation between them. The last stage is an almost closed woodland with crown cover of 50% to 90%, made up of 8–12-m-tall trees casting considerable shade so that the ground layer is much reduced. This form is called *cerradão*. Clearly, the dividing line between these physiognomies is somewhat arbitrary, but researchers usually agree surprisingly well on the classification. Other formations commonly associated with the cerrado landscape will be referred to by their local names (e.g., *veredas, campo de murundus*). The Brazilian nomenclature will be used throughout the book because it is currently well accepted internationally, unambiguous, and appropriate. As a general rule, whenever a given "type" of vegetation physiognomy is referred to by its Brazilian name in some part of the book, the reader will be directed to chapter 6 for a detailed description of that particular physiognomy.

ACKNOWLEDGMENTS

We are grateful to Ana Rabetti and Ana Carvalho, from the Biology Library of the Universidade Estadual da Campinas, for their most valuable help with literature compilation. Augusto C. Franco, William A. Hoffmann, and Ary T. Oliveira-Filho offered useful suggestions on the manuscript.

REFERENCES

Dias, B. F. S. 1992. Cerrados: Uma caracterização. In B. F. S. Dias, ed., *Alternativas de Desenvolvimento dos Cerrados: Manejo e Conservação dos Recursos Naturais Renováveis*, pp. 11–25. Brasília: Fundação Pró-Natureza.

Eiten, G. 1972. The cerrado vegetation of Brazil. *Bot. Rev.* 38:201–341.

Ferri, M. G. 1977. Ecologia dos cerrados. In M. G. Ferri, ed., *IV Simpósio sobre o Cerrado*, pp. 15–36. São Paulo: Editora da Universidade de São Paulo.

Goodland, R. 1971. A physiognomic analysis of the "cerrado" vegetation of central Brazil. *J. Ecol.* 59:411–419.

Goodland, R. and M. G. Ferri 1979. *Ecologia do Cerrado*. São Paulo: Editora da Universidade de São Paulo.

Labouriau, L. G. 1966. Revisão da situação da ecologia vegetal nos cerrados. *An. Acad. Bras. Ciênc.* 38:5–38.

McDade, L. A. and K. S. Bawa. 1994. Appendix I: Patterns of research productivity, 1951–1991. In L. A. McDade, K. S. Bawa, H. A. Hespenheide, and G. S. Hartshorn, eds., *La Selva: Ecology and Natural History of a Neotropical Rain Forest*, pp. 341–344. Chicago: University of Chicago Press.

Myers, N., R. A. Mittermeier, C. G. Mittermeier, G. A. B. Fonseca, and J. Kent. 2000. Biodiversity hotspots for conservation priorities. *Nature* 403:853–858.

Nadkarni, N. M. 2000. Scope of past work. In N. M. Nadkarni and N. T. Wheelwright, eds., *Monteverde: Ecology and Conservation of a Tropical Cloud Forest*, pp. 11–13. Oxford: Oxford University Press.

Ratter, J. A., P. W. Richards, G. Argent, and D. R. Gifford. 1973. Observations on the vegetation of northeast Mato Grosso: I. The woody vegetation types of the Xavantina-Cachimbo Expedition area. *Phil. Trans. Royal Soc. London B* 266:499–492.

Sarmiento, G. 1984. *The Ecology of Neotropical Savannas*. Cambridge, MA: Harvard University Press.

Sinclair, A. R. E. and P. Arcese, eds. 1995. *Serengeti II: Dynamics, Management, and Conservation of an Ecosystem*. Chicago: University of Chicago Press.

Sinclair, A. R. E. and M. Norton-Griffiths, eds. 1979. *Serengeti: Dynamics of an Ecosystem*. Chicago: University of Chicago Press.

Solbrig, O. T., E. Medina, and J. F. Silva, eds. 1996. *Biodiversity and Savanna Ecosystems Processes: A Global Perspective*. Berlin: Springer-Verlag.

Warming, E. 1892. *Lagoa Santa: Et bidrag til den biologiske plantegeographi*. Copenhagen: K. danske vidensk Selsk., 6.

Werner, P. A., ed. 1991. *Savanna Ecology and Management*. Oxford: Blackwell Scientific.

Part I

Historical Framework and the Abiotic Environment

2

Relation of Soils and Geomorphic Surfaces in the Brazilian Cerrado

Paulo E. F. Motta, Nilton Curi, and Donald P. Franzmeier

THE CERRADO REGION IS LOCATED BETWEEN THE EQUATORIAL zone and 23° south latitude. It is bordered by the Amazon forest to the north, by the Atlantic forest to the south and southeast, and by the *caatinga* (deciduous xerophytic vegetation) of the semiarid region to the northeast. Also included in the cerrado region is the nonflooded part of the western *pantanal* (wet plains; see chapter 6). During its evolutional process, the areal extent of the cerrado expanded and contracted in response to climatic fluctuations. During dry periods, the cerrado expanded at the expense of forest (Ab'Saber 1963). During wet periods, forest expanded at the expense of cerrado except in places that were depleted of plant nutrients and that presented some water deficiency (Resende 1976). Once established, the cerrado tends to maintain itself with more tenacity than other vegetation formations because the climate and soil factors that favor it are not extreme (Ker and Resende 1996). In contrast, other vegetation types are favored by more severe conditions. For example, the xerophytic *caatinga* is maintained by the very pronounced water deficiency in a semiarid climate. The *pantanal*, an extensive, low-lying waterlogged plain with hydrophytic grassland in the central-western region, is maintained by a severe oxygen deficiency. The cerrado region has great climatic diversity because of its wide latitudinal and altitudinal ranges. In addition to its 15° range in latitude, the cerrado varies in altitude from 100 m in the *pantanal* to 1,500 m in some of the more elevated tablelands of the Central Plateau.

SOIL FORMATION PROCESSES AND TROPICAL SOILS

In this chapter we present soil characterization data, classify the soils according to the Brazilian soil classification system (Embrapa 1999) and U.S. Soil Taxonomy (Soil Survey Staff 1999), and discuss how soil properties affect plant growth. The next section provides background for the subsequent sections of the chapter.

Soil Formation Processes

The relationship of soils to their environment is explained by the equation,

$$s = f(cl, o, r, p, t, \ldots),$$

which shows that any soil property (s) is a function of regional climate (cl), organisms (o), landscape position or relief (r), geologic parent material (p), time (t), and possibly additional factors (...). Many soils of the cerrado region formed from weatherable minerals (p) on old (t) land surfaces conducive to leaching because of their landscape position (r) in a warm climate (cl) where organisms (o) were very active. Together, the individual factors all contribute to the formation of highly weathered tropical soils in much, but not all, of the cerrado. They are called Latosols in the Brazilian soil classification system, and Oxisols in the U.S. (comprehensive) system.

The 10 most abundant elements in the earth's crust are $O > Si > Al > Fe > Ca > Mg = Na > K > Ti > P$ (Sposito 1989). The fate of these elements during soil formation provides an overview of soil formation. Minerals and rocks from which soils form are made up mainly of the first eight elements of the list, and clay minerals are composed mainly of the first three. Oxygen is unique among the 10 elements. It has a negative charge and is much larger than the others—so large that most of the other, positively charged, elements fit within a "stack" of Os and balance their negative charge.

During soil formation, parent rocks weather and release weathering products that are leached from the soil or remain in the soil and combine to form clay minerals, many with a negative charge. Most base cations (Ca, Mg, Na, and K) are leached from the soil if they are not held by negative charges on clay minerals. Some of the Si released in weathering is leached, and some remains to form clay minerals. Al weathering products are mainly insoluble and remain in the soil. In freely drained soils, Fe also tends to remain in the soil as Fe-oxide minerals such as goethite and hematite. In wet soils, Fe-oxides are reduced and dissolved, and soluble Fe^{2+} is leached. In summary, the mobility of elements in the soil follows the sequence, Ca > Na > Mg > K >> Si >> Fe > Al. The elements at the beginning of the sequence are major plant nutrients and are subject to leaching. Because they are so highly weathered, Latosols tend to be infertile and rich in Al and Fe. By this process, Fe-oxides accumulate in soils because other materials are lost, which could be called a *passive* accumulation of Fe.

Iron can also accumulate in soils by *active* processes. When the water table is high and soils are saturated, oxygen is not available to accept electrons produced by microbial respiration, so they are accepted by Fe^{3+}, resulting in reduction to Fe^{2+} which can move within the soil profile and landscape. When the water table is low, oxygen becomes available, and Fe^{2+} is oxidized to Fe^{3+} and precipitates as iron oxide minerals to form Fe-rich soil materials in subsurface horizons. When first formed in soils, this material is soft. When it dries, it hardens irreversibly, meaning that it does not soften up when the soil becomes moist. Previously, both the hard and soft materials were called *laterite*. In order to distinguish between the two forms, the soft material was called *plinthite*, and the hard material was called *ironstone* in early versions of Soil Taxonomy (Soil Survey Staff 1999). In the Brazilian Soil Classification (Embrapa 1999), these materials are called *plinthite* and *petroplinthite*, respectively. Adjectival forms of these words are used in the names of many soil classes. Depending on the size of the original Fe concentrations in the soil, plinthite may harden into small (sand- and gravel-size), large (gravel and cobbles), or even continuous masses of petroplinthite when the soil dries.

Various kinds of clay minerals form during soil formation. They are made up of sheets composed of Si and O and of Al and O. One way to describe different clay minerals is by the number of Si sheets and Al sheets in their structure. Thus, 2:1 clay minerals have two Si sheets and one Al sheet. Examples are mica, smectite, vermiculite, and illite. In the structure of these minerals, Al^{3+} may substitute for Si^{4+}, which leaves an extra negative charge on the clay surface to which cations such as Ca^{2+} are attracted. This Ca is called exchangeable Ca, because it can exchange with other cations in the soil solution, and the total charge on the mineral is called cation exchange capacity, CEC. Soils on young land surfaces tend to be rich in 2:1 clay minerals.

Kaolinite, a 1:1 clay mineral, and gibbsite ($Al(OH)_3$), a 0:1 clay mineral with no Si and little or no CEC, are abundant in Latosols, especially kaolinite. In the course of soil formation, base cations are leached and clays lose CEC. The two processes complement each other with the result that Latosols have very low contents of exchangeable base cations and are thus infertile. When base cations are removed from negative sites, they are first replaced with H^+ which makes the soil acid, but later acid Al-compounds replace H^+.

Soil Characterization

Tables 2.1–2.3 present characterization data for the main soils of the cerrado. The discussion below explains how the properties reported in these tables relate to soil formation processes, soil classification, and soil fertility.

Color. Three attributes of color are represented in a Munsell designation such as 5YR 4/8. *Hue* (5YR) represents the spectral colors (Y = yellow, R = red). Soil hues grade from yellowish to reddish in the sequence 2.5Y, 10YR, 7.5YR, 5YR, 2.5YR, 10R. *Value* (4) represents the relative darkness, from black ≈ 2, to light or pale ≈ 8. *Chroma* (8) represents the purity of the hue. Chroma = 0 is a black-white transition, and chroma ≈ 8 is relatively pure red, yellow, etc. Soil color has several important interpretations. Low chroma (≤ 2) indicates soil wetness and lack of Fe-oxides. Hue, with higher chromas, indicates the kind of Fe-oxide minerals present and is used to subdivide Latosols. Hematite is reddish, and goethite is yellowish. Yellow Latosols have 10YR and 7.5YR hues, and goethite is dominant. Red-Yellow Latosols have 5YR hue, and neither mineral dominates the color. Red Latosols have 2.5YR and redder hues, and hematite is dominant.

s, r (silt, clay). Represents soil texture, the relative contents of sand, silt, and clay. Sand = 1,000 − s − r. Other factors being similar, more weathered soils contain more clay than less weathered ones. Most Latosols are rich in clay.

C (organic carbon). C oxidizes readily in tropical soils, but the C content in subsoils is high relative to well-drained soils of temperate areas, probably because of ant and termite activity.

pH. Soil pH is a measure of soil acidity. pH is measured in both water and KCl solution. In KCl, K^+ replaces H^+ and other cations, and Cl^- replaces mainly OH^-. If the soil has more cation exchange capacity, CEC, than anion exchange, AEC, more H^+ is replaced than OH^-, and the pH is *lower* in KCl than in water. Then, ΔpH, $pH_{H_2O} - pH_{KCl}$ is positive. On the other hand, a negative ΔpH indicates that AEC is larger than CEC and that the soil has a net positive charge. Such a soil could adsorb more NO_3^- than K^+ or NH_4^+, for example.

T (cation-exchange-capacity, CEC). Total negative charge in soil measured at pH 7. It originates mainly in clay particles and organic matter.

S (sum of bases). Amount of CEC that is balanced by base cations (Ca^{2+}, Mg^{2+}, K^+, Na^+). $T - S$ = acidity (H^+ or Al-compounds) on exchange sites. Soils lose base cations and soil fertility during weathering.

V (base saturation). $V = (S/T) \times 100$. The lower the value, the more leached (and weathered) the soil and the less its supply of plant-available Ca, Mg, and K. For reference, V ranges up to 100%.

m (Al saturation). The percentage of negative sites balanced by positively charged Al-compounds. Soils are considered to be high in Al (allic) if the extractable Al content is > 0.5 $cmol_c$/kg soil and $m \geq 50\%$. Al may be toxic to some plant roots growing in these soils. If roots are stunted they are limited in their ability to take up water, so plants may show drought symptoms.

Fe_2O_3 (content of Fe-oxides, mostly as goethite and hematite). These minerals may also be a source of positive charge in soils.

TiO_2 (Ti-containing minerals are very resistant to weathering). Generally, the higher the content, the more weathered the soil.

K_i (molar SiO_2/Al_2O_3 ratio of the clay fraction). K_i decreases with the degree of weathering of the soil. Latosols must have $K_i < 2.2$ and usually < 2.0.

K_r (molar $SiO_2/(Al_2O_3 + Fe_2O_3)$ ratio of the clay fraction). $K_r > 0.75$ indicates that the clay fraction has significant kaolinite content, and $K_r < 0.75$ indicates that it consists mainly of oxides.

Plant-Soil Relations

Latosols tend to have good physical but poor chemical properties relative to plant growth. The good physical properties are mainly due to high aggregate stability. Aggregates of clay (largely kaolinite and gibbsite) are stabilized by high contents of Fe- and Al-oxides, by organic matter, or both. Strong aggregate stability allows water and air to move through the soil readily and permits roots to penetrate with little resistance. Stable aggregates are also less subject to erosion than unstable ones.

Latosols are low in plant nutrients, especially P and Ca, and many are low in micronutrients. In many cases the Al content is so high that it is toxic to plant roots. Large applications of lime and P fertilizer are needed to make these soils productive for agricultural crops. Lime ($CaCO_3$) neutralizes some of the acidity, decreases available Al levels, and increases the amount of Ca^{2+} on exchange sites and thus available to plants. Large applications of P are required because much of the fertilizer P is tied up

Table 2.1 Color, Physical, and Chemical Attributes of Selected Horizons of the Soils of the First Geomorphic Surface Horizon

Horizon[a]	Depth (cm)	Color moist	s	r	Org. C	pH H$_2$O	pH KCl	S	T	V (%)	m (%)	Fe$_2$O$_3$	TiO$_2$	K$_i$	K$_r$
			(g/kg)	(g/kg)	(g/kg)			(cmol$_c$/kg)	(cmol$_c$/kg)			(g/kg)	(g/kg)		
Typic Acric RED LATOSOL															
Ap	0–17	2.5YR 3/5	100	690	19.5	4.7	4.6	2.1	9.3	23	13	130	9.5	0.72	0.54
Bw2	106–150	2.5YR 4/6	100	730	5.8	4.9	6.0	0.4	2.2	18	0	133	11.2	0.72	0.55
Typic Dystrophic RED-YELLOW LATOSOL															
A	0–18	5YR 4/4	110	820	15.2	4.7	3.9	1.0	12.4	8	47	96	12.0	0.60	0.51
Bw	70–100	5YR 5/8	100	790	11.4	5.0	4.7	0.9	6.6	14	10	103	12.6	0.61	0.51
Typic Acric RED-YELLOW LATOSOL															
A	0–17	5YR 3/3	130	750	22.7	4.5	4.4	0.5	9.2	5	71	114	10.8	0.53	0.43
Bw2	93–170	5YR 4/8	80	820	9.2	5.3	5.8	0.3	3.4	9	0	120	13.3	0.54	0.43
Endopetroplinthic Dystrophic YELLOW LATOSOL															
Ap	0–19	10YR 4/3	150	660	17.5	5.2	4.1	4.0	7.8	51	0	122	14.3	0.62	0.48
Bw2	119–155	7.5YR 5/6	90	790	6.5	5.0	4.1	0.4	2.7	15	0	130	17.8	0.63	0.48
Bwcf	155–189	7.5YR 5/6	100	770	5.7	4.7	4.6	0.4	2.2	18	0	119	17.5	0.67	0.53
Petroplinthic Acric YELLOW LATOSOL															
Acf	0–25	7.5YR 4/2	70	310	13.7	4.6	4.1	0.6	7.2	8	67	156	10.6	0.71	0.42
Bwcf1	78–135	7.5YR 5/6	90	620	8.0	4.7	4.6	0.4	4.0	10	33	156	11.0	0.71	0.50
Bwcf2	135–220	7.5YR 5/8	120	610	4.8	4.1	5.6	0.3	2.3	13	0	174	11.0	0.73	0.49
Typic Dystrophic HAPLIC PLINTHOSOL															
AB	5–17	10YR 7/2	130	700	19.0	5.2	5.4	0.5	4.8	10	0	54	13.3	0.31	0.28
Bf	17–38	2.5Y 7/4	150	720	9.7	5.4	6.6	0.5	2.4	21	0	107	14.9	0.35	0.29
Bgf2	75–120	2.5Y 7/4	160	690	3.4	6.1	7.3	0.3	0.9	33	0	118	15.3	0.40	0.33

Source: Embrapa (2001).

Abbreviations: s = silt; r = clay; S = sum of bases; T = cation-exchange-capacity; V = base saturation; m = Al saturation.

[a] p = pedoturbation, w = intensive weathering, c = indurated concretions, f = plinthite, g = gley (Embrapa, 1988).

Table 2.2 Color, Physical, and Chemical Attributes of Selected Horizons of the Soils of the Second Geomorphic Surface Horizon

Horizon[a]	Depth (cm)	Color moist	s	r (g/kg)	Org. C (g/kg)	pH H$_2$O	pH KCl	S	T (cmol$_c$/kg)	V (%)	m (%)	Fe$_2$O$_3$	TiO$_2$ (g/kg)	K_i	K_r
Typic Acric RED LATOSOL															
A	0–22	5YR 4/4	150	390	11.1	4.8	4.5	0.4	3.9	10	33	57	5.2	0.64	0.53
Bw2	105–160	2.5YR 4/6	160	440	3.4	5.6	5.8	0.4	1.4	29	0	68	6.0	0.62	0.51
Typic Acriferric RED LATOSOL															
A	0–15	2.5YR 3/6	190	590	21.5	4.8	4.3	0.5	9.5	5	44	255	25.9	0.63	0.35
Bw	65–100	10R 3/6	180	610	10.9	5.1	5.2	0.4	4.9	8	0	263	26.5	0.58	0.33
Typic Dystrophic RED-YELLOW LATOSOL															
Ap2	8–24	7.5YR 4/4	90	410	8.7	5.0	3.8	0.6	4.1	15	33	46	4.7	0.60	0.52
Bw1	56–140	5YR 5/8	110	450	3.1	5.2	5.1	0.3	1.6	19	0	60	5.5	0.65	0.55
Endopetroplinthic Acriferric YELLOW LATOSOL															
Ap	0–24	10YR 4/4	90	570	14.4	4.7	4.5	0.4	6.8	6	60	181	21.3	0.61	0.40
Bw2	88–119	7.5YR 5/6	90	650	6.4	5.5	5.1	0.3	2.3	13	0	183	21.9	0.59	0.40
Bwcf	119–160	7.5YR 5/6	100	710	4.7	5.3	6.2	0.4	1.5	27	0	181	18.3	0.61	0.42

Source: Embrapa (2001).

Abbreviations: s = silt; r = clay; S = sum of bases; T = cation-exchange-capacity; V = base saturation; m = Al saturation.

[a] p = pedoturbation, w = intensive weathering, c = indurated concretions, f = plinthite (Embrapa, 1988).

Table 2.3 Color, Physical, and Chemical Attributes of Selected Horizons of the Soils of the Third Geomorphic Surface Horizon

Horizon[a]	Depth (cm)	Color moist	s	r	Org. C	pH H$_2$O	pH KCl	S	T	V (%)	m (%)	Fe$_2$O$_3$	TiO$_2$	K_i	K_r
			(g/kg)	(g/kg)	(g/kg)			(cmol$_c$/kg)	(cmol$_c$/kg)			(g/kg)	(g/kg)		
Typic Orthic ARGILUVIC CHERNOSOL															
Ap	0–2.5	5YR 3/2	250	480	19.8	5.8	5.0	18.2	25.5	71	0	160	37.0	1.64	0.99
Bt	37–90	2.5YR 3/6	190	600	15.0	6.1	5.5	14.2	17.6	81	0	157	27.0	1.53	1.02
Typic Eutrophic RED ARGISOL															
A1	0–10	10YR 2/1	290	350	47.3	5.7	4.9	16.1	24.7	65	0	139	35.6	2.00	1.09
Bt2	70–115	3.5YR 2.5/4	190	540	8.2	6.7	5.6	8.6	10.9	79	0	146	26.7	1.68	1.11
Typic Eutrophic RED-YELLOW ARGISOL															
A	0–15	10YR 3/4	450	370	15.6	5.8	4.6	6.1	10.7	57	0	65	7.0	1.94	1.51
Bt	50–70	5YR 5/6	340	530	4.9	5.8	4.9	4.2	6.6	64	0	95	7.0	1.78	1.36
Typic Tb Dystrophic HAPLIC CAMBISOL															
A	0–10	10YR 3/3	250	340	15.9	4.6	3.9	1.8	7.8	23	63	55	4.2	2.13	1.54
Bi	35–65	7.5YR 6/6	270	440	5.6	4.7	4.0	0.6	5.2	12	84	72	4.0	2.01	1.51
Typic Tb Eutrophic HAPLIC CAMBISOL															
A	0–12	10YR 3/2	350	210	38.6	5.6	5.2	13.3	19.5	68	0	83	11.1	2.02	1.59
Bi	31–55	5YR 5/6	360	160	3.0	5.7	4.8	2.2	3.9	56	0	94	11.6	1.86	1.62
Typic Dystrophic LITHOLIC NEOSOL															
A	0–20	10YR 4/4	360	420	15.0	4.6	3.9	1.0	7.3	14	75	75	12.0	2.04	1.46
Typic Tb Dystrophic FLUVIC NEOSOL															
A	0–20	10YR 3/2.5	410	450	29.9	4.9	4.2	4.8	14.9	32	25	58	6.4	1.68	1.42
C3	70–120	Variegated	220	260	2.9	5.5	4.1	1.1	3.6	31	48	38	4.5	1.73	1.42
Typic Tb Eutrophic FLUVIC NEOSOL															
A	0–20	10YR 4/2	620	250	14.5	4.9	4.5	5.2	6.8	76	19	37	3.7	1.86	1.59
C2	40–60	10YR 4/2	390	140	5.7	5.5	4.8	3.0	4.9	61	0	21	2.5	1.90	1.64

Source: Embrapa (2001).

Abbreviations: s = silt; r = clay; S = sum of bases; T = cation-exchange capacity; V = base saturation; m = Al saturation.

[a]p = pedoturbation, t = clay accumulation, i = incipient development (Embrapa, 1988).

by Fe- and Al-oxides. Organic matter helps to hold the meager supply of plant nutrients in Latosols.

GEOMORPHIC SURFACES AND SOILS

The Central Brazil region constitutes a classic example of polycyclic landscape evolution, with both young (Pleistocene) forms and well-preserved remnants of much older surfaces (Lepsch and Buol 1988). Overall, three major geomorphic surfaces have been identified by Feuer (1956) in the area of the Federal District (FD). He called them the first, second, and third surfaces (see figs. 2.1, 2.2). A geomorphic surface is a portion of the landscape specifically defined in space and time (Ruhe 1969). The geomorphic surfaces consist of *plains*, the generally level or rolling surfaces, and *bevels*, erosion surfaces that cut and descend from a plain (Bates and Jackson 1987).

First Geomorphic Surface

The first surface (Surface I) corresponds to the peneplane formed during the arid South American erosion cycle (Braun 1971), and is often called the South American Surface. This cycle lasted long enough to affect almost all of the Brazilian landscape (King 1956; Suguio and Bigarella 1979). Subsequent moister climatic conditions propitiated the deepening of the weathering mantle. After the epirogenic upliftings of the Medium Tertiary (King 1956), and the consequent lowering of the base level of erosion, dissection of this surface was initiated.

In the region south of the Federal District, the high tablelands (900 to 1,100 m altitude) with slopes of less than 3% are remnants of the South American surface. This surface is covered with a thick layer of Tertiary sediments (Radambrasil 1983). We know little about the origin and mode of deposition of these sediments. In part of the region, the edges of the remnants of this surface are covered by a thick layer of hard iron-rich fragments (petroplinthite). Highly resistant to erosion, this layer effectively protects and maintains the remnants of said surface. In other places, the plateau is protected by quartzitic mountain crests. Where there is no such protection, the tablelands are eroded rapidly by parallel slope retreat.

Soils on Surface I

The distribution of the soils on Surface I depends on the size of the tableland remnants. In wide remnants, the soil distribution is similar to

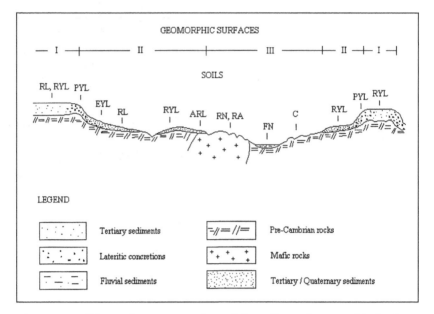

Figure 2.1 Schematic representation of the soils on Geomorphic Surfaces I, II, and III, south of the Federal District. RL = Red Latosol; RYL = Red-Yellow Latosol; PYL = Petroplinthic Yellow Latosol; EYL = Endopetroplinthic Yellow Latosol; ARL = Acriferric Red Latosol; RN = Red Nitosol; RA = Red Argisol; FN = Fluvic Neosol; C = Cambisol.

that described by Macedo and Bryant (1987) for areas of the Federal District. From the center to the border there is a sequential occurrence of: Red Latosols (RL), Red-Yellow Latosols (RYL), and Petroplinthic Yellow Latosols (PYL) (see figs. 2.1 and 2.3a). They are all in the Oxisol order in Soil Taxonomy (Soil Survey Staff 1999). The PYL soil constitutes the major part of the borders with the second geomorphic surface. In the narrower remnants, RL does not occur, and the PYL soil, rich in ironstone fragments, is more widespread and in some places is the main soil of the area.

These soils developed from fine sediments of unknown origin apparently not related to the underlying strata (Brasil 1962; Braun 1971). The soils differ from each other in the hydric regime or natural soil drainage along the gentle slopes. The Red Latosols on the higher areas near the center of the tableland remnant have good internal drainage as shown by their red color. The red color is due to hematite, an iron oxide mineral in the clay fraction which indicates an oxidizing environment (Cornell and

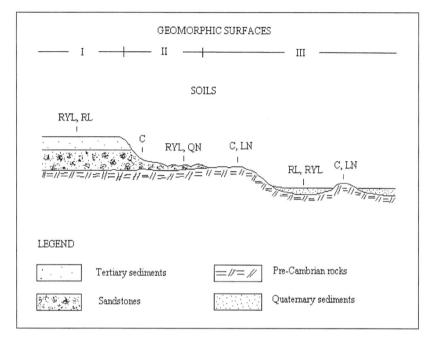

Figure 2.2   Schematic representation of the soils on Geomorphic Surfaces I, II, and III in northwest Minas Gerais state. RYL = Red-Yellow Latosol; RL = Red Latosol; C = Cambisol; QN = Quartzarenic Neosol; LN = Litholic Neosol.

Schwertmann 1996). The Red-Yellow Latosols and the Petroplinthic Yellow Latosols both have the red-yellow colors of the iron oxide mineral goethite, which indicates moister soil conditions than in redder soils. The red-yellow and yellow soils are adjacent to soils with seepage zones and iron oxide concretions on the edges of the high tablelands (Macedo and Bryant 1987). Apparently Fe was reduced in the red-yellow soils, transported in solution to the edge of the tableland remnant, and oxidized there to form Fe-rich concretions.

Hydromorphic soils occur near the borders of the tablelands where there are occasional springs associated with a drainage net. Hydromorphic soils are also common in scattered low-lying areas in the interior of the tablelands. These areas are generally flat with many microelevations or swells averaging 1 to 1.5 m high and 5 to 10 m in diameter (see fig. 2.4A). According to Corrêa (1989) these swells are paleotermite mounds. Between the swells there are gently concave swales through which water flows much of the year. This topography, regionally called *covoal* (termite mounds), or *murundu field* (chapter 6), has been described in various

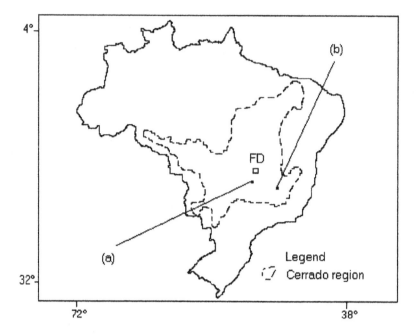

Figure 2.3   Map of Brazil, showing the schematic localization of the area (a) south of the Federal District (FD) and (b) northwest of the state of Minas Gerais.

regions of the Central Plateau (Embrapa 1982; Corrêa 1989; Motta and Kämpf 1992; Resende et al. 1999). The depth of the water table in the soils fluctuates seasonally, resulting in the reduction, transport, and oxidation of Fe to form plinthite. The main soils in this landscape are Plinthosols (Aquox) and, in smaller proportion, Plinthic Red-Yellow Latosols. Both formed under hydrophytic grassland. Hydromorphic soils also occur in the *veredas*, small valleys with distinctive hydrophytic vegetation (Lima 1996) characterized by a tree-shrub set in which the *buriti* palm (*Mauritia vinifera* Mart.) predominates, and a grass zone in areas in which water seeps to the surface (see chapter 6).

In part of the region, Surface I has a prominent border in which the soils contain much gravel, cobbles, and boulders of petroplinthite (see figs. 2.4B and 2.4C). Except for the high content of coarse fragments and the fact that clay contents increase with depth (table 2.1 and field observations), the B horizons are similar to latosollic B horizons of other soils of the high tablelands. In general, the surface on which these soils occur is relatively high in the tableland landscape. Apparently, nearby surfaces

Figure 2.4 (A) Undulating topography on Surface I due to *covoal*, or *murundus* (termite mounds). (B) Surface II in the foreground and surface I in the background, south of the Federal District. (C) Petroplinthite (ironstone) on the soil surface near the border of Surface I. (D) View of the region south of the Federal District, showing the bevel and escarpment between Surface I and Surface II. (E) Isolated elevation on a gentle undulated area of the second geomorphic surface. (F) Gully erosion on the borders of Surface II. (G) Red Nitosols and Red Argisols on Surface III. (H) Aspect of relief and vegetation on the erosion segment of the third geomorphic surface.

were lowered by erosion because the soils on them lacked the protective cover of petroplinthite. In other parts of the region, the bevel leading to the second surface consists of escarpments 100 to 200 m high (see fig. 2.4D), probably due to the effect of tectonic movement of small geographic expression (Cline and Buol 1973).

In the region of the Minas Gerais Triangle, the Surface I is 300 to 400 m lower than the corresponding surface in the Federal District, and the distribution of soils does not follow a definite pattern. In the Triangle, there is a predominance of high-clay Red Latosols and Red-Yellow Latosols on flat and gently undulating topography.

The soils of Surface I are very highly weathered and have very low natural fertility and a limited reserve of nutrients (table 2.1). The K_i ratio for all horizons is well below 2.0, the upper limit for Latosols, and the K_r ratio of most horizons is less than 0.75, which suggests that the clay fraction consists mainly of Fe- and Al-oxide minerals. This mineralogy is confirmed by the low or negative ΔpH ($pH_{H_2O} - pH_{KCl}$) values in most horizons. The rest of the clay is probably kaolinite. In addition, some soils have more than 50% aluminum saturation in the surface horizon (table 2.1). The low nutrient supply and reserve capacity in these soils is illustrated by sum of bases, S, in table 2.1. In many subsoil horizons, the sum of Ca, Mg, and K is only 0.3 to 0.5 $cmol_c kg^{-1}$, a negligible quantity of these plant nutrients.

Semideciduous tropical cerrado is the main form of native vegetation in the flat segment as well as in the borders of the first geomorphic surface, although there is occurrence of semiperennial tropical cerrado, and, in more restricted areas, semideciduous and semiperennial tropical forest, beyond hydrophytic grassland.

Second Geomorphic Surface

A cycle of pediplanation in the mid-Tertiary, called the "Velhas" erosion cycle by King (1956), initiated the dissection of the Surface I and the formation of the second geomorphic surface (Surface II). South of the Federal District, this surface is a plain that slopes downward in the main direction of water flow (see fig. 2.4E) from the borders of the first surface.

Soils on Surface II

Soils on the Surface II show more influence of the underlying bedrock than soils on Surface I. South of the Federal District, soils over Precambrian schists (Araxá Group) are clayey, those over mafic granulite (Anápolis-

Itauçu Granulitic Complex) are very clayey, and soils on quartzite (Araxá Group) are medium-textured. Soils formed largely from Tertiary sediments from the first surface are also very clayey.

On Surface II, the relief is mainly gently undulating with small flat areas. The native vegetation is again cerrado, with small areas of semideciduous forest. In the region of the Minas Gerais Triangle, there is a predominance of *cerradão* (woodland savanna; chapter 6). In the Central Plateau, the predominant soils are Red Latosols and Red-Yellow Latosols, but there are small areas of Acriferric Red Latosols (ARL) mainly south of the Federal District (Cline and Buol 1973; Mothci 1977; Embrapa 2001), where they formed from mafic rocks (Anápolis-Itauçu Granulitic Complex). On the higher parts of this surface, just below the borders of Surface I, some soils have strata rich in concretions (concentration masses) of petroplinthite. These concretions show no edges and are mixed with the latosollic (oxic) material at variable depths in profiles of Endopetroplinthic Yellow Latosols (EYL) (fig. 2.1 and table 2.2). The deposition of these concretions as stonelines, their mixing with the latosollic material, and the absence of morphological evidence of reducing conditions necessary for the formation of ironstone concretions suggest that these concretionary deposits originated from the borders of the first surface. The EYL occur in a zone adjacent to the escarpment base, while the Red Latosols or Acriferric Red Latosols are in a lower position, hundreds of meters from the EYL zone.

In the northwest of the state of Minas Gerais (fig. 2.2), this surface cut Cretaceous sandstones (Areado Formation). The relief is dominantly gently undulating, and the soils are medium-textured Red-Yellow Latosols, and Cambisols (Inceptisols) or sandy-textured Neosols (Quartzarenic Neosols, QN, Entisols). Here the drainage net is more dense and the *veredas* more common. In the Minas Gerais Triangle, medium-textured Red Latosols predominate, also developed from Cretaceous sandstones (Bauru Group) (Embrapa 1982).

The mixture of this preweathered material with the less altered material that originated from weathering of the subjacent rocks constitutes the parent material of the soils of the Surface II. The relative proportion of these two materials influences the distribution and attributes of the soils.

Surface II is being dissected by contemporary erosion processes creating the third surface; in many places the two surfaces are intertwined. In the hilly borders with many remnants of Surface II, the soils are very susceptible to gully erosion. Where native vegetation has been removed to facilitate farming, grazing, or mining projects, gully erosions may progress rapidly (see fig. 2.4F).

Although Surface I is much older than Surface II, the K_i index, which expresses the degree of loss of silica from soils and their degree of weathering, did not show substantial differences (tables 2.1, 2.2), in agreement with the observations of Cline and Buol (1973). The index might even be smaller for soils on the Surface II; this might be explained, beyond the contribution of the preweathered material from Surface I, by the more freely drained conditions on Surface II, which accelerate the processes of weathering and leaching. The K_r ratios are also small, and ΔpH is small or negative in several horizons of soils on Surface II (table 2.2). Similar to soils on Surface I, soils on Surface II have low natural fertility, and all of them have medium (20% to 40%) to high (>40%) levels of Al saturation in the plow layer (table 2.2). In subsoils, the cation exchange capacity is so low that the soil can hold neither base cations, which are needed by plants, or Al, which is detrimental to plants.

Third Geomorphic Surface

The third surface (Surface III) was formed when geologic erosion cut through Surfaces I and II, forming a new surface including an erosion segment with sloping relief and a depositional segment with gentle relief. South of the Federal District, this surface is still in the initial stage of development and includes little but the erosion segment. In other areas, however, the depositional segment is more extensive.

In the northwest of the state of Minas Gerais (see figs. 2.2 and 2.3B), removal of sandy material from Surface II exposed Precambrian pelitic (clay-rich) rocks of the Bambuí Group. The weathering products of these rocks were subsequently reworked. Where the process of planation of the former surface is extensive, the new surface is called an "Exhumed Pre-Cretaceous Surface" (Cetec 1981). In the major part of the region, however, removal of the former surface is less extensive, and there is predominance of well-dissected landforms, resulting in many residual hills within the deposition zone.

The depositional segment of Surface III is most extensive along the valleys of the main rivers, which start in the Central Plateau. These rivers are in the Amazon, São Francisco and Paraná basins. This segment is well expressed in the São Francisco Depression, where the cover deposits lie between 400 and 600 m altitude and were dated as Pleistocene by Penteado and Ranzani (1973). The deposits vary from a few centimeters to several meters thick. Local textural variations are linked to the erosive reworking of Pleistocene deposits and to a greater or lesser contribution of fluvial material. Regional variations are related to the proportions of sandy detritus from Cretaceous formations and clayey detritus from Precambrian formations. Sandier covers occur in areas where the escarpments which border the depression are located mainly in sandstones (Areado Formation). Clayey covers are observed in areas where the escarpments are in Precambrian pelitic rocks (Bambuí Group).

Soils on Surface III

Shallow soils predominate on the erosional segment of Surface III, mainly Cambisols (C, Inceptisols) and, in smaller proportions, Red-Yellow Argisols (RYA), Red Argisols (RA) and Red Nitosols (RN), the last three occurring in the Alfisol order in Soil Taxonomy (Soil Survey Staff 1999), beyond the Litholic Neosols (LN, Entisols). The RN soils occur in association with mafic rocks outcroppings (see fig. 2.4G). Because of the close relationship between soils and geology, the soils of the erosional segment of Surface III are much more variable than those of the other surfaces, especially in relation to base saturation and texture (table 2.3). For example, soils may be allic (low base saturation, Al-rich), dystrophic (low base saturation), or eutrophic (high base saturation), and texture varies from medium to clayey. In general, soils that have a textural B horizon (argillic horizon) also have high clay contents, > 450 g kg^{-1}, and medium to high base saturation (table 2.3). The majority of Cambisols and Litholic Neosols, on the other hand, are characterized by low natural fertility and high aluminum saturation. Cambisols generally have a high content of gravels and cobbles. Mainly, they support *campo cerrado* (open cerrado).

On the depositional segment in the São Francisco Basin the main soils are clayey Red Latosols and Red-Yellow Latosols, although Cambisols and Litholic Neosols are important in residual elevations. Vegetation follows the great variability of soils. The native vegetation includes forest, *cerradão*, *cerrado*, *campo cerrado*, and tropical grassland (physiognomic descriptions in chapter 6). The RN, RA, and RYA soils have relatively good natural fertility, and the native vegetation is forest (see fig. 2.4H). Riparian forests occur in the fluvial plains over Fluvic Neosols (FN, Fluvents) in the wider valleys, or over Hydromorphic soils along the smaller water flows. The relief on Surface III varies from gently undulating to mountainous.

In contrast to soils on older surfaces, soils on Surface III do not exhibit net positive charge. Additionally, the K_i indices, although low enough to be characteristic of low-activity clays, tend to be higher than those for soils on the other surfaces. K_r ratios are also higher. All these trends confirm that these soils are less weathered than those on older surfaces.

CONCLUDING REMARKS

The Brazilian cerrado region consists of three main geomorphic surfaces, each having a set of soils with distinctive attributes. On the oldest surface (Surface I), the much longer exposure of the Tertiary sediments to weathering and leaching has overcome the influence of the subjacent rocks on the kinds and distribution of soils. On that surface the main soil differences are due to the water regime. The seasonal saturation of some soils results in redistribution of Fe in the soil profile and landscape, and formation of petroplinthite. This rocklike material retards erosion, especially at the edge of tablelands, thereby helping keep the tablelands intact. Thus, petroplinthite has significance for both soils and geomorphology. The index values based on laboratory data, K_i, K_r, and ΔpH, confirm that these soils are highly weathered. On the youngest surface (Surface III), the relationship between the distribution and properties of soils and their parent materials is more evident. On the intermediate surface (Surface II), the distribution of soils as well as their attributes are related to the degree of mixing of the Tertiary material from Surface I and the products of decomposition of the subjacent rocky material. The index values are similar to those on Surface I, however, showing that these soils are also highly weathered.

REFERENCES

Ab'Saber, A. N. 1963. Contribuição a geomorfologia da área dos cerrados. In M.G. Ferri, ed., *Simpósio sobre o Cerrado*, pp. 119–128. São Paulo: Editora da Universidade de São Paulo.

Bates, R. L. and J. A. Jackson. 1987. *Glossary of Geology*. 3rd edition. Am. Geol. Institute, Alexandria, VA.

Brasil. 1962. *Levantamento de Reconhecimento de Solos da Região sob Influência do Reservatório de Furnas*. Rio de Janeiro: Serviço Nacional de Pesquisas Agronômicas, Ministério da Agricultura.

Braun, O. P. G. 1971. Contribuição à geomorfologia do Brasil Central. *Rev. Bras. Geogr.* 32:3–39.

Cetec. 1981. *Segundo Plano de Desenvolvimento Integrado do Noroeste Mineiro: Recursos Naturais*. Belo Horizonte: Fundação Centro Tecnológico de Minas Gerais.

Cline, M. G. and S. W. Buol. 1973. *Solos do Planalto Central do Brasil*. Ithaca: Cornell University Press.

Cornell, R. M. and U. Schwertmann. 1996. *The Iron Oxides: Structure, Properties, Reactions, Occurrence and Uses*. Weinheim: VCH.

Corrêa, G. F. 1989. *Les Microreliefs "Murundus" et Leur Environnement*

Pédologique dans l'Ouest du Minas Gerais, Région du Plateau Central Brésilien. Thèse de doctorat, Université de Nancy, Vandoeuvres-les-Nancy, France.

Embrapa. 1982. *Levantamento de Reconhecimento dos Solos e Aptidão Agrícola das Terras do Triângulo Mineiro.* Rio de Janeiro: Embrapa, Serviço Nacional de Levantamento e Conservação de Solos.

Embrapa. 1988. *Definição e Notação de Horizontes e Camadas do Solo.* Segunda edição. Rio de Janeiro: Embrapa, Serviço Nacional de Levantamento e Conservação de Solos.

Embrapa. 1999. *Sistema Brasileiro de Classificação de Solos.* Rio de Janeiro: Embrapa, Centro Nacional de Pesquisa de Solos.

Embrapa. 2001. In press. *Levantamento de Reconhecimento de Alta Intensidade dos Solos do Município de Silvânia, GO.* Rio de Janeiro: Embrapa, Serviço Nacional de Levantamento e Conservação de Solos.

Feuer, R. 1956. "An Exploratory Investigation of the Soils and Agricultural Potential of the Soils of the Future Federal District in the Central Plateau of Brazil." Ph.D. thesis, Cornell University, Ithaca, USA.

Ker, J. C. and M. Resende. 1996. Recursos edáficos dos cerrados: Ocorrência e potencial. In: Biodiversidade e produção sustentável de alimentos e fibras nos cerrados. In R.C. Pereira and L.C.B. Nasser, eds., *Anais do VIII Simpósio Sobre o Cerrado*, pp. 15–19. Brasília: Embrapa, Centro de Pesquisa Agropecuária dos Cerrados.

King, L. C. 1956. A geomorfologia do Brasil Oriental. *Rev. Bras. Geogr.* 18:147–265.

Lepsch, I. F. and S. W. Buol. 1988. Oxisol-landscape relationships in Brazil. In F. H. Beinroth, M. N. Camargo, and H. Eswaran, eds., *Proceedings of the International Soil Classification Workshop*, 8, pp. 174–189.

Lima, S. C. 1996. "As Veredas do Ribeirão Panga no Triângulo Mineiro e a Evolução da Paisagem." Doctor in Science thesis, Universidade de São Paulo, São Paulo, Brasil.

Macedo, J. and R. B. Bryant. 1987. Morphology, mineralogy and genesis of a hydrosequence of Oxisols in Brazil. *Soil Sci. Soc. Am. J.* 51:690–698.

Mothci, E. P. 1977. "Características e Gênese de um Seqüência de Oxisols no Planalto Central Brasileiro." Master's thesis, Universidade Federal do Rio Grande do Sul, Porto Alegre, Brasil.

Motta, P. E. F. and N. Kämpf. 1992. Iron oxide properties as support to soil morphological features for prediction of moisture regimes in Oxisols of Central Brazil. *Z. Pflanzenernähr* 155:385–390.

Penteado, M. M. and G. Ranzani. 1973. *Relatório de viagem ao Vale do Rio São Francisco: Geomorfologia.* São Paulo: Editora da Universidade de São Paulo.

Radambrasil. 1983. *Folha SE 22 Goiânia: Geologia, Geomorfologia, Pedologia, Vegetação e Uso Potencial da Terra.* Rio de Janeiro: Ministério das Minas e Energia.

Resende, M. 1976. *Mineralogy, Chemistry, Morphology and Geomorphology of Some Soils of the Central Plateau of Brazil*. Ph.D. thesis, Purdue University, West Lafayette, USA.

Resende, M., N. Curi, S. B. Rezende, and G. F. Corrêa. 1999. *Pedologia: Base para Distinção de Ambientes*. Viçosa, Brasil: Núcleo de Estudos de Planejamento de Uso da Terra.

Ruhe, R. 1969. *Quaternary Landscapes in Iowa*. Ames: Iowa State University Press.

Soil Survey Staff. 1999. *Soil Taxonomy*. 2nd edition. U.S. Department of Agriculture Handbook 436. Washington, D.C.: U.S. Government Printing Office.

Sposito, G. 1989. *The Chemistry of Soils*. New York: Oxford University Press.

Suguio, K. and J. J. Bigarella. 1979. Ambiente fluvial. In J. J. Bigarella, K. Suguio, and R. D. Becker, eds., *Ambientes de Sedimentação: Sua Interpretação e Importância*. Curitiba: Editora Universidade do Paraná.

3

Late Quaternary History and Evolution of the Cerrados as Revealed by Palynological Records

Marie-Pierre Ledru

WHETHER CERRADOS ARE ANTHROPOGENIC OR NATURAL FORMA-tions has been a matter of strong debate over the last century. The fact is that cerrados and forests can occur in the same region, at the same latitude, under the same climatic conditions (chapters 2, 6). These observations generated two types of hypotheses about the origin of the cerrado. The first favors the human-induced origin of the vegetation and is based on observations of fire-adapted species, which suggest that cerrados would result from the development of dry forests under the influence of fire (Lund 1835; Loefgren 1897; Aubréville 1961; Schnell 1961; Eiten 1972). The second hypothesis supports the natural origin of the cerrados based on the occurrence of cerrados in areas only recently colonized by humans, and on the discovery of extinct giant mammals living in open forest landscapes at the end of the Pleistocene (Warming 1918 in Warming and Ferri 1973; Azevedo 1965; Cartelle 1991; Guerin 1991; Vilhena Vialou et al. 1995). Palynological research started in Brazil at the end of the 1980s and has provided an excellent tool for the elucidation of patterns and causes of vegetation and climate change. Pollen grains accumulated in lakes or bogs are well preserved in sedimentary deposits and can show how the vegetation changed through time and space. The construction of pollen reference collections and publication of a pollen atlas for the cerrados (Salgado-Labouriau 1973) has facilitated the research. In this chapter I will first give a definition of the cerrado in terms of modern pollen rain and subsequently review the published fossil pollen diagrams as a basis for examination of the evolution of the cerrados during the Late Quaternary.

POLLEN ANALYTICAL METHODS

Samples for pollen analysis derived from recent surface or core sediments are treated according to the standard palynological techniques defined in Faegri and Iversen (1989). These techniques involve concentrating the pollen and spores content of each sample. The inorganic component is removed with acids; humic acids with KOH and palynomorphs are separated from the sediment residue using a solution of density 2 and mounted in glycerine for light microscopical analysis. Pollen and spores are identified by comparison with pollen reference collections and published pollen atlases and are counted to calculate their frequencies. Pollen taxa are grouped and expressed as percentages of a total pollen sum. Aquatics, wetland taxa, and spores are usually excluded from the total pollen sum and expressed as percentages thereof. Pollen diagrams represent the counts for each taxon along a depth scale for fossil samples. Counts can be also expressed as a ratio of Arboreal Pollen/Non-Arboreal Pollen (AP/NAP) or as different ecological groups such as cerrado elements or semideciduous forest elements. Changes in fossil pollen frequencies are interpreted to reflect environmental and climatic changes. Charcoal particles deposited in the sediments can also be counted. They are expressed either as a total of the pollen sum or as particle concentration per unit volume or weight of sample. Radiocarbon dates obtained from the same sediment core and lithology of the core are also reported on fossil pollen diagrams.

Detailed pollen counts of several pollen records used in this review are archived in the NOAA–World Data Center A for Paleoclimatology (Latin American Pollen Database).

MODERN POLLEN RAIN AND CERRADOS INDICATORS

Pollen taxa that are representative for a specific vegetation type are called indicator taxa. Pollen indicator taxa must show good preservation, dispersion, and production to have any chance of being found in fossil sediments. In addition, the source plants must be well represented in the vegetation, and a comparison of modern pollen rain with phytosociological surveys is often needed to define their value. Once an association of indicator species is defined, they can be related to present-day climatic parameters such as the length of the dry season or mean winter tempera-

ture. This approach can be used to provide a quantitative climatic interpretation of the fossil pollen spectra (see table 3.1). The problem with the
cerrados is that, apart from the wind-pollinated Poaceae and Cyperaceae,
most of the pollen taxa recorded are insect-pollinated.

The first study of modern pollen rain for the cerrados was published
by Salgado-Labouriau (1973). Pollen traps were collected monthly in a
cerrado located in the state of Goiás (Central Plateau). Results showed
mainly a dominance of Poaceae (Gramineae) with 74%, followed by the
Fabaceae-Caesalpiniaceae-Mimosaceae (Leguminosae) with 3.2%. Ledru
(1991) collected surface samples in six different types of cerrado in Mato
Grosso, Distrito Federal, and São Paulo (see figs. 3.1–3.3; table 3.2).
Pollen counts were undertaken in relation to phytosociological data in
order to define cerrados indicators well represented in both pollen rain
and vegetation. Two taxa can be distinguished and considered good cerrado indicators, *Byrsonima* and *Didymopanax*. Well represented as a tree,
Curatella is often considered a cerrado indicator but does not produce
much pollen. *Byrsonima* and *Didymopanax* are actually overrepresented
in terms of pollen proportions (i.e., percent of pollen is greater than percent of trees; *Byrsonima* occurs as small trees with a Diameter at Breast
Height (DBH) < 10 cm and therefore is not counted in the botanical surveys). *Byrsonima* is present in different proportions in the six types of cerrado studied, with frequencies of less than 5%. *Qualea* and *Caryocar* are
underrepresented in the pollen counts compared to the associated basal
area of these trees (fig. 3.3). Grasses are not taken into account in the phytosociological surveys but are also abundant and representative of the
cerrados (chapter 7), with ca. 50% Poaceae in *campo cerrado*, and less
than 20% Poaceae in *cerradão* (see also chapter 6). *Vellozia* (Velloziaceae) is a good indicator for *campo cerrado*; *Borreria* (Rubiaceae) is
also well represented although not recorded in the cerrado of Cuiabá
Salgadeira (Mato Grosso), and the *cerradão* of Bauru (São Paulo). The
differences in taxa distribution complicate the definition of a quantified

Table 3.1   Relation Between Climatic Parameters
and Vegetation in Central Brazil

	Climate	
Vegetation	*Dry season*	*Mean winter temperature*
Cerrado	4–5 months	$\geq 15°C$
Semideciduous forest	2–3 months	$> 10°C$ and $< 15°C$
Araucaria forest	None	$\leq 10°C$

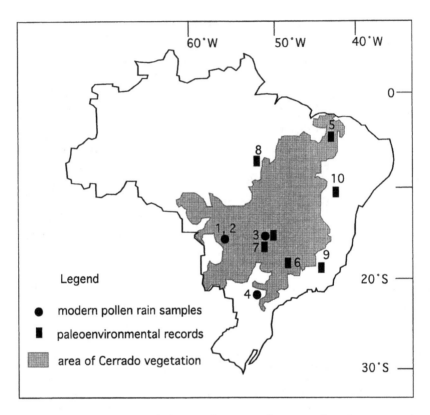

Figure 3.1   Map of Brazil showing location of sites cited in the text: modern and paleoenvironmental records. 1. Cuiabá Salgadeira; 2. Cuiabá Rio Claro; 3. Brasília and Aguas Emendadas; 4. Bauru; 5 Lagoa do Caço; 6. Lagoa dos Olhos; 7. Cromínia; 8. Carajás; 9. Lagoa do Pires; 10. Icatu River Valley.

pattern of indicator taxa. In lakes or in the *veredas* (palm swamp forest; chapter 6), where cores for paleoenvironmental studies are generally recovered, surface samples show high levels of *Mauritia* (Buriti palm) and Cyperaceae (Barberi 1994; Salgado-Labouriau et al. 1997). This further complicates the possibility of recognizing the regional vegetation.

PALEORECORDS LOCATED IN THE CERRADO AREA: POLLEN AND CHARCOAL RESULTS

Pollen records obtained from lake or peat bog sediments provide a detailed paleoenvironmental reconstruction of the changes that affected the cerrados during the Late Quaternary.

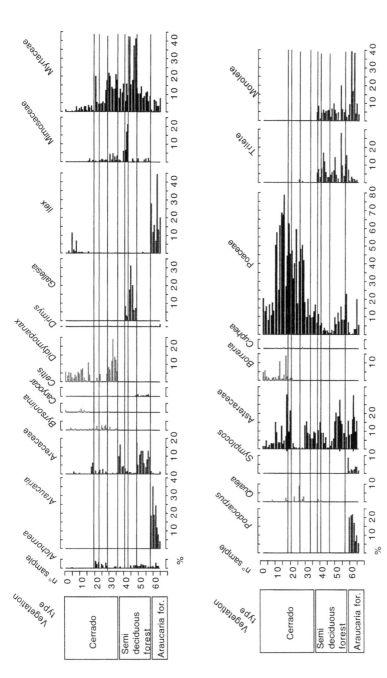

Figure 3.2 Modern pollen rain in the cerrado environment and other Brazilian forests. The cerrado pollen counts were collected in different types of cerrado and are presented in the following order: samples 1 to 21 come from Brasília (Distrito Federal, DF); 22 to 24 from Cuiabá Salgadeira (Mato Grosso, MT); 25 to 29 from Cuiabá Rio Claro (Mato Grosso, MT); and 30 to 37 from Bauru (São Paulo, SP). See also table 3.2.

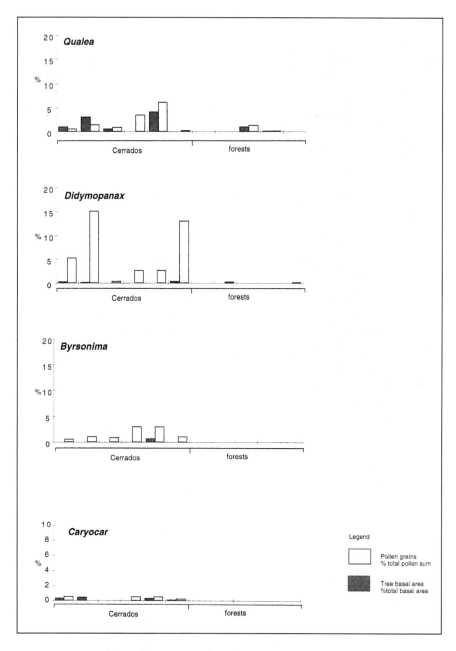

Figure 3.3 Relation between tree basal area (Diameter at Breast Height Percent) and pollen percent for four cerrado taxa. The sites are in the same order as in figure 3.2. Every pollen sample represents a mean value of the pollen counts presented in figure 3.2 to allow comparison with tree basal area.

Table 3.2 Location and Structural Vegetation Attributes of Surface Sample Sites Located in Different Cerrado Areas of Brazil

Sampling site	Elevation (m)	Latitude	Density (trees/ha)	Basal area (m²/ha)	Source[a]
Brasília (DF)					
cerradão	1030	15°35' S	2231	20.9214	1
Cerrado sensu stricto	1125	15°35' S	911	9.65	1
Campo cerrado	1175	15°35' S	203	1.6686	1
Cuiabá (MT)					
cerradão	350	15°21' S	1546	16.116	2, 3
Cerrado sensu stricto	350	15°21' S	1888	21.044	2, 3
Bauru (SP):					
Cerrado sensu stricto	570	22°19' S	8198	40.8793	4

[a]Key to sources: 1 = Ribeiro et al. 1985; 2 = Oliveira-Filho 1984; 3 = Oliveira-Filho and Martins 1986; 4 = Cavassan and Martins 1984.
Note: See chapter 6 for descriptions of cerrado physiognomies; and figure 3.2.

Four palynological records show changes in the vegetation composition: Aguas Emendadas (15° S 47°35' W, 1,040 m elevation; Barberi 1994; Barberi et al. 1995); Cromínia (17°17' S 49°25' W, 730 m elevation; Vicentini 1993; Salgado-Labouriau et al. 1997); Lagoa do Caço (25°8' S, 43°25' W, 80 m elevation; Sifeddine et al. 1999; Ledru et al. 2001); and Lagoa dos Olhos (19°38' S 43°54' W, 730 m elevation; De Oliveira 1992; see figs. 3.1, 3.4A–D).

Pollen records from the cerrados date back to about 32,000 YBP (years Before Present). The indicator taxa *Byrsonima*, *Didymopanax*, and *Curatella* appear throughout the fossil spectra with frequencies of less than 5%, which is far lower than in the modern spectra for the latter two genera. In the Cromínia record (ca. 32,000 YBP), *Byrsonima* and *Mauritia* are recorded at frequencies of less than 2%, and between 10% and 30%, respectively. Arboreal Pollen (AP) frequency is high and the climate can be defined as moist and warm, probably with seasonality in precipitation. Microscopic charcoal particles are also abundant. This observation reveals that fires had occurred in the cerrados without human influence, for the presence of people in South America is not confirmed before 12,000 YBP (Cooke 1998; see also chapter 4). This pre-full-glacial moist and warm phase is found in several other tropical South American records (De Oliveira 1992; Ledru et al. 1996). In tropical northeastern Australia high microscopic charcoal frequencies are recorded from ca. 38,000 YBP, when moist forest was replaced by sclerophyll vegetation.

(A) Aguas Emendadas

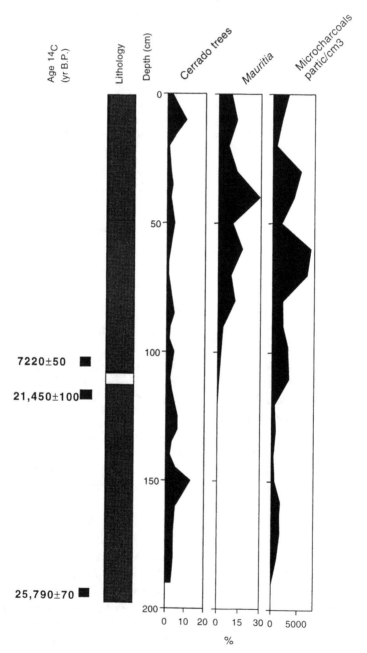

Figure 3.4 Summary pollen diagrams of the records located within the cerrado region, (A) Aguas Emendadas; (B) Cromínia; (C) Lagoa dos Olhos; (D) Lagoa do Caçó (data from Barberi et al. 1995; Salgado-Labouriau et al. 1997; De Oliveira 1992; Ledru, in preparation). Soil codes under (C).

(B) Cromínia

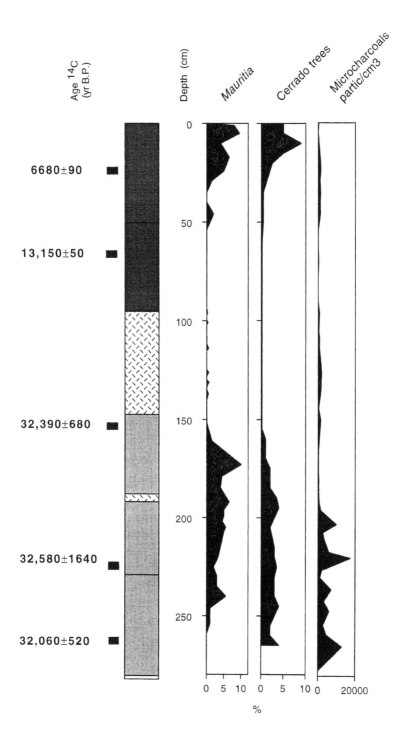

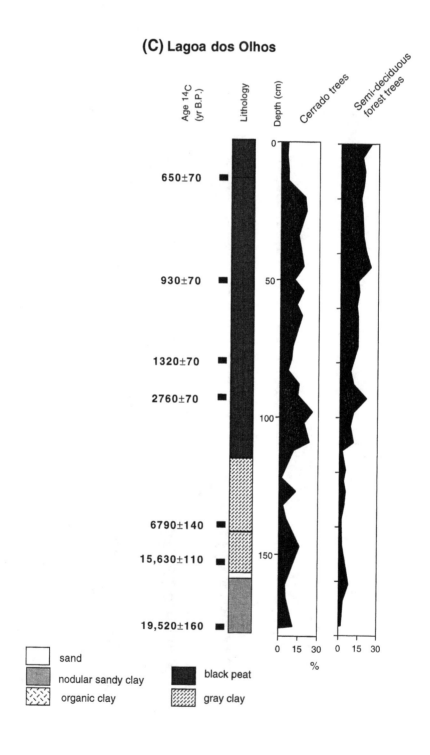

(C) Lagoa dos Olhos

(D) Lagoa do Caçó

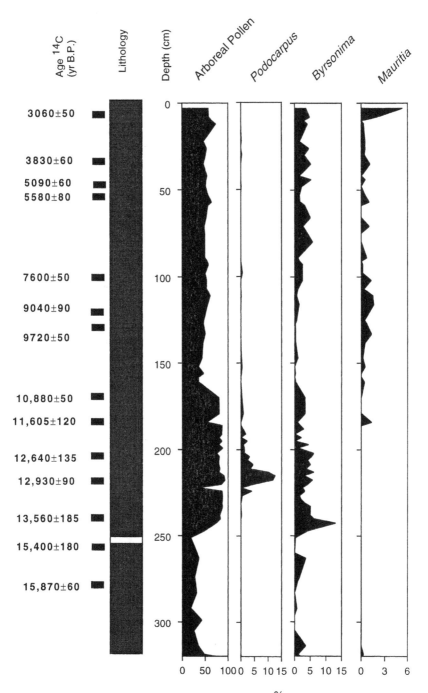

This was probably related to human influence that induced the expansion of the fire-adapted vegetation type (Kershaw 1986; Kershaw et al. 1997). Brazilian records instead show that fire-adapted vegetation can be the result of climate influence alone, although older records are needed in lowland South America to better compare with Australian vegetation history.

A particularly dry and cold climatic phase is recorded in the tropics between 20,000 and 18,000 YBP, corresponding to the last glacial maximum. It was detected in light of the absence of sediment accumulation or palynologically sterile sediment during this period of time (Ledru et al. 1998a). Where pollen is present, this cold and dry phase started at ca. 25,000 YBP and is characterized by a decrease in swamp forest taxa, absence of *Mauritia*, low frequencies of Arboreal Pollen, and dominance of Poaceae, Asteraceae, and Cyperaceae. The lakes were replaced by marshes. Absence of cerrado indicators and microscopic charcoal particles show that climatic conditions did not allow the development of the cerrado at this time.

When sedimentation rates increased after ca. 18,000 YBP at the onset of the late glacial, the absence of assemblages with modern analogues does not allow clear definition of the environment. At Lagoa dos Olhos an association of *Podocarpus* and *Caryocar* is recorded, while at Lagoa do Caçó *Podocarpus* is recorded along with high frequencies of *Byrsonima*. These pollen assemblages characterize cool climatic conditions if compared with modern *Podocarpus* associations in the montane region of Minas Gerais (A.T. Oliveira-Filho, personal communication). The absence of charcoal particles or *Mauritia* pollen grains during the whole lateglacial time period also suggests dry conditions.

After ca. 10,000 YBP and until 7,000 YBP the landscape was more open. The virtual absence of *Mauritia* from the records suggests that palm swamp forests remain restricted, and a gap in sedimentation is often recorded attesting to a dry climate without seasonality. This could be due to changes in orbital parameters that attenuated the "monsoon effect" and reduced the overall precipitation (Martin et al. 1997; Ledru et al. 1998b).

The cerrado of Lagoa do Caçó indicates a different pattern of evolution. *Byrsonima* is recorded during the whole sequence, indicating that cerrado-like vegetation, probably as a *campo cerrado* type, was maintained. No microscopic charcoal fragments are recorded before the beginning of the Holocene, when cerrado tree frequencies started to increase, and the presence of *Mauritia* attests to the establishment of a seasonal climate and increase in temperatures. These differences are probably due to the northern location of the lake, close to the Amazon Basin, and to the influence of the Meteorological Equator. In other cerrado pollen records, when sedimentation restarts after 7,000 YBP, the palm swamp forest is well established, and the climate is warm (mean winter temperatures above 15°C) and seasonal (4 to 5 months dry season). Microscopic charcoal particles are abundant, Poaceae increases, and Asteraceae decreases. This indicates influence of both human and climate on the landscape.

A dry early Holocene is also documented in the Colombian savannas, where an expansion of grassland taxa is recorded. After 6,000 YBP the expansion of gallery forest taxa, mainly the swamp forest taxa *Mauritia* and *Mauritiella*, and the abundance of microscopic charcoal particles, indicate wetter climatic conditions for the Late Holocene and full development of the savanna (Behling and Hooghiemstra 1998; Behling and Hooghiemstra 1999).

Biomass was sufficiently high to induce repeated fires during the Late Holocene in Central Brazil. The increase in seasonality that started at ca. 7,000 YBP was necessary to allow cerrado vegetation to grow on the Central Plateau. It started at different times during the Holocene according to the latitude of the respective record.

PALEORECORDS LOCATED OUTSIDE THE CERRADO AREA SHOWING CERRADO INDICATORS

In other records located in tropical forest areas (Amazonia, montane forest, Atlantic forest), a local increase of cerrado-type indicators confirms that these regions were linked with central regions during extreme climatic conditions.

A pollen record located within the Amazonian Basin (Carajás, 50°25' W 6°20' S, 700–800 m elevation), shows three main changes in vegetation composition during the last 60,000 YBP (Absy et al. 1991; Sifeddine et al. 2001; see fig. 3.5). Cerrado taxa such as *Byrsonima* and *Ilex* are present today on the plateau around the lakes. These taxa are well represented, together with high frequencies of rainforest taxa during past moist climatic phases. Open landscape elements associated with *campo cerrado* vegetation, Poaceae, *Borreria*, and *Cuphea*, increase in frequency during dry climatic periods when moisture levels were insufficient to maintain the forests. *Podocarpus* frequencies increase together with the *campo cerrado* species, attesting to an open landscape with *Podocarpus* on the lake margins similar to the situation found in the record from the Maranhão region during the late-glacial period (Ledru et al. 2001). A cerrado-type vegetation could grow on the southern and eastern periphery of the Amazonian

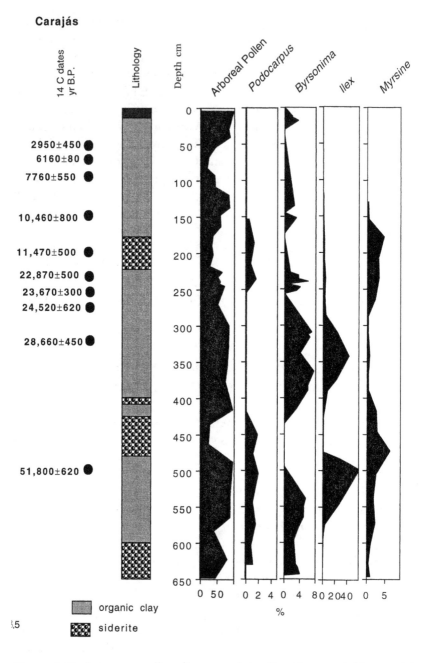

Figure 3.5 Summary pollen diagram of the Carajás record (Amazonia) (data from Absy 1991).

basin, and within the dry corridor during drier and cooler climates (Bush 1994; van der Hammen and Absy 1994). This cerrado vegetation is maintained today on the Plateau of the Serra do Carajás due to specific edaphic conditions.

In the Lagoa do Pires record (17°57' S, 42°13' W, 390 m elevation), located in a semi-deciduous forest region, an increase of cerrado elements (mainly represented by *Curatella*) is registered at the beginning of the Holocene, when the climate was drier in this region (Behling 1995).

A pollen record in the Icatu River valley (10°24' S 43°13' W, ca 400 m elevation) shows an increase of montane and moist forest elements at the beginning of the Holocene, preceding the development of cerrado-*caatinga* vegetation after 9,000 YBP (De Oliveira et al 1999). This moist early Holocene also detected in the Salitre record was interpreted as related to stronger polar advections (Ledru 1993). Fluctuations in *Mauritia* pollen frequencies are detected throughout the Holocene.

CONCLUSION

The floristic composition of the cerrado vegetation varied through time as it does today according to location (chapter 6). The earliest record of cerrado-type vegetation dates back to 32,000 YBP and is located on the Central Brazilian Plateau. Vegetation resembling present-day cerrados does not occur prior to 7,000 YBP in central Brazil and 10,000 YBP in northern Brazil. Their presence in both regions is most likely a consequence of a progressive increase of seasonality, concomitant with an increase in temperature. Both factors, in addition to man's influence, contributed to increased fire frequencies, although cerrado vegetation was probably fire-adapted before people arrived in South America. Local increases of cerrado-type vegetation are recorded in moist forest areas when climatic conditions became drier and/or colder. During these periods forest taxa remained connected through a network of gallery forests. This contributed to the differences in physiognomy observed today (chapter 6). Cerrados recognized in Amazonia or in São Paulo might be the result of this ancient connection.

REFERENCES

Absy, M. L., A. Cleef, M. Fournier, L. Martin, M. Servant, A. Sifeddine, M. Ferreira da Silva, F. Soubiès, K. Suguio, B. Turcq, and T. van der Hammen.

1991. Mise en évidence de quatre phases d'ouverture de la forêt amazonienne dans le sud-est de l'Amazonie au cours des 60000 dernières années: Première comparaison avec d'autres régions tropicales. *C. R. Acad. Sc. Paris* 312:673–678.

Aubréville, A. 1961. *Etude Ecologique des Principales Formations Végétales du Brésil et Contribution à la Connaissance des Forêts de l'Amazonie Brésilienne.* Nogent: Centre Technique Forestier Tropical.

Barberi, M. 1994. "Paleovegetação e Paleoclima no Quaternario Tardio da Vereda de Aguas Emendadas, DF." Master's thesis, Universidade de Brasília, Brasília, Brasil.

Barberi, M., M. L. Salgado-Labouriau, K. Suguio, L. Martin, B. Turcq., and J. M. Flexor. 1995. Análise palinológica da vereda de Águas Emendadas (DF). *Vth Congresso da Associação Brasileira de Estudos do Quaternário,* Universidade Federal Fluminense, Niteroí, Brasil.

Behling, H. 1995. A high resolution Holocene pollen record from Lago do Pires, SE Brazil: Vegetation, climate and fire history. *J. Paleolimnol.* 14:253–268.

Behling, H. and H. Hooghiemstra. 1998. Late Quaternary palaeoecology and palaeoclimatology from pollen records of the savannas of the Llanos Orientales in Colombia. *Palaeogeogr. Palaeoclimatol. Palaeoecol.* 139: 251–267.

Behling, H. and H. Hooghiemstra. 1999. Environmental history of the Colombian savannas of the Llanos Orientales since the Last Glacial Maximum from lake records El Pinal and Carimagua. *J. Paleolimnol.* 21:461–476.

Bush, M. 1994. Amazonian speciation: A necessarily complex model. *J. Biogeogr.* 21:5–17.

Cartelle, C. 1991. Um novo Mylodontinae (Edentata, Xenarthra) do pleistoceno final da região intertropical brasileira. *An. Acad. Bras. Ciênc.* 63:161–170.

Cavassan, O., O. Cesar, and F. R. Martins. 1984. Fitossociologia da vegetação arbórea da reserva Estadual de Bauru, Estado de São Paulo. *Rev. Bras. Bot.* 7:91–106.

Cooke, R. 1998. Human settlement of central America and northernmost South America (14,000–8000 BP). *Quat. Int.* 49/5:177–190.

De Oliveira, P. E. 1992. "A Palynological Record of Late Quaternary Vegetational and Climatic Change in Southeastern Brazil." Ph.D. thesis, Ohio State University, Columbus, Ohio, U.S.A.

De Oliveira, P.E., A. M. F. Barreto, and K. Suguio. 1999. Late Pleistocene/Holocene climatic and vegetational history of the Brazilian caatinga: The fossil dunes of the middle São Francisco River. *Palaeogeogr. Palaeoclimatol. Palaeoecol.* 152:319–337.

Eiten, G. 1972. The cerrado vegetation of Brazil. *Bot. Rev.* 38:1–341.

Faegri, K. and J. Iversen. 1989. *Textbook of Pollen Analysis.* 4th edition. Chichester: John Wiley and Sons.

Guérin, C. 1991. La faune de vertébrés du Pléistocène supérieur de l'aire archéologique de São Raimundo Nonato (Piauí, Brésil). *C. R. Acad. Sc. Paris* 313:567–572.

Kershaw, A. P. 1986. Climatic change and Aboriginal burning in north-east Queensland during the last two glacial/interglacial cycles. *Nature* 322:47–49.

Kershaw, A. P., M. B. Bush, G. S. Hope, K.-F. Weiss, J. G. Goldammer, and R. Sanford. 1997. The contribution of humans to past biomass burning in the tropics. In J. S. Clark, H. Cachier, J. G. Goldammer, and B. Stocks, eds., *Sediment Records of Biomass Burning and Global Change*, pp. 413–442. Berlin: Springer-Verlag.

Ledru, M.-P. 1991. "Etude de la Pluie Pollinique Actuelle des Forêts du Brésil Central: Climat, Végétation, Application à l'Etude de l'Evolution Paléoclimatique des 30000 Dernières Années." Ph.D. thesis, Musée National d'Histoire Naturelle, Paris, France.

Ledru, M.-P. 1993. Late Quaternary environmental and climatic changes in central Brazil. *Quat. Res.* 39:90–98.

Ledru, M-P., P. I. Soares Braga, F. Soubiès, M. Fournier, L. Martin, K. Suguio, and B. Turcq. 1996. The last 50,000 years in the Neotropics (southern Brazil): Evolution of vegetation and climate. *Palaeogeogr. Palaeoclimatol. Palaeoecol.* 123:239–257.

Ledru, M.-P., J. Bertaux, A. Sifeddine, and K. Suguio. 1998a. Absence of Last Glacial Maximum records in lowland tropical forest. *Quat. Res.* 49: 233–237.

Ledru, M.-P., M. L. Salgado Labouriau, and M. L. Lorscheitter. 1998b. Vegetation dynamics in Southern and Central Brazil during the last 10,000 yr B.P. *Rev. Palaeobot. Palynol.* 99:131–142.

Ledru M.-P., R. Campello Cordeiro, J. M. D. Landim, L. Martin, P. Mourguiart, A. Sifeddine, and B. Turcq. 2001. Late-glacial cooling in Amazonia inferred from pollen at Lago do Caço, northern Brazil, *Quat. Res.* 55:47–56.

Loefgren, A. and G. Edwall. 1897–1905. Flora Paulista. *Bol. Com. Geogr. Geolog. São Paulo* 1–4:12–15.

Lund, P. W. 1835. Bemaerkninger over Vegetationen paa de indre Hogsletter af Brasilien, isaer I plantehistorike henseende. *Kgl. Danske Videnskab. Selsk. Skrifer.* 6:145–188.

Martin, L., J. Bertaux, T. Correge, M.-P. Ledru, P. Mourguiart, A. Sifeddine, F. Soubies, D. Wirrmann, K. Suguio, and B. Turcq. 1997. Astronomical forcing of contrasting rainfall changes in tropical South America between 12,400 and 8800 cal yr B.P. *Quat. Res.* 47:117–122.

Oliveira-Filho A. T. 1984. "Estudo Florístico e Fitossociológico em um Cerrado na Chapada dos Guimarães, Mato Grosso: Uma Análise de Gradientes." Master's thesis, Universidade Estadual de Campinas, Campinas, Brazil.

Oliveira-Filho A. T. and F. R. Martins. 1986. Distribuição, caracterização e composição florística das formações vegetais da região da Salgadeira na Chapada dos Guimarães (MT). *Rev. Bras. Bot.* 9:207–223.

Ribeiro, J. F., J. C. Souza Silva, and G. J. Batmanian. 1985. Fitossociologia de tipos fisionômicos de cerrado em Planaltina DF. *Rev. Bras. Bot.* 8:131–142.

Salgado-Labouriau, M. L. 1973. *Contribuição à Palinologia dos Cerrados.* Acadêmia Brasileira de Ciências: Rio de Janeiro.

Salgado-Labouriau, M. L., V. Casseti, K. R. C. F. Vicentini, L. Martin, F. Soubies, K. Suguio, and B. Turcq. 1997. Late Quaternary vegetational and climatic changes in cerrado and palm swamp from central Brazil *Palaeogeogr. Palaeoclimatol. Palaeoecol.* 128:215–226.

Schnell, R. 1961. Le problème des homologies phytogéographiques entre l'Afrique et l'Amérique tropicale. *Mem. Mus. Hist. Nat.* 11:137–241.

Sifeddine, A., R. C. Cordeiro, M.-P. Ledru, B. Turcq, A. L. S Albuquerque, L. Martin, J. M. Landim Dominguez, H. Pasenau, G. Ceccantini, and J. J. Abrão. 1999. Late glacial multi-proxy paleo-environmental reconstruction in Caço Lake (Maranhão state, Brazil). *INQUA Congress, Durban, South Africa,* Abstract p. 167.

Sifeddine, A., L. Martin, B. Turcq, C. Volkemer-Ribeiro, F. Soubies, R. C. Cordeiro, and K. Suguio. 2001. Variations of the Amazonian rainforest environment: A sedimentological record covering 30,000 years. *Palaeogeogr. Palaeoclimatol. Palaeoecol.* 168:221–235.

Van der Hammen, T. and M. L. Absy. 1994. Amazonia during the last glacial. *Palaeogeogr. Palaeoclimatol. Palaeoecol.* 109:247–261.

Vicentini, K. R. C. F. 1993. "Análise Palinológica de uma Vereda em Cromínia, Goiás." Master's thesis, Universidade de Brasília, Brasília, Brazil.

Vilhena Vialou A., T. Aubry, M. Benabdelhadi, C. Cartelle, L. Figuti, M. Fontugne, M. E. Solari, and D. Vialou. 1995. Découverte de Mylodontinae dans un habitat préhistorique daté du Mato Grosso (Brésil): L' abri rupestre de Santa Elina. *C. R. Acad. Sc. Paris* 320:655–661.

Warming, E. and M. G. Ferri. 1973. *Lagoa Santa e a Vegetação de Cerrados Brasileiros.* : São Paulo e Belo Horizonte: Editora da Universidade São Paulo e Editora Itatiaia.

4

The Fire Factor

Heloisa S. Miranda, Mercedes M. C. Bustamante,
and Antonio C. Miranda

FIRE IS A COMMON FEATURE OF THE CERRADOS, AS IT IS FOR MOST
savanna ecosystems. Fires set by man or lightning are common and have
been for thousands of years. Vicentini (1993), in a paleoclimatic and paleo-
vegetational study, has registered the occurrence of fire 32,400 Years
Before Present (YBP) in the region of Cromínia (GO); De Oliveira (1992)
registered the presence of charcoal particles dated from 13,700 YBP in lake
sediments in cerrado of the southeastern Brazil; and Coutinho (1981) has
reported the occurrence of charcoal pieces dated from 8,600 YBP from a
campo cerrado soil horizon lying at 2 m depth. Although Gidon and Delibrias (1986) date the presence of man in Brazil to 32,000 YBP, according to
Prous (1992) and Cooke (1998) there is no evidence of human presence in
central Brazil before 12,000–11,000 YBP. Therefore, the particles of charcoal and burned wood dated from 11,000 YBP could, at least in part, be
caused by the early inhabitants of the cerrado region (Salgado-Laboriau
and Vicentini 1994). The indigenous people of the cerrado region used fire
for hunting, stimulation of fruit production, control of undesirable species,
and tribal war (Coutinho 1990a; Mistry 1998). Nowadays, the principal
cause of fire in the cerrado is agricultural, its purpose either to transform
cerrado into crop fields or to manage natural (more open cerrado forms)
or planted pasture (Coutinho 1990a; see chapter 5).

Although fire is considered one of the determinants of the cerrado
vegetation, the rapid occupation of the cerrado region has changed the
natural fire regime (season and frequency of burning) with consequences
for the vegetation structure and composition.

In this chapter we present a review of cerrado fire ecology, with
emphasis on fuel dynamics, fire behavior, nutrient fluxes, and changes in

the structure and composition of the vegetation. A review of fire effects on population dynamics of woody plants is presented in chapter 9.

THE CERRADO FIRES

Cerrado fires, like most savanna fires, are characterized as surface fires, which consume the fine fuel of the herbaceous layer. Luke and McArthur (1978) define fine fuel as live and dead grasses, and leaves and stems with diameter smaller than 6 mm. Depending upon the physiognomic form (chapter 6) and the time since the last fire, the total fine fuel load, up to a height of 2 m, may vary from 0.6 kg/m^2 to 1.2 kg/m^2. The fine fuel of the herbaceous layer represents 97% of the fuel load for *campo sujo*, 90% for cerrado *sensu stricto*, and 85% for *cerradão* (Miranda 2000; see descriptions of physiognomies in chapter 6). These values are similar to those presented by Castro and Kauffman (1998) for cerrado *sensu stricto*, by San José and Medina (1977) for other South American savannas, and by Kelmann et al. (1987) for African savannas.

The vegetation of the herbaceous layer represents 94% of the fuel consumed during the fires. Most of the fine fuel in the woody layer is not consumed during the fires (see table 4.1). This may be a consequence of the high water content of the live fuel, of the fast rate of spread of the fire front (Kauffman et al. 1994; Miranda et al. 1996a; Castro and Kauffman 1998), and of the height of the flames during the fires. Flame height for savanna fires ranges from 0.8 m to 2.8 m (Frost and Robertson 1987). Castro Neves (unpublished data) determined a reduction of 4% in the canopy cover of a *cerradão* immediately after a prescribed fire. The abscission of the damaged leaves resulted in a reduction of 38% in the canopy cover in the 15 days after the fire, suggesting that most of the live leaves are not consumed but damaged by the hot air flow during the fire.

The fuel consumption in the different physiognomic forms (table 4.1) reflects the difference in fine fuel composition and in fire regime. In the *campo sujo* most of the fine fuel is composed of grasses (live and dead, or dormant) that are not in close contact with the wet soil surface and are well exposed to the wind and solar radiation, quickly losing moisture to the environment.

In the denser forms of cerrado (i.e., *cerradão*) the composition of the fine fuel reflects the fire regime much more than the open form, with the dead leaves of the litter layer representing most of the fuel load after long periods of protection from fire. In this case the dead fuel is in close contact with the soil surface. The microclimate may affect the rate of fuel moisture

Table 4.1   Range of Fine Fuel Load and Fuel Consumption
During Prescribed Fires in the Cerrado Vegetation
at the Reserva Ecológica do IBGE, Brasília, DF

	Campo sujo	Cerrado sensu stricto	Cerradão
Before fire (kg/m^2)			
Herbaceous layer	0.64–0.96	0.59–1.11	0.50–0.75
Woody layer	0.03–0.03	0.06–0.14	0.08–0.14
Total	0.67–0.99	0.65–1.25	0.58–0.89
Dead fuel (%)	69–75	76–61	50 – 69
Live fuel (%)	31–25	24–39	50 – 31
After fire (kg/m^2)			
Herbaceous layer	0.05–0.06	0.09–0.00	0.08–0.04
Woody layer	0.01–0.01	0.06–0.04	0.06–0.14
Total	0.06–0.07	0.15–0.04	0.14–0.18
Fuel consumption (%)	91–93	77–97	75–76

Source: H. S. Miranda, unpublished data.
Note: See chapter 6 for descriptions of cerrado physiognomies.

loss due to shading of the fine fuel by the trees and shrubs, leaving patches of vegetation unburned (Miranda et al. 1993; Kauffman et al. 1994).

As the fire front advances, air temperatures rise sharply. Miranda et al. (1993, 1996a), in studies of vertical distribution of temperatures (1 cm, 60 cm and 160 cm height) during cerrado fires, have registered maximum values in the range of 85°C to 840°C (regardless of the physiognomic form), with the highest temperatures occurring most of the time at 60 cm above ground. The range of temperatures reported is similar to that of savanna fires, where a considerable range of temperatures has been recorded, from 70–800°C at ground level, or just above it, to 200–800°C at about 1 m (Frost and Robertson 1987). The great variability in the temperature may reflect the varying compositions and spatial distributions of the fine fuel; fuel water content; days since last rain; and weather conditions at the moment of the fire (Miranda et al. 1993). The duration of temperatures above 60°C varied from 90 to 270 seconds at 1 cm height, from 90 to 200 seconds at 60 cm height, and from 20 to 70 seconds at 160 cm height. Although 60°C is considered the lethal temperature for plant tissue, Kayll (1968) has shown that lethal temperature varies in relation to the exposure time, with leaves withstanding 49°C for 2 hours, 60°C for 31 seconds, or 64°C for 3 seconds. The duration of temperatures above 60°C in cerrado fires was long enough to kill the leaves exposed to the hot air flow.

Wright (1970) shows that the death of plant tissue depends primarily on moisture content and is an exponential function of temperature and time. Consequently, the heat tolerance of a tree (the ability of a tree's organs to withstand high temperatures) along with its fire resistance (mainly determined by its size, bark thickness, and foliage distribution) may vary with season as a result of the seasonal changes in plant water content. Guedes (1993) and Rocha e Silva and Miranda (1996) have shown that the short duration of the heat pulse and the good insulating effect of thick bark, characteristic of the cerrado trees, provide protection to the cambium so that the increase in cambium temperature during the fires is small. These authors determined a minimum bark thickness of 6–8 mm for effective protection of the cambium tissue. However, in the lower branches, where bark is not thick enough to produce an effective insulation, the cambium may reach high temperatures, remaining over 60°C long enough to cause the death of the cambium tissue, and consequently the death of the branches, altering the structure of the tree canopy.

As a consequence of the short time of residence of the fire front, the increase in soil temperature is small. At 1 cm depth the highest temperatures range from 29°C to 55°C. Soil temperature changes are negligible at and below 5 cm depth, with a maximum increase of 3°C regardless of the physiognomic form of cerrado being burned. The maximum temperatures are registered 10 min after the fire at 1 cm depth and after 4 h at 10 cm depth (Coutinho 1978; Castro Neves and Miranda 1996). The rise in soil temperature during cerrado fires may have little effect on soil organic matter, microbial population, and buried seeds, and also is likely to have little effect on the loss of nutrients from the soil pool.

The reduction of the vegetation cover and the deposition of an ash layer over the soil surface result in a postfire alteration in soil microclimate. Castro Neves and Miranda (1996) found that the albedo (ρ) between 10:00 A.M. and 2:30 P.M. is reduced from 0.11 to 0.03 after a *campo sujo* fire, where 94% of the vegetation was consumed. This decrease in ρ represents a 10% increase in the energy absorbed. One month after the fire, ρ returned to 54% of the prefire value. Soil heat flux (G) changed from 55 W/m² before the fire to 75 W/m² after the fire, representing 7% of the incident solar radiation (Castro Neves and Miranda 1996). The alteration in ρ and G results in an increase in the amplitude of soil temperature after the fire on the order of 30°C at 1 cm depth, and 10°C at 5 cm depth, with no alteration at 10 cm depth (Dias 1994; Castro Neves and Miranda 1996). Although these alterations in soil microclimate may have some effect on plant colonization and soil microorganisms (Frost and Robertson 1987), they are of short duration as a consequence of the fast recovery of the vegetation of the herbaceous layer. Andrade (1998) has shown that 80% of the fuel load of the herbaceous layer of a *campo sujo* is recovered one year after a fire. Neto et al. (1998) have determined that after 2 years the biomass of the herbaceous layer of *campo sujo* has completely recovered from fire.

Bustamante et al. (1998) showed that soil water content (0–10 cm depth) was lower in a burned cerrado *sensu stricto* area than in an adjacent unburned area. The difference in soil water content between the two areas lasted for 15 months, probably as a consequence of plant cover removal and alteration of the vegetation composition. The frequent fires in the area reduced the woody plant density, favoring the recolonization of grasses that used water of the superficial soil layer.

NUTRIENT CYCLING

Nutrient cycling is a very important aspect of fire ecology, especially in the cerrado region where the native vegetation presents a low nutrient content (Haridasan 2000), with low decomposition rate of the litter (Silva 1983), and the soils are poor (see chapter 2). As a consequence of the many different physiognomic forms, floristic composition, soil characteristics, differences in fire regime, and differences in the sampling methods used, there is a great variability in the data regarding the nutrient content of the cerrado vegetation (André-Alvarez 1979; Batmanian 1983; Pivello and Coutinho 1992; Kauffman et al. 1994; Castro 1996).

During a fire the nutrients may be lost by volatilization or as particles deposited in the soil as ash, or remain in the unburned vegetation. The studies of loss of nutrients from the cerrado vegetation during fires suggest that the loss is greatest in *campo limpo* and declines along the gradient from *campo sujo* to *cerradão* (see table 4.2; chapter 6). Silva (1990) studied the partition of biomass and nutrients in the tree layer of a cerrado *sensu stricto*. The woody components of the vegetation are the major pool of nutrients (see table 4.3). In general, the large woody parts of the vegetation do not burn during cerrado fires. The nutrient stock in the leaves of cerrado trees (Silva 1990) is smaller than the stock of the green vegetation of the herbaceous layer (Batmanian 1983, see table 4.4). The difference in nutrient stock and the higher fuel consumption during fires in more open areas of cerrado may explain the decline in nutrient loss with the increase in the density of woody plants.

In general 300 to 400 kg/ha of ash are deposited on the soil surface after a cerrado fire (Coutinho 1990b). Some of the nutrients deposited in

Table 4.2 Loss of Nutrients from Vegetation During Fires
in Different Physiognomic Forms of Cerrado

| Physiognomy | Loss of nutrients (%) | | | | | | Source |
	N	P	K	Ca	Mg	S	
Campo limpo	97	50	60	58	—	16	Kauffman et al. (1994)
Campo limpo	85	—	—	—	—	88	Castro (1996)
Campo limpo	82	72	50	40	—	17	Kauffman et al. (1994)
Campo sujo	81	—	—	—	—	51	Castro (1996)
Campo cerrado	93–97	45–61	29–62	22–71	19–62	43–81	Pivello and Coutinho (1992)
Campo cerrado	66	47	53	67	—	35	Kauffman et al. (1994)
Cerrado sensu stricto	49	47	46	60	—	34	Kauffman et al. (1994)
Cerrado sensu stricto	25	—	—	—	—	44	Castro (1996)
Cerradão	18	—	—	—	—	30	Castro (1996)

Note: See chapter 6 for descriptions of cerrado physiognomies.

the soil are quickly absorbed. Cavalcanti (1978) observed that, immediately after a fire, there was an increase in the concentration of nutrients to a depth of 5 cm, with a significant reduction in the next 3 months. The author observed little alteration in nutrient concentration at greater depths. After a fire in cerrado *sensu stricto*, Batmanian (1983) measured an increase in the concentration of K, Na, Ca and Mg to a depth of 60 cm; no alteration in the concentrations of N and P was observed. The high concentrations of K, Na, Ca, and Mg lasted for 3 months. Batmanian (1983) and Cavalcanti (1978) suggest that most of the nutrients liberated during the fires are absorbed by the superficial roots of the plants of the herbaceous layer. In fact, Dunin et al. (1997), in a study comparing the evapotranspiration of a burned *campo sujo* (3 months after the fire) with an unburned *campo sujo* (1 year since last fire), concluded that almost all the water used by the vegetation of the burned area is removed from the first 1.5 m of the soil. According to Coutinho (1990a), this is the region where the alteration of postfire nutrient concentration has been observed and where the greatest concentration of fine roots is found (Castro 1996).

Table 4.3 Partition of Above-Ground Biomass and Nutrients
in Different Components of the Tree Layer of Cerrado *Sensu Stricto*

| | Partition of biomass and nutrients | | | | | |
	Trunk	Branches	Stems	Leaves	Fruits	Total (kg/ha)
Biomass (kg/ha)	6591	4280	9416	1049	64	21,400
Biomass (%)	30.8	20.0	44.0	4.9	0.3	
Nutrient (kg/ha)						
P	1.3	0.8	1.3	0.8	0.3	4.5
K	6.7	4.8	8.9	6.0	3.3	29.7
Ca	6.5	3.7	9.3	2.9	0.5	22.9
Mg	3.6	1.9	4.0	1.2	0.3	11.0
Al	4.9	2.9	5.9	2.0	0.4	

Source: Silva 1990.
Note: Values represent the mean for 35 woody species and a tree density of 1333 trees/ha.

As discussed before, the rise in soil temperature is small during cerrado fires and is restricted to the first centimeters. Therefore, it may have little effect on the loss of nutrients from the soil pool. Raison (1979) reported a loss of 25% of N in the soil after 2 h at 200°C. No alteration in the N concentration in the 0–20 cm soil layer was recorded by Kauffman et al. (1994) and Kozovits et al. (1996). Further studies of N and S concentration, to a depth of 2 m (Castro 1996), likewise showed no change after cerrado fires. Similar results were reported by Montes and San José (1993) for another neotropical savanna. The loss of nutrients in

Table 4.4 Nutrient Content of the Herbaceous
Layer During the Wet and Dry Seasons in an
Unburned Cerrado *Sensu Stricto*

| | Nutrient content (kg/ha) | | | |
| | Grasses | | Nongrasses | |
Nutrient	Wet season	Dry season	Wet season	Dry season
P	0.9	0.6	0.5	0.3
K	5.9	3.8	6.0	2.5
Ca	0.5	0.5	2.5	1.8
Mg	0.9	1.2	1.1	0.7
Al	3.7	3.0	0.6	0.5

Source: Batmanian 1983.

the system pool (vegetation + soil) is therefore a consequence of the burning of the above-ground biomass representing 3.8% of the system pool (Kauffman et al. 1994).

During cerrado fires the maximum temperatures are around 800°C (Miranda et al. 1993, 1996a), and most of the nutrients are lost by volatilization. Considering that Ca and Mg have high volatilization temperatures, 1240°C and 1107°C, respectively (Wright and Bailey 1982), Coutinho (1990a) assumes that they are lost by particle transport. Kauffman et al. (1994) estimated that about 33% of N, 22% of P and 74% of S are lost by volatilization during cerrado fires. Castro (1996) presented similar values for N (35%) and S (91%).

Although a large proportion of the nutrients is lost from a determined area during a fire, some will return to the ecosystem as dry or wet deposition. Coutinho (1979) reported that for a cerrado area there is an annual total deposition of 2.5 kg/ha of K, 3.4 kg/ha of Na, 5.6 kg/ha of Ca, 0.9 kg/ha of Mg, and 2.8 kg/ha of PO_4. Considering Coutinho's (1979) data on the input of nutrients, Pivello and Coutinho (1992) estimated that the replacement time for P and S lost during burning was far less than 1 year; in the range of 1 to 3.4 years for Ca; 1.6 to 4.1 years for K; and 1 to 5.3 years for Mg. They concluded that an interval of 3 years between burnings was initially considered adequate to stimulate the recycling of the elements retained in the dead plant material and to avoid a critical nutrient impoverishment in the ecosystem. The time interval between burnings suggested by Pivello and Coutinho (1992) was confirmed by Kauffman et al. (1994).

In addition to the return of nutrients through dry and wet deposition, one has also to consider the transfer of nutrients through the decomposition of the scorched leaves that are prematurely dropped after the fire. Silva (1983) determined that in the litter accumulated during 1 year in a cerrado *sensu stricto* area there is 4.8 kg/ha of K, 3.6 kg/ha of Ca, 3.0 kg/ha of Mg, and 0.8 kg/ha of P, and that after 300 days of decomposition there is a reduction of 70%, 55%, and 35% in the initial concentrations of K, Mg, and Ca, respectively. Considering that senescent leaves have a lower nutrient concentration than mature green leaves, the premature drop of scorched leaves may play an important role in the recycling of nutrients caused by fires, even considering the low decomposition rate for the cerrado litter.

Alterations in the carbon cycle have also been observed for cerrado areas submitted to prescribed fires. Burned areas present higher soil CO_2 fluxes than unburned areas, and this effect lasts several months after the fire (Poth et al. 1995). The higher fluxes might be due to the increase of soil organic matter availability in response to the increase of soil pH. The CO_2 fluxes to the atmosphere over *campo sujo* areas under different fire regimes were also studied (Santos 1999; Silva 1999). From June to August the *campo sujo* fixed more CO_2 than was released through respiration. Maximum assimilation rates varied from 2.5 to 0.03 μmol CO_2 m^{-2} s^{-1}. In September the *campo sujo* became a source of CO_2 to the atmosphere, with a maximum emission rate of 1.5 μmol CO_2 m^{-2} s^{-1}. A prescribed fire in late September resulted in an increase of the CO_2 emission to 4.0 μmol CO_2 m^{-2} s^{-1}. In November and December the *campo sujo* again became a sink for CO_2, with the assimilation rate increasing to 15.0 μmol CO_2 m^{-2} s^{-1}. During the seven months of measurements the *campo sujo* accumulated 0.55 t C ha^{-1} (25% of the amount of carbon accumulated in one year by the cerrado *sensu stricto* as determined by Miranda et al. 1996b, 1997).

Burned areas are also a source of trace gases to the atmosphere. Poth et al. (1995) measured soil fluxes of NO, N_2O and CH_4 from cerrado sites that had been burned within the previous 2 days, 30 days, and 1 year, and from a control site last burned in 1976. NO and N_2O fluxes responded to fire with the highest fluxes observed from newly burned sites after addition of water. NO fluxes immediately after burning are among the highest observed for any ecosystem studied to date. However, these rates declined with time after burning, returning to control levels 1 year after the fire. The authors concluded that cerrado is a minor source of N_2O and a sink of atmospheric CH_4.

FIRE EFFECTS ON THE VEGETATION

The flora of the herbaceous/undershrub stratum is highly resistant to fire (Coutinho 1990a). Some plants are annuals, growing in the rainy season, and many species exhibit subterranean organs such as rizomes, bulbs, and xylopodia (Rawitscher and Rachid 1946; chapter 7) that are well insulated by soil. A few days after the fire, the organs sprout with full vigor (Coutinho 1990a).

Many plant species appear to depend upon fire for sexual reproduction. Intense flowering can be observed a few days or weeks after cerrado fires for many species of the herbaceous layer. Oliveira et al. (1996) observed 44 species of terrestrial orchids flowering after fires in areas of cerrado *sensu stricto*, *campo sujo*, and *campo limpo*, with some of the species flowering in the first two weeks after fire. Intense flowering of *Habenaria armata* was observed just after an accidental fire, while in the four preceding years, when the vegetation was protected, no flowering individuals could be observed. Coutinho (1976) observed that a great number of species depend on fire to flower, responding with intense flowering to burns occurring any season of the year. In experiments with four species of the herbaceous/undershrub stratum (*Lantana montevidensis, Stylosanthes capitata, Vernonia grandiflora* and *Wedelia glauca*), Coutinho (1976) showed that burning, cutting the plants close to the soil, exposing them to a period of drought, or causing the death of their epigeous parts all resulted in a high percentage of flowering. He concluded that the effect of fire on the induction of flowering is not a result of thermal action or fertilization by the ashes. In a comparative study on the effects of fire and clipping on the flowering of 50 species of the flora of the herbaceous layer of a *campo sujo*, Cesar (1980) concluded that, for most of the species studied, flowering was independent of the season of burning and resulted in a similar phenological response in both treatments. Fire-induced flowering has been frequently reported, especially for grasses and geophytic lilies and orchids (Gill 1981). The causes of intense flowering may be related both to the increase in productivity after fire and to the damage caused by fire to the above-ground plant parts, possibly stimulating the production of flower primordia (Whelan 1995).

Regeneration after fire for savanna vegetation through germination of soil stored seeds has been reviewed by Frost and Robertson (1987), but few studies report on seed dispersal in relation to fire (Whelan 1986). For the herbaceous vegetation of cerrado, Coutinho (1977) observed that *Anemopaegma arvense, Jacaranda decurrens, Gomphrena macrocephala* and *Nautonia nummularia* dispersed their seeds shortly after fire. This suggests that fire may be beneficial to such species, since it promotes or facilitates the dispersal of their anemocoric seeds. Seed germination of *Echinolaena inflexa* (a C3 grass common in all cerrado forms) was higher after a mid-dry season quadrennial fire in a *campo sujo* than in an area protected from fire for 21 years. However, recolonization by vegetative growth was higher than by seeds in both areas (Miranda 1996). After the fire, the density of *E. inflexa* was twice the density determined for the unburned area. Parron (1992) observed no difference in the density of *E. inflexa* in a *campo sujo* burned annually for 3 years, at the beginning of the dry season, and an adjacent area protected from fire for 3 years. These results may reflect the interaction between fire regime and the reproductive strategies of *E. inflexa*. San José and Farinas (1991), in a long-term monitoring of density and species composition in a *Trachypogon* savanna, showed that the dominant species, *Trachypogon plumosus*, was replaced by *Axonopus canescens* when fire was suppressed. The alteration in species density was associated with differences in reproductive strategies: *T. plumosus* presented vegetative reproduction; *A. canescens,* sexual reproduction. For *A. canescens*, fire suppression may increase the probability of seedling survival.

Most of the woody species of the cerrado present strong suberization of the trunk and branches, resulting in an effective thermal insulation of the internal living tissues of those organs during fires. Nevertheless, plants differ greatly in their tolerance to fire and in their capacity to recover subsequently.

Most of the work on the response of savanna woody vegetation to fire is related to mortality, regeneration through seedlings, or resprouting from epicormic meristems or lignotubers (Frost and Robertson 1987). Landim and Hay (1996) reported that fire damaged 79% of the fruits of *Kielmeyera coriacea*, irrespective of tree height (1 to 3 m), but there were no differences in flower and bud initiation in the next reproductive period. Hoffmann (1998) observed that fruits and seeds of *Miconia albicans, Myrsine guianensis, Roupala montana, Periandra mediterranea, Rourea induta*, and *Piptocarpha rotundifolia* were damaged by a late dry season biennial fire, with a negative impact on sexual reproduction. All species but *P. rotundifolia* exhibited overall reductions in seed production in the years following fire (see chapter 9). An increase in the reproductive success for *Byrsonima crassa* after a mid-dry-season fire was reported by Silva et al. (1996).

Although most of the cerrado trees are well insulated by thick bark, the small individuals may not have produced an effective insulation between fires, being more susceptible to the effects of the high temperature of the flames (Guedes 1993). Consequently, frequent fires reduce the density of woody vegetation through the mortality of the smaller individuals (Frost and Robertson 1987) and through the alteration of the regeneration rate of the woody species (Hoffmann 1998; Matos 1994; Miyanish and Kellman 1986).

In a study of the effects of a biennial fire regime on the regeneration of *Blepharocalyx salicifolius* in cerrado *sensu stricto*, Matos (1994) found twice the number of individuals (seedlings and juveniles) in an area protected from fire for 18 years than in an area that was burned biennially. The mortality caused by the biennial fires was greater than 90% for seedlings and less than 50% for juveniles. The author estimated that the critical size for survival and resprout after fire was 50 cm in height with 0.6 cm in basal diameter. Hoffmann (1998), investigating the postburn reproduction of woody plants in areas subjected to biennial fires, observed that fire caused a high mortality in seedlings of *Miconia albicans* (100%), *Myrsine guianensis* (86%), *Roupala montana* (64%), *Periandra mediterranea* (50%), and *Rourea induta* (33%). Root suckers of *M. guianensis, R. montana,* and *R. induta* had a higher survival rate, perhaps a consequence of their stem diameters (1.7 mm to 2.4 mm, two to four times greater than the seedlings) and connection with the mother plant (chapter 9).

The effect of two annual fires on small individuals (from 20 cm to 100 cm in height and diameter greater than 1.5 cm, at 30 cm from the soil) was investigated by Armando (1994) for nine woody species: *Aspidosperma dasycarpon, Blepharocalyx salicifolius, Caryocar brasiliense, Dalbergia miscolobium, Hymenaea stigonocarpa, Stryphnodendron adstringens, Sclerolobium paniculatum, Siphoneugena densiflora,* and *Virola sebifera.* The two consecutive fires resulted in a reduction of 10 cm in the mean height of the plant community and in a mortality of 4%. Only four species presented reduction in the number of individuals: *D. miscolobium* (12%), *S. adstringens* (14%), *S. densiflora* (14%), and *S. paniculatum* (15%).

Sato (1996) determined that, for woody vegetation submitted to a biennial fire regime in the middle of the dry season, after 18 years of protection, the highest mortality rate occurred among the individuals with height between 0.3 m and 2.0 m. After the first fire they accounted for 40% of the mortality, and for 72% after the second fire. Ramos (1990) observed that young trees and shrubs up to 128 cm tall, and with diameter smaller than 3 cm (measured at 30 cm above ground), are seriously damaged by biennial fires.

Mortality rates related to fire season have been reported by Sato and Miranda (1996) and Sato et al. (1998) for cerrado vegetation. The authors considered only the individuals with stem diameter equal to or greater than 5.0 cm, at 30 cm from soil surface. After 18 years of protection from fire, three biennial fires at the middle of the dry season resulted in mortality rates of 12%, 6%, and 12%, with a final total reduction in the number of individuals of 27% after the third fire. In an experimental plot burned at the end of the dry season, the mortality rates were 12%, 13%, and 19%, with a reduction of 38% in the number of individuals. Williams (1995) presents similar values for mortality of tropical savanna trees in Australia, and higher rates are presented by Rutherford (1981) and Frost and Robertson (1987) for species of African savanna. Most of the trees that died in the second fire (≈60%) had suffered top kill during the first fire after a long period of protection. Some of the mortality following the second and third fires may be an indirect effect of fire. Cardinot (1998), studying the sprouting of *Kielmeyera coriacea* and *Roupala montana* after fires, in the same experimental plots, reported that the mortality of some trees is a consequence of herbivory and nutrient shortage. The higher mortality rates for late dry season fires may be related to the phenology of many species of the cerrado vegetation that launch new leaves, flowers, and fruits during the dry season (Bucci 1997).

Similar results have been reported by Rocha e Silva (1999) for *campo sujo*. After protection from fire for 18 years, three biennial fires, at the middle of the dry season, resulted in tree and shrub mortality rates of 5%, 8% and 10%, reducing the number of individuals by 20%. In an experimental plot burned for four years at the middle of the dry season, mortality rates were 10% and 12%, with a reduction of 20% of the number of individuals, suggesting that two quadrennial burns produce the same mortality as three biennial burns. Of the 30 woody species present in the experimental plots, only 7 did not suffer alteration in the number of individuals after the fires: *Byrsonima verbascifolia*, *Caryocar brasiliensis*, *Eremanthus mollis*, *Eriotheca pubescens*, *Qualea parviflora*, *Syagrus comosa*, and *Syagrus flexuosa*.

The alteration in the regeneration rates of woody species and the high mortality rate determined in these studies suggest that the biennial fire regime is changing the physiognomies of cerrado *sensu stricto* and *campo sujo* to an even more open form, with grasses as the major component of the herbaceous layer. This alteration, in turn, favors the occurrence of more intense and frequent fires.

REFERENCES

Andrade, S. M. A. 1998. "Dinâmica do Combustível Fino e Produção Primária do Estrato Rasteiro de Áreas de Campo Sujo de Cerrado Submetidas a Diferentes Regimes de Queima." Master's thesis, Universidade de Brasília, Brasília, Brazil.

André-Alvarez, M. 1979. "Teor de Nutrientes Minerais na Fitomassa do Estrato Herbáceo Subarbustivo do Cerrado de Emas (Pirassununga, Est. de São Paulo)." Master's thesis, Universidade de São Paulo, São Paulo, Brazil.

Armando, M. S. 1994. "O Impacto do Fogo na Rebrota de Algumas Espécies de Árvores do Cerrado." Master's thesis, Universidade de Brasília, Brasília, Brazil.

Batmanian, G. J. 1983. "Efeitos do Fogo na Produção Primária e a Acumulação de Nutrientes no Estrato Rasteiro de um Cerrado." Master's thesis, Universidade de Brasília, Brasília, Brazil.

Bucci, F. F. B. 1997. "Floração de Algumas Espécies de Melastomataceae do Distrito Federal: Uso de Dados de Herbário para Obter Padrões Sazonais." Master's thesis, Universidade de Brasília, Brasília, Brazil.

Bustamante, M. M. C., F. B. Nardoto, A. A. Castro, C. R. Garofalo, G. B. Nardoto, and M. R. S. Silva. 1998. Effect of prescribed fires on the inorganic-N concentration in soil of cerrado areas and on the assimilation of inorganic-N by woody plants. In D. X. Viegas, ed., *Proceedings of the 14th Conference on Fire and Forest Meteorology* 2:1361–1379. Coimbra: University of Coimbra, Portugal.

Cardinot, G. K. 1998. "Efeitos de Diferentes Regimes de Queima nos Padrões de Rebrotamento de Kielmeyera coriacea Mart. e Roupala montana Aubl:, Duas Espécies Típicas do Cerrado." Master's thesis, Universidade de Brasília, Brasília, Brazil.

Castro, E. A. 1996. "Biomass, Nutrient Pools and Response to Fire in the Brazilian Cerrado." Master's thesis, Oregon State University, Corvallis, U.S.

Castro, E. A. and J. B. Kauffman. 1998. Ecosystem structure in the Brazilian Cerrado: A vegetation gradient of aboveground biomass, root mass and consumption by fire. *J. Trop. Ecol.* 14:263–283.

Castro Neves, B. M. and H. S. Miranda. 1996. Efeitos do fogo no regime térmico do solo de um campo sujo de Cerrado. In H. S. Miranda, C. H. Saito, and B. F. S. Dias, (eds.), *Impactos de Queimadas em Áreas de Cerrado e Restinga*, pp.20–30. Brasília: ECL/Universidade de Brasília.

Cavalcanti, L. H. 1978. "Efeito das Cinzas Resultantes da Queimada sobre a Produtividade do Estrato Herbáceo Subarbustivo do Cerrado de Emas." Ph.D. thesis, Universidade de São Paulo, São Paulo, Brazil.

Cesar, L. 1980. "Efeitos da Queima e Corte sobre a Vegetação de um Campo Sujo na Fazenda Água Limpa, Brasília–DF." Master's thesis, Universidade de Brasília, Brasília, Brazil.

Cooke, R. 1998. Human settlement of central America and northernmost South America (14,000–8000 BP). *Quat. Int.* 49/5:177–190.

Coutinho, L. M. 1976. "Contribuição ao Conhecimento do Papel Ecológico das Queimadas na Floração de Espécies do Cerrado." Livre Docente's thesis, Universidade de São Paulo, São Paulo, Brazil.

Coutinho, L. M. 1977. Aspectos ecológicos do fogo no cerrado: II. As queimadas e a dispersão de sementes de algumas espécies anemocóricas do estrato herbáceo subarbustivo. *Bol. Bot.,USP* 5:57–64.

Coutinho, L. M. 1978. Aspectos ecológicos do fogo no cerrado: I. A temperatura do solo durante as queimadas. *Rev. Bras. Bot.* 1:93–97.

Coutinho, L. M. 1979. Aspectos ecológicos do fogo no cerrado: III. A precipitação atmosférica de nutrientes minerais. *Rev. Bras. Bot.* 2:97–101.

Coutinho, L. M. 1981. Aspectos ecológicos do fogo no cerrado: Notas sobre a ocorrência e datação de carvões vegetais encontrados no interior do solo, em Emas, Pirassunga, S.P. *Rev. Bras. Bot.* 4:115–117.

Coutinho, L. M. 1990a. Fire in the ecology of the Brazilian Cerrado. In J. G. Goldammer, ed., *Fire in the Tropical Biota: Ecosystem Processes and Global Challenges*, pp.82–105. Berlin: Springer Verlag.

Coutinho, L. M. 1990b. O Cerrado: A ecologia do fogo. *Ciência Hoje* 12:22–30.

De Oliveira, P. E. 1992. "A palynological record of Late Quarternary vegetation and climatic change in southeastern Brazil." Ph. D. thesis, Ohio State University, Columbus, Ohio, U.S.A.

Dias, B. F. S. 1992. Cerrados: Uma caracterização. In B. F. S. Dias, ed., *Alternativas de Desenvolvimento dos Cerrados: Manejo e Conservação dos Recursos Naturais Renováveis*. Brasília: IBAMA, Ministério do Meio Ambiente.

Dias, I. F. O. 1994. "Efeitos da Queima no Regime Térmico do Solo e na Produção Primária de um Campo Sujo de Cerrado." Master's thesis, Universidade de Brasília, Brasília, Brazil.

Dunin, F. X., H. S. Miranda, A. C. Miranda, and J. Lloyd. 1997. Evapotranspiration responses to burning of campo sujo savanna in Central Brazil. *Proceedings of the Bushfire'97*, pp. 146–151. Australia: CSIRO.

Frost, P. H. G. and F. Robertson. 1987. The ecological effects of fire in savannas. In B. H. Walker, ed., *Determinants of Tropical Savannas*, pp. 93–141. Oxford: IRL Press Limited.

Gidon, N. and G. Delibrias. 1986. Carbon-14 dates point to man in the Americas 32,000 years ago. *Nature* 321:769–771.

Gill, A. M. 1981. Adaptative response of Australian vascular plant species to fires. In A. M. Gill, R. H. Groves, and I. Noble, eds., *Fire and the Australian Biota*, pp. 243–272. Canberra: Australian Academy of Science.

Guedes, D. M. 1993. "Resistência das Árvores do Cerrado ao Fogo: Papel da Casca como Isolante Térmico." Master's thesis, Brasília, Universidade de Brasília, Brazil.

Haridasan, M. 2000. Nutrição mineral de plantas nativas do Cerrado. *Rev. Bras. Fisiol. Vegetal* 12:54–64.

Hoffmann, W. 1998. Post-burn reproduction of woody plants in a neotropical savanna: The relative importance of sexual and vegetative reproduction. *J. Appl. Ecol.* 35:422–433.

Kauffman, J. B., D. L. Cummings, and D. E. Ward. 1994. Relationships of fire, biomass and nutrient dynamics along a vegetation gradient in the Brazilian cerrado. *J. Ecol.* 82:519–531.

Kayll, A. J. 1968. Heat tolerance of tree seedlings. *Proceedings of the Tall Timbers Fire Ecology Conference.* 3:88–105.

Kelmann, M., K. Miyanishi, and P. Hiebert. 1987. Nutrient retention by savanna ecosystem. *J. Ecol.* 73:953–962.

Kozovits, A. R., M. M. C. Bustamante, L. F. Silva, G. T. Duarte, A. A. Castro, and J. R. Magalhães. 1996. Nitrato de amônio no solo e sua assimilação por espécies lenhosas em uma área de cerrado submetida a queimadas prescritas. In H. S. Miranda, C. H. Saito and B. F. S. Dias, eds., *Impactos de Queimadas em Áreas de Cerrado e Restinga*, pp. 137–147. Brasília: ECL/Universidade de Brasília.

Landim, M. F. and J. D. Hay. 1996. Impacto do fogo sobre alguns aspectos da biologia reprodutiva de *Kielmeyera coriacea* Mart. *Rev. Bras. Biol.* 56:127–134.

Luke, R. H. and A. G McArthur. 1978. *Bushfire in Australia*. Canberra: Australian Government Publishing Service.

Matos, R. B. M. 1994. "Efeito do Fogo sobre Regenerantes de Blepharocalyx salicifolius (H.B.K.) Berg. (Myrtaceae) em Cerrado Aberto, Brasília, DF." Master's thesis, Universidade de Brasília, Brasília, Brazil.

Miranda, A. C., H. S Miranda, I. F. O. Dias, and B. F. S Dias. 1993. Soil and air temperatures during prescribed cerrado fires in Central Brazil. *J. Trop. Ecol.* 9:313–320.

Miranda, A. C., H. S Miranda, J. Grace, J. Lloyd, J. McIntyre, P. Meier, P. Riggan, R. Lockwood, and J. Brass. 1996b. Fluxes of CO_2 over a Cerrado *sensu stricto* in Central Brazil. In J. H. Gash, C. A. Nobre, J. M. Roberts, and R. L. Victoria, eds., *Amazonian Deforestation and Climate*, pp. 353–363. Chichester: John Wiley and Sons.

Miranda, A. C., H. S Miranda, J. Lloyd, J. Grace, J. A. Francey, J. McIntyre, P. Meier, P. Riggan, R. Lockwood, and J. Brass. 1997. Fluxes of carbon, water and energy over Brazilian cerrado: An analysis using eddy covariance and stable isotopes. *Plant Cell Environ.* 20:315–328.

Miranda, H. S. 2000. Queimadas de Cerrado: Caracterização e impactos na vegetação. In *Plano de Prevenção e Combate aos Incêndios Florestais do DF*, pp.133–149. Brasília: Secretaria de Meio Ambiente e Recursos Hídricos.

Miranda, H. S., E. P. Rocha e Silva, and A. C. Miranda. 1996a. Comportamento do fogo em queimadas de campo sujo. In H. S. Miranda, C. H. Saito and B. F. S. Dias, eds., *Impactos de Queimadas em Áreas de Cerrado e Restinga*, pp. 1–10. Brasília: ECL/Universidade de Brasília.

Miranda, M. I. 1997. "Colonização de Campo Sujo de Cerrado por Echinolaena inflexa (Poaceae)." Master's thesis, Universidade de Brasília, Brasília, Brazil.

Mistry, J. 1998. Fire in the cerrado (savannas) of Brazil: An ecological review. *Prog. Phys. Geog.* 22:425–448.

Miyanish, K. and M. Kellman. 1986. The role of fire in the recruitment of two neotropical savanna shrubs, *Miconia albicans* and *Clidemia sericea*. *Biotropica* 18:224–230.

Montes, R. and J. J. San José. 1993. Ion movement in the well drained neotropical savanna. *Curr. Top. Bot. Res.* 1:391–418.

Neto, W. N., S. M. A. Andrade, and H. S. Miranda. 1998. The dynamics of the herbaceous layer following prescribed burning: a four year study in the Brazilian savanna. In D. X. Viegas, ed., *Proceedings of the 14th Conference on Fire and Forest Meteorology* 2:1785–1792. Coimbra: University of Coimbra, Portugal.

Oliveira, R. S., J. A. N. Batista, C. E. B. Proença, and L. Bianchetti. 1996. Influência do fogo na floração de espécies de *Orchidaceae* em cerrado. In H. S. Miranda, C. H. Saito, and B. F. S. Dias, eds., *Impactos de Queimadas em Áreas de Cerrado e Restinga*, pp.61–67. Brasília: ECL/Universidade de Brasília.

Parron, L. M. 1992. "Dinâmica de Crescimento, Sobrevivência, Produção de Sementes, Repartição de Biomassa Aérea e Densidade das Gramíneas Echinolaena inflexa e Trachypogon folifolius, numa Comunidade de Campo Sujo, Com e Sem Queima." Master's thesis, Universidade de Brasília, Brasília, Brazil.

Pivello, V. R. and L. M. Coutinho. 1992. Transfer of macro-nutrients to the atmosphere during experimental burnings in an open cerrado (Brazilian savanna). *J. Trop. Ecol.* 8:487–497.

Poth, M., I. C. Anderson, H. S. Miranda, A. C. Miranda, and P. J. Riggan. 1995. The magnitude and persistence of soil NO, N_2O, CH_4, and CO_2 fluxes from burned tropical savanna in Brazil. *Global Geoch. Cycles* 9:503–513.

Prous, A. 1992. *Arqueologia Brasileira.* Brasília: Editora Universidade de Brasília.

Raison, R. J. 1979. Modification of the soil environment by vegetation fires, with particular reference to nitrogen transformations: A review. *Plant and Soil* 51:73–108.

Ramos, A. E. 1990. "Efeitos da Queima sobre a Vegetação Lenhosa do Cerrado." Master's thesis, Universidade de Brasília, Brasília, Brazil.

Rawitscher, F. and M. Rachid. 1946. Tronco subterrâneos de plantas brasileiras. *An. Acad. Bras. Ciênc.* 18:261–280.

Rocha e Silva, E. P. and H. S. Miranda. 1996. Temperatura do câmbio de espécies lenhosas do cerrado durante queimadas prescritas. In R. C. Pereira and L. C. B Nasser, eds., *Anais do VII Simpósio sobre o Cerrado/1st International Symposium on Tropical Savannas*, pp.253–257. Brasília: Empresa Brasileira de Pesquisa Agropecuaria.

Rocha e Silva, E. P. 1999. "Efeito do Regime de Queima na Taxa de Mortalidade e Estrutura da Vegetação Lenhosa de Campo Sujo de Cerrado." Master's thesis, Universidade de Brasília, Brasília, Brazil.

Rutherford, M. C. 1981. Survival, regeneration and leaf biomass changes in a woody plants following spring burns in *Burkea africana-Ochna pulchra* savanna. *Bothalia* 13:531–552.

San José, J. J. and M. R. Farinas. 1991. Temporal changes in the structure of a *Trachypogon* savanna protected for 25 years. *Acta Oecol.* 12:237–247.

San José, J. J. and E. Medina. 1977. Producción de metéria organica en la sabana de Trachypogon, Calabozo, Venezuela. *Bol. Soc. Ven.Cienc. Nat.* 134:75–100.

Salgado-Laboriau, M. L. and K. R. F. C. Vicentini. 1994. Fire in the Cerrado 32,000 years ago. *Curr. Res. Pleist.* 11:85–87.

Santos, A. J. B. 1999. "Fluxos de Energia, Carbono e Água em Áreas de Campo Sujo." Master's thesis, Universidade de Brasília, Brasília, Brazil.

Sato, M. N. 1996. "Taxa de Mortalidade da Vegetação Lenhosa do Cerrado Submetida a Diferentes Regimes de Queima." Master's thesis, Universidade de Brasília, Brasília, Brazil.

Sato, M. N. and H. S. Miranda. 1996. Mortalidade de plantas lenhosas do cerrado *sensu stricto* submetidas a diferentes regimes de queima. In H. S. Miranda, C. H. Saito, and B. F. S. Dias, eds., *Impactos de Queimadas em Áreas de Cerrado e Restinga*, pp.102–111. Brasília: ECL/Universidade de Brasília.

Sato, M. N., A. A. Garda, and H. S. Miranda. 1998. Effects of fire on the mortality of woody vegetation in Central Brazil. In D. X. Viegas, ed., *Proceedings of the 14th Conference on Fire and Forest Meteorology* 2:1777–1784. Coimbra: University of Coimbra, Portugal.

Silva, D. M. S., J. D. Hay, and H. C. Morais. 1996. Sucesso reprodutivo de *Byrsonima crassa* (Malpighiaceae) após uma queimada em um cerrado de Brasília–DF. In H. S. Miranda, C. H. Saito, and B. F. S. Dias, eds., *Impactos de Queimadas em Áreas de Cerrado e Restinga*, pp. 122–127. Brasília: ECL/Universidade de Brasília.

Silva, F. C. 1990. "Compartilhamento de Nutrientes em Diferentes Componentes da Biomassa Aérea em Espécies Arbóreas de um Cerrado." Master's thesis, Universidade de Brasília, Brasília, Brazil.

Silva, G. T. 1999. "Fluxos de CO_2 em um Campo Sujo Submetido a Queimada Prescrita." Master's thesis, Universidade de Brasília, Brasília, Brazil.

Silva, I. S. 1983. "Alguns Aspectos da Ciclagem de Nutrientes em uma Área de Cerrado (Brasília: DF): Chuva, Produção e Decomposição de Liter." Master's thesis, Universidade de Brasília, Brasília, Brazil.

Vicentini, K. R. C. F. 1993. "Análise palinológica de uma vereda em Cromínia-GO." Master's thesis, Universidade de Brasília, Brasília, Brazil.

Whelan, R. J. 1986. Seed dispersal in relation to fire. In D. R. Murray, ed., *Seed Dispersal*, pp. 237–271. Sydney: Academic Press.

Whelan, R. J. 1995. *The Ecology of Fire*. Cambridge, Mass: Cambridge University Press.

Williams, R. J. 1995. Tree mortality in relation to fire intensity in a tropical savanna of the Kakadu region, Northern Territory, Australia. *CalmSci. Suppl.* 4:77–82.

Wright, S. J. 1970. A method to determine heat-caused mortality in bunchgrass. *Ecology* 51:582–587.

Wright, S. J. and A. W. Bailey. 1982. *Fire Ecology*. New York: John Wiley and Sons.

5

Past and Current Human Occupation, and Land Use

Carlos A. Klink and Adriana G. Moreira

BIT BY BIT, THE CERRADO LOSES GROUND

"Drive south from Rondonópolis, and for mile after mile the flat table-land stretches away to the far horizon, a limitless green prairie carpeted with swelling crops. The monotony of the landscape is broken only by the artifacts of modern agribusiness: a crop-dusting plane swoops low over the prairie to release its chemical cloud, while the occasional farmhouses have giant harvesting machines lined up in the yard outside. It could be the mid-western United States. In fact, it is the very heart of tropical South America, its central watershed, in the Brazilian State of Mato Grosso." That is how a major newspaper (*The Economist* 1999) has recently described the cerrado landscape in central Brazil.

Over the past four decades, the Cerrado Biome has become Brazil's largest source of soybeans and pastureland, and a significant producer of rice, corn, and cotton. In contrast to the small farms in other parts of Brazil, a very different kind of farming, capital-intensive, large-scale, mechanized, and scientific, has developed in the poor soils but cheap land of the cerrado.

From a narrow, revenue generation perspective, the benefits of commercial agriculture in the cerrados are clear. Soybeans and soy products are among the largest of Brazil's export commodities, and the cerrados support the largest cattle herd in the country. Even so, the development of modern agriculture in the cerrado region has exacerbated social

69

inequality at a high environmental cost: landscape fragmentation, loss of biodiversity, biological invasion, soil erosion, water pollution, land degradation, and heavy use of chemicals (Klink et al. 1993, 1995; Davidson et al. 1995; Conservation International et al. 1999; see chapter 18).

Transformation of the cerrado landscapes continues at a fast pace. Here we describe these changes and discuss their implications for sustainable use and conservation. We start with a brief historical overview of past and current human occupation and then explain the main driving forces behind agricultural expansion and the recent transformation of the cerrados. We end with a look at future land use in the cerrado region, offering recommendations for progressive management and conservation.

HUMAN OCCUPATION

Presettlement Inhabitants

Archaeological evidence indicates that 9,000 YBP (Years Before Present) a hunter-gatherer culture flourished in the open habitats of the cerrado. Excavations near the town of Itaparica (state of Bahia), and later from other sites (especially in the town of Serranópolis, state of Goiás), revealed chipped stone tools and rock paintings from cave campsites (Schmitz 1992; Barbosa and Nascimento 1994). Stratified prehistoric deposits uncovered in Serranópolis revealed cerrado tree fruits and wood, and faunal remains, to 8,800 and 6,500 YBP (Schmitz 1992).

The first known cerrado humans, known as the Itaparica Tradition (Schmitz 1992), were foragers who used simple tools and subsisted on native plants and game, such as deer, armadillos, and lizards. Paintings on cave walls typically depict animals but rarely show human figures (Schmitz 1992). The existence of cerrado cave dwellers, and related cultures in the northeastern *caatinga* dry lands and in the Amazon rainforest, are changing our understanding of South American paleoindians (Guidon 1991; Roosevelt et al. 1996)

The Itaparica Tradition persisted until 6,500 years YBP, when it was replaced by "specialized" hunters and gatherers subsisting primarily on small animals and terrestrial mollusks (Barbosa and Nascimento 1994). Why the Itaparica Tradition was replaced by what became the modern indigenous population is unknown. It has been hypothesized that the regional climate may have become wetter and permitted dispersion to other areas in Brazil (Barbosa and Nascimento 1994; see chapter 3).

The last millennium has been characterized by sedentary indigenous populations that hunted and made use of utensils and agriculture until the arrival of the first Europeans. In all, 230 indigenous societies speaking 170 distinct languages have been identified in Brazil. Culturally distinct societies have occupied the cerrado region, including the Bororo, Karajá, Parakanã, Kayapó, Canela, Krahô, Xavante and Xerente.

At the time of Anglo-European colonization in the 16th century, the Xavante and the Xerente societies occupied extensive areas in the cerrado (Maybury-Lewis 1988). They foraged on native plant fruits and roots, fished, and hunted. They also used cerrado plants for home and utensil construction. The Xavante set fire to the native grasslands to facilitate hunting of cerrado game (Leeuwenberg and Salimon 1999). The number of indigenous people in the cerrado decreased dramatically since the contacts with the European colonizers. Initially they were decimated by disease or enslavement; more recently, the advance of the agricultural frontier has displaced many from their native lands.

Postsettlement Occupation

The idea of central Brazil as a region to be conquered and transformed has been embedded in the Brazilian society since colonial times. Cerrado exploration started with the Portuguese who searched for precious minerals and Indians for enslavement in the sixteenth century. The first permanent settlements were established in the early eighteenth century and were associated with gold mining. Some farming developed among these communities, and economic activity shifted to cattle production with the exhaustion of the mines (Klink et al. 1993, 1995).

It was only after the Paraguay War (1864–1870) that cerrado occupation was promoted by the Brazilian authorities, concerned with the defense and maintenance of the border, challenged by low population density. The government encouraged occupation of the border province of southern Mato Grosso by providing incentives to grow tea (Almeida and Lima 1956). The existence of large areas of native grassland also made the area attractive for cattle raising and the development of large ranches.

The occupation of the core area of the cerrado region was delayed until the 1900s. The first major economic boom in the cerrados came during the period 1920–1930, when coffee growing and processing industries were flourishing in the state of São Paulo, which hence became the major market for cerrado cattle (Hees et al. 1987). Later, the Getulio Vargas government (1930–1945) actively promoted a colonization of southern Goiás, providing land, subsidies, and technical assistance, thus encouraging farmers to settle on and clear fertile forested lands.

Infrastructure

Both the distance of the cerrado from the major Brazilian coastal urban areas and the lack of a transport system have posed major obstacles for cerrado occupancy and development. Construction of the first railway, which linked São Paulo to Mato Grosso, was initiated in 1905 but not completed until 1947 (Lucarelli et al. 1989). After 1946, roads replaced railways as the main link between the Brazilian regions. In conjunction with the construction of Brazil's new capital, Brasília, in the cerrado heartland in the late 1950s, roads crossing the cerrado were built to connect Brasília with São Paulo, Rio de Janeiro, and Belo Horizonte in the southeast, and Belém in the Amazon region (Lucarelli et al. 1989). The construction of Brasília and highways linking the new capital with the main Brazilian cities made way for the cerrado occupation that began in the mid-1960s. At the time of this writing, the construction of railways is again becoming fashionable. Most noticeable is Ferronorte, a newly built railway linking the cerrado to Brazil's largest port, Santos, in the state of São Paulo.

Population Growth and Urbanization

The population of the cerrado region grew by 73% between 1950 and 1960, mostly due to employment opportunities associated with the construction of Brasília (Lucarelli et al. 1989). The population growth is not limited to this decade, however, since from 1870 to 1960 the regional population grew at a rate twice that of Brazil, as a result of internal migration. Preliminary data of the national demographic survey done in 2000 (IBGE 2001) indicate that the cerrado population may have reached 18 million inhabitants in 2000 (see table 5.1) There has been a strong trend towards urbanization since the 1940s, particularly in the southern part of the cerrado. As of 2000, almost 30% of the cerrado inhabitants lived in eight cities: Brasília, Goiânia, Teresina, Campo Grande, Uberlândia, Cuiabá, Montes Claros, and Uberaba. Internal migration is the main cause of population growth in urban areas. The population of Palmas, the state capital of Tocantins, has grown 12.2% since 1996 (IBGE 2001), attracted by the ongoing construction of the city.

Population growth and agricultural development had important implications for cerrado land use. Until the late 1950s, the contribution of the cerrados to Brazil's agricultural output was very low, with the extent of farmland and agricultural output contributing <10% of the national total. This state of affairs changed dramatically after the 1960s,

Table 5.1 The Population
of Cerrado from 1872 to 2000

Year	Population
1872	221,000
1890	320,000
1900	373,000
1920	759,000
1940	1,259,000
1950	1,737,000
1960	3,007,000
1970	5,167,000
1980	7,545,000
1991	12,600,000
2000	18,000,000

Sources: Klink et al. 1993, 1995, and
IBGE 2001, preliminary data.

when the cerrados became Brazil's major producer and exporter of important cash crops.

LAND USE

Changes in land use in the cerrado are a function of the technological innovations, capital investments, energy, and knowledge applied with the objective of promoting the expansion of intensive agriculture. Until 40 years ago, the region was used primarily for extensive cattle raising. At the time of this writing it is estimated that 35% of its natural cover has been totally converted into planted pastures with African grasses and cash crops, mainly soybeans and corn (see table 5.2). It is estimated that 60% of the cerrado area is used by humans directly (Conservation International et al. 1999). In 1970, 202,000 km^2 of land (an area 2.2 times the size of Portugal) was used for intensive crops and planted pastures in the region. The area cleared had grown 3.3 times since then, and in 1996 it was equivalent to 672,000 km^2 (table 5.2), an area the size of Texas in the U.S.

The total area, and sometimes the annual rate, of native vegetation clearing is greater in the cerrado than in the Brazilian Amazon rainforest. For example, between 1970 and 1975 the average annual rate of land clearing in the cerrado region was 40,600 km^2 per year, 1.8 times the estimated deforestation rate of the Amazon rainforest during the period 1978–1988 (see fig. 5.1). Projections for the year 2000 show that the total

Table 5.2 Changes in Land Cover and Land Use
in the Cerrado, 1970–1996

	1970	1985	1995/1996[a]
Total cleared land (km^2)	202,000	508,000	672,000
Crops (km^2)	41,000	95,000	103,000
Planted pastures (km^2)	87,000	309,000	453,000
Cleared but not in use	74,000	104,000	116,000[b]
Proportion of total cerrado area cleared	10.6%	26.7%	33.6%

[a]Dates of last national agriculture survey.
[b]Estimated from Klink et al. (1995).

cleared land in the cerrados can reach 800,000 to 880,000 km^2, roughly 1.6 times the size of France (Klink et al. 1995). During the 25-year period shown in table 5.2, the area under crops had grown 250%, planted pasture 520%, and land that had been cleared but had not been cultivated, 150%. This "uncultivated" land represents productive land that had been cleared in the past but either had never been used or had been abandoned.

Several environmental and economical conditions favored these transformations. Although the rainfall distribution within the year is uneven, the mean total annual rainfall (1500 mm) is considered sufficient for crop production. Temperatures are warm year-round, and sunshine does not restrict photosynthesis. Level topography and deep well-drained soils propitiate mechanization, and the cerrado savannas and woodlands are less expensively and more easily cleared for farming or cattle ranching than tropical rainforest.

PROMOTION OF AGRICULTURAL EXPANSION

The growth of agriculture in the Cerrado Biome is the outcome of a combination of factors, including the growth of the demand for agricultural products in Brazil and abroad, public investments in infrastructure, technological advances in the agronomic sciences, and the implementation of policies for regional development, particularly during the rapid growth period of 1968–1980. The strong performance of the Brazilian economy, associated with a national development policy aimed at integrating the "empty" spaces of the cerrado and Amazon regions into the capitalist economy of the richer southern and southeastern regions of the country, created the right atmosphere for investments (Mueller 1990; Klink et al.

(A)

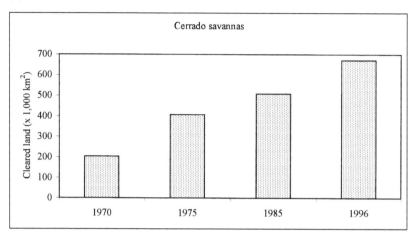

(B)

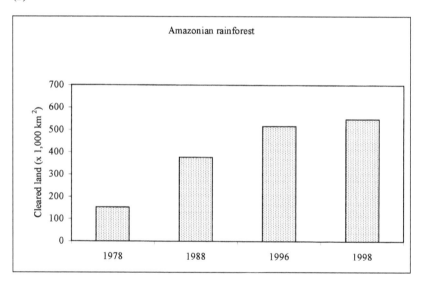

Figure 5.1   The total cleared area of (A) natural cerrado vegetation, and (B) Amazon rainforest (data for the cerrado from IBGE 1999 and Klink et al. 1993, 1995; data for the Amazon from INPE 2000).

1995). The economic stagnation and crises of the 1980s changed these prospects, but did not totally eliminate the programs and policies that resulted in the clearing of new lands for agricultural purposes.

The main causes for the expansion of the agricultural frontier in the cerrado can be divided into two broad categories: (1) policies aimed at expanding the agricultural output of the country as a whole; and (2) policies specifically targeting the cerrado, mainly developmental programs, use of new technologies, and subsidies.

Subsidized Loans and Inflation

Credit was one of the main instruments utilized by the Brazilian government to promote agricultural development. Probably nowhere are the environmental, economic and social contradictions of intervention in the economic process as visible as in the credit policy of Brazil in the 1960–1980 period. Subsidized credit has had a direct impact on the profitability expectation of farmers with access to loans, as well as an indirect but powerful impact on land prices. The greatest incentive was provided by a policy of low-interest farm loans. Because these loans were at fixed rates of 13% to 15%, they were further subsidized in that they ignored rapidly rising inflation rates (averaging 40% annually during that time). Even lower rates were imposed for fertilizers, pesticides and equipment. As a result, credit to the rural sector grew between 1969 and 1979 at a rate 188% faster than total output (199% for agriculture, 164% for livestock production) (Klink et al. 1993, 1995). The growth was particularly high during the 1969–1976 period, when agricultural credit rose at a yearly 24% compound interest rate (World Bank 1982). After 1977, credit flows, especially long-term investment credit, slowed.

Loans were not evenly distributed among crop types. Over 75% of production loans were concentrated in six crops: soybeans, rice, coffee, wheat, maize, and sugarcane. Soybeans alone received 20% of the credit available for Brazilian farmers (Klink et al. 1993). Since loans were allocated based on the size of the planted area, they encouraged extensive and inefficient agriculture (Goedert 1990). Credit was concentrated in the south and southeast regions of Brazil, which received almost twice as much credit per hectare as the cerrado. The cerrado, in turn, received 70% more credit than the Amazon and northeast, where the small producers of food crops are concentrated. This readily available credit increased the demand for land in the cerrado and drove up land prices (Klink et al. 1993).

POLOCENTRO and PRODECER

The failure of the settlement initiatives in the Amazon region (Mahar 1989) and the desire to hoist the cerrado economy (Klink et al. 1993) led to the creation of the Program for the Development of the Cerrado (POLOCENTRO) in 1975. Its goals were to settle farmers in places with good farming potential (twelve areas were selected), to improve infra-structure (mainly the construction of secondary roads and electricity), and to develop agricultural research and technology. The program's original target was to farm 60% of the exploited area and to give preference to foodstuffs. Farmers received subsidized loans, and credit lines were at rel-atively low fixed interest rates with no monetary correction. Rising infla-tion and considerable grace periods effectively transformed these loans into donations (Klink et al. 1995).

This program had a major impact on cerrado agriculture. Between 1975 and 1982, 3,373 agriculture projects were approved, totaling U.S.$577 million. Medium and large farmers benefited most (Klink et al. 1993). Eighty-one percent of the farms were 200 ha or larger in size and accounted for 88% of the total funds allocated; farms larger than 1,000 ha accounted for 39% of all projects and received more than 60% of the credit. An estimated 2.4 million ha of native land had been transformed between 1975 and 1980 alone. The program's original target to give pref-erence to foodstuffs was never realized. Instead it induced the expansion of commercial agriculture in the cerrado region. Most of the land was used for cattle ranching, and soybean became the main crop.

After the inauguration of the "New Republic" in 1985, many devel-opmental programs, including the POLOCENTRO, were closed in Brazil. However, in the late 1970s a new program, the Brazil-Japan Cooperative Program for the Development of the Cerrado (PRODECER), was initi-ated. It selects experienced farmers from the south and southeast of Brazil for settlement in the cerrado. It is financed by loans from both the Brazil-ian and the Japanese governments. In contrast with the POLOCENTRO, loans are granted at real (not fixed) interest rates. PRODECER is still active. At the time of this writing, two projects are being implemented, in Balsas (7°30' S, 46°20' W) and Porto Nacional (10°08'S, 48°15' W), and each will establish 40,000 hectares of new agricultural land, mainly for soybeans.

New Technologies

The development of appropriate technologies to enable farmers to deal with the nutrient-poor, acidic soils has also helped promote the agri-cultural development in the cerrado. These include technologies for soil fertilization, such as application of phosphate fertilizer and lime to cor-rect both nutrient deficiency and acidity; *Rhizobium*-based nitrogen fix-ation; the development of crop varieties; heavy use of herbicides and pes-ticides; and modern machinery (chapter 2). The performance of the cerrado Agricultural Center of the Brazilian Institute for Agricultural Research (EMBRAPA) has been impressive in almost every respect (Paterniani and Malavolta 1999). However, the technologies that were developed were strongly biased toward medium and large capital-intensive farmers, and cash crops, especially soybeans (Klink et al. 1993).

Because the expansion of agriculture in the cerrado is based on the incorporation of technology requiring increased use of machinery in agricultural operations, the need for manpower decreases in both time and space. As the number of tractors increases, relative employment growth decelerates. Between 1970 and 1985, the period of greatest agricultural expansion, employment grew 2.7% per year, whereas the farmed area expanded 5.4%, planted pastures grew 8.4%, the bovine herd increased 5.5%, and the stock of tractors, 13.6% (Klink et al. 1993, 1995).

Minimum Price Policy

The price support policy in Brazil has been in effect since the 1930s to guarantee a minimum price for agricultural products. It assumed a special significance for the cerrado in the 1980s. Responding to pressures from the World Bank and the International Monetary Fund, farm credit lines were restricted, reduced, or eliminated. Consequently, production costs increased substantially in the cerrado. The government then started to purchase large amounts of cerrado products, particularly soybeans, rice, and corn (Mueller and Pufal 1999). This favored farmers from remote areas in the cerrado, because they benefited indirectly from the nationwide unified fuel price. The net result of the price support policy was that it encouraged the expansion of commercial farming in areas that could not have supported profitable production without subsidies, and, consequently, the deforestation of new land.

AGRICULTURAL EXPANSION

Integration of cerrado farming and ranching into the national economy is a recent phenomenon. Agricultural activities, however, are not evenly distributed, and intraregional differences exist. In the state of Mato Grosso do Sul, southern Mato Grosso, central, southwestern and southeastern Goiás, the Federal District, the Triângulo Mineiro, and western Minas Gerais, modern, consolidated farming activities are well established. The basic infrastructure is well developed, along with access to the more dynamic markets of the country. The "modern" subregion is responsible for most of the soybean, corn, coffee, and bean production in the cerrado. It also produces a large share of the regional rice and cassava, as well as most of the bovine herd. In the remaining cerrado areas, agriculture is still developing, the road network and commercialization facilities are precarious, but further deforestation is expected as agriculture expands.

The expansion of cash crops, particularly soybean and corn, has been considerable. Soybean production, virtually nonexistent in the cerrado in the early 1970s, is currently 15 million metric tons (see fig. 5.2A). Production advanced slowly until 1989, suffered a fall in 1990 due to the collapse of the agricultural policy, and grew rapidly after that. The expansion during the 1990s was due to the restructuring of agricultural policies, high international prices for soybean and soy products, and productivity increases. The latter is evident in the disparity between yields and acreage in the 1990s (fig. 5.2A).

Transport and commercialization difficulties are hampering the expansion of soybean in the northern cerrado. This may change if the planned export river and railway corridors to the north are built; these will combine the Carajá and Norte-Sul railroads, inland waterways and highways, going to the port of São Luis in the northern state of Maranhão.

Corn production is currently expanding in some parts of the cerrado region, having increased from 0.5 million metric tons in 1960 to 5.5 million metric tons in 1995 (see fig. 5.2B). The "modern" subregion is responsible for more than 85% of the total increase in planted area of the cerrado. Cerrado corn production is part of the south central agri-industrial complex, which has experienced dramatic productivity increases over the last fifteen years. For example, between 1985 and 1994, production almost doubled, whereas the farmed area increased only 20% (Klink et al. 1995).

Rice has been an important crop in the cerrado, its production peaking in 1980 (fig. 5.2B). Since 1980, rice production has declined to the levels of the 1970s. The plunge in rice production has resulted primarily from planting soybean in recently cleared areas, instead of planting rice first as was done in the past.

Cerrado produces 40% of the Brazilian soybeans and 22% of corn. Despite the recent declines in production, the share of rice in the national market is still 12%. A recent trend is the increase of cotton production in the cerrado. As a strategy to reduce risks by diversifying crop production, soybean farmers are also investing in cotton, which represents 33% of the national production (Mueller and Pufal 1999).

Coffee production in 1990 was close to 250,000 metric tons. Nevertheless, the marked decline in the international price of coffee, the retreat

(A)

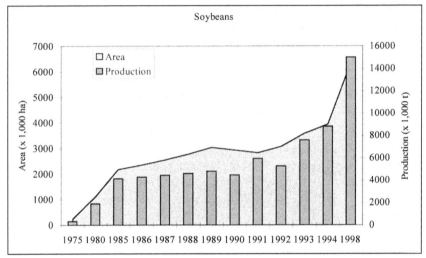

(B)

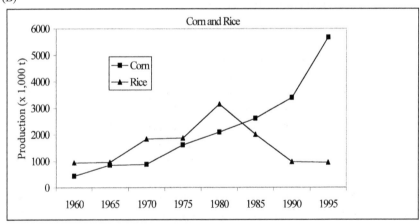

Figure 5.2 (A) Expansion of soybeans in the cerrado. Area is given in 1,000 hectares and production in 1,000 metric tons. (B) Growth of corn and rice in the cerrado. Production is given in 1,000 metric tons (data from IBGE 1999; Klink et al. 1993, 1995; Mueller and Pufal 1999).

of government support, and the virtual collapse of the International Coffee Agreement led to waning interest on the part of the farmers, which in turn led to the eradication of some coffee plantations in the Triângulo Mineiro (state of Minas Gerais), the main coffee producing zone in the region. Bean (*Phaseolus*) production has had a modest and irregular expansion in the cerrado. Most of the bean production comes from the "modern" subregion, where productivity is substantially higher than in the remaining cerrado areas because of the use of irrigated cropping systems, particularly in the state of Goiás (Klink et al. 1995).

Livestock

The cerrado is an important cattle ranching region. Ranching varies from relatively modern and efficient farms to extensive operations with rudimentary methods and low productivity. Modern techniques have been most readily adopted in the areas closest to markets with better access to technical assistance, basic and support infrastructure, and relatively sophisticated meat packing facilities.

The bovine herd in the cerrado region increased by 21.4 million head between 1970 and 1985: that is, from 16.6 million to almost 38 million (see fig. 5.3). The average yearly growth was initially high (3.6%) but has decreased since 1995, partly due to the relative saturation in areas with better transport, processing, and commercialization infrastructure. Total number of cattle in the cerrado is more than 51 million, representing 33% of the national herd. The recent frontier expansion areas may have a certain growth potential but are constrained by deficient transport networks. Ranching in these areas is done mainly on extensive native pastures.

The increase in the number of bovines is a direct consequence of the increase in the area of planted pastures. Planted pastures are by far the most important land use in the cerrado, representing 67% of the total cleared land (table 5.2), an area the size of Sweden and Denmark combined. To establish planted pastures, the savannas are clear-cut and burned, and then seeded with grasses of African origin, such as *Andropogon gayanus*, *Brachiaria brizantha*, *B. decumbens*, *Hyparrhenia rufa*, and *Melinis minutiflora*. Legumes, like *Centrosema* and *Stylosanthes*, are used as a source of protein (Barcellos 1996; Billoz and Palma 1996; Sano et al. 1999; see chapter 7).

Charcoal

Farming in the cerrado requires the removal of trees and roots, which usually are piled up and burned. Today it is becoming common to sell the firewood from deforestation for charcoal production to offset the costs of land clearing (Klink et al. 1995). This is usually done by itinerant, family-based charcoal producers. In the past, the use of cerrado vegetation for charcoal production was associated with the installation of large steel

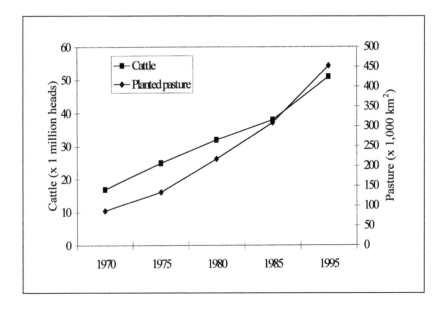

Figure 5.3   Expansion of the number of cattle and planted pastures in the cerrado. Cattle is given in 1,000 heads and pasture in 1,000 ha (data from IBGE 1999; Klink et al. 1993, 1995).

factories in the state of Minas Gerais in the 1940s and 1950s. As the natural cerrado vegetation around the plants vanished, and the transport costs rose in the late 1980s, the steel plants started to reforest extensively cleared cerrado land with *Eucalyptus* trees for charcoal production (Klink et al. 1995).

ENVIRONMENTAL IMPACTS

The perception of abundant land has driven most of the land use changes in the cerrado in the last 40 years. Although the extent of environmental modification is less well documented than the economical transformation, it is clear that the net impact has been negative. Soil mechanization of large tracts of monoculture expose great areas of bare soil, resulting in erosion and soil compaction. Soil losses on a 5% slope under bare fallow can be as high as 130 metric tons per hectare annually (Goedert 1990).

Agricultural expansion has led to an increase in burning, and areas still covered with natural vegetation are now burned almost every year

(Klink et al. 1993; chapter 4). Fire controls the proportion of woody and herbaceous plants in the cerrado. Because of their negative effect on tree and shrub seedlings, fires tend to favor herbaceous plants at the expense of woody plants. Protection against fire for sufficiently long periods of time favors the appearance of more wooded physiognomies in the cerrado (Moreira 2000; chapters 4, 8, 9).

The cerrado is one of the richest terrestrial ecoregions ("hotspots") on Earth (Mittermeier et al. 1999). It has a unique fauna and the largest diversity of all savanna floras in the world (ca. 10,000 species) (Ratter et al. 1997; Conservation International et al. 1999; Mittermeier et al. 1999; chapter 18). Large-scale transformation of the cerrado landscape is endangering its biodiversity with habitat fragmentation and even animal extinction. Three amphibians, 15 reptiles, and 33 bird species are now threatened with extinction in the cerrado (Conservation International et al. 1999).

The development of a modern agriculture in the cerrado has not been able to maintain and increase the ecosystem yield capacity without degradation. For instance, it is estimated that 50% of the planted pastures with African grasses in the cerrado (an area of 250,000 km^2) is degraded, to the detriment of its production capacity (Barcellos 1996). Extensive cultivation of African grasses in the cerrado has increased concerns over biotic invasions. African grasses proliferate, persist, and spread into a new range in which they cause detriment to the environment (Berardi 1994; Klink 1994). Given their current scale, African grasses are major agents of change (chapter 7).

One of the most widespread African species is the molassa or fat grass (*Melinis minutiflora*), known for the disruption it brings to biodiversity and ecosystem functioning in other parts of the world (Berardi 1994). Although it has been superseded by more productive African species, it is extensively found in disturbed areas, roadsides, abandoned plantations, and nature reserves in the cerrado (Klink et al. 1995). It can attain extremely high biomass that, when dried, becomes a highly combustible fuel that initiates a grass-fire interaction capable of preventing the regrowth of natural vegetation (Berardi 1994). In places where M. *minutiflora* reaches high cover, the diversity of the local flora is considerably lower than in native areas (chapter 7). Compared to natural savanna fires, fire temperature in M. *minutiflora* was far higher and had a much longer residence time, with flames over 6 meters high (Berardi 1994).

Large expanses of natural cerrado vegetation have been transformed from a mixture of trees and grasses into planted pastures and crops. Many cerrado trees and shrubs are deep-rooted and uptake soil water at 8 meters depth or more (Oliveira 1999; see also chapter 10). Simulations of the effects of the conversion of natural cerrado into open grasslands on regional climate have shown reduced precipitation by approximately 10%, an increase in the frequency of dry periods within the wet season, changes in albedo, and increased mean surface air temperature by 0.5°C (Hoffmann and Jackson 2000).

Considerable amounts of carbon are stored in roots and soil organic matter in the cerrado. Up to 70% of the alive biomass in cerrado vegetation is underground (Castro and Kauffmann 1998), and up to 640 tons of soil organic carbon has been found to a depth of 620 cm under natural vegetation (Abdala et al. 1998). Given the extension of cerrado vegetation already transformed into planted pastures and agriculture, it is possible that a significant change in both root biomass and the regional carbon sink has already occurred.

STRATEGIES FOR FUTURE CERRADO LAND USE

The agricultural development in the cerrado has been selective. Subsidies have favored commercial crops, and credit concessions have provided a strong incentive to open new land. Increases in production are due more to the increased area of land under cultivation than to gains in productivity. The impact of many of the agricultural development policy initiatives pursued since 1960 has also been inequitable. Most of the programs have favored wealthy farmers and large landholders. The use of technology-intensive agriculture generated only modest gains in farm employment and explains the decline of the rural population in the areas of dynamic agriculture in the southern cerrado (Cunha et al. 1994). Moreover, the environmental impact has been negative.

The demand for food, fibers, and other agricultural products exerts much pressure on the natural resources of the cerrado. Therefore it is necessary to identify the best agricultural growth options: whether to encourage intensification on areas already cleared, or to continue the expansion of the agricultural frontier. Both options have ecological costs. Horizontal expansion leads to a larger transformed area, destruction of habitats and biodiversity, and natural ecosystems disturbances, while farming intensification may increase soil degradation, chemical contamination, and water pollution. In the future, both types of agricultural systems shall coexist in the cerrado. The prospects for commercial farming are promising: the devaluation of Brazil's currency in January 1999 should help farm exporters; the increase of farm credit by the Banco do Brasil should stimulate cash crops; the transport costs should be lowered when the Fer-ronorte railroad is fully running and loading terminals are built; and the use of new technologies (e.g., genetically modified crops) is already on the horizon.

The modernization of agriculture should permit more intensive land use in the "modern" subregion of the Cerrado Biome, in which case no new area needs to be cleared. This tendency has been perceived in some parts of Brazil, as well as in other countries. Nevertheless, as isolated areas of the cerrado become more accessible by roads and railways, expansion of the agriculture frontier should be expected, especially for planted pastures.

In theory cerrado land is appropriate for sustainable agricultural activities, but this requires adequate public action. Usually the fragility of the natural ecosystems does not rank among farmers' top priorities, even where cash crops or cattle raising is inappropriate. The cerrado has an enormous environmental heterogeneity, including ecosystems that are rel-atively stable and resistant to changes, and others that are extremely sen-sitive to anthropogenic modifications.

Sustainable use of the cerrados will only be achieved if the strategies that direct its use are clearly defined. For example, the use of the no-till (direct drilling) technique that reduces soil erosion during soybean farm-ing (Landers 1996), or the use of organic farming along with pig raising (Mueller and Pufal 1999), are methods already in use for large scale farm-ing in the cerrado.

Based on the existing knowledge, many authors have suggested that an agri-environmental zoning of cerrado is desirable and possible (Goed-ert 1990; Cunha et al. 1994; Klink et al. 1995; Mueller and Pufal 1999). For instance, Mueller and Pufal (1999), based on an initial proposition made by Cunha et al. (1994), suggest that four major categories can be used for this zoning: areas used preferentially for crops, areas for crops and cattle raising, areas for crops and afforestation, and areas for con-servation, particularly those with fragile ecosystems. This last category could be as large as a third of the cerrado area, or 630,000 km^2 (Mueller and Pufal 1999).

This chapter has shown that policies were often formulated with lit-tle attention to their implications for cerrado land use and, as a result, have increased rates of land degradation, encouraged inefficient forms of development, and caused social conflict. The time is ripe for policies that take into consideration the interests of all cerrado land users (including small farmers and indigenous people) and the environmental services pro-vided by the cerrado (Klink and Moreira 2000). These policies should rely on mechanisms that couple market forces with the natural economic and ecological capacity of cerrado ecosystems.

ACKNOWLEDGMENTS

We thank Paulo S. Oliveira and an anonymous referee for a thorough review that greatly improved the chapter.

REFERENCES

Abdala, G., L. Caldas, M. Haridasan, and G. Eiten. 1998. Above and below-ground organic matter and root: Shoot ratio in a cerrado in central Brazil. *Braz. J. Ecol.* 2:11–23.

Almeida, F. F. M and M. A. Lima. 1956. The west central plateau and Mato Grosso "Pantanal." In Fundação Instituto Brasileiro de Geografia e Estatística, ed., *18th International Geographical Congress Excursion Guidebook no. 1.* Rio de Janeiro: Instituto Brasileiro de Geografia e Estatística.

Barbosa, A. S. and I. V. Nascimento. 1994. Processos culturais associados à vegetação. In M. N. Pinto, ed., *Cerrado: Caracterização, Ocupação e Perspectivas,* pp. 155–170. Brasília: Universidade de Brasília.

Barcellos, A. O. 1996. Sistemas extensivos e semi-intensivos de produção: Pecuária bovina de corte nos cerrados. In R. C. Pereira and L. C. B. Nasser, eds., *Biodiversidade e Produção Sustentável de Alimentos e Fibras nos Cerrados,* pp. 130–136. VIII Simpósio sobre o cerrado. Brasília: Empresa Brasileira de Pesquisa Agropecuária.

Berardi, A. 1994. "Effects of the African grass Melinis minutiflora on Plant Community Composition and Fire Characteristics of a Central Brazilian Savanna." Master's thesis, University College, University of London, London, UK.

Billaz, R. and V. Palma. 1996. La expansion de la agricultura y de la ganadería en las sabanas tropicales de America del Sur. In R. C. Pereira and L. C. B. Nasser, eds., *Biodiversidade e Produção Sustentável de Alimentos e Fibras nos Cerrados (VIII Simpósio sobre o Cerrado),* pp. 484–491. Brasília: Empresa Brasileira de Pesquisa Agropecuária.

Castro, E. A. and J. B. Kauffman. 1998. Ecosystem structure in the Brazilian cerrado: A vegetation gradient of aboveground biomass, root mass and consumption by fire. *J. Trop. Ecol.* 14:263–284.

Conservation International, Funatura, Fundação Biodiversitas, and Universidade de Brasília. 1999. *Ações Prioritárias para a Conservação da Biodiversidade do Cerrado e Pantanal.* Brasília: www.bdt.org.br/workshop/cerrado/br.

Cunha, A. S., C. C. Mueller, E. R. A. Alves, and J. E. Silva. 1994. *Uma Avali-*

ação da Sustentabilidade da Agricultura nos Cerrados. Estudos de Política Agrícola no. 11. Brasília: Instituto de Pesquisa Econômica e Aplicada and Programa das Nações Unidas para o Desenvolvimento (UNDP).

Davidson, E. A., D. C. Nepstad, C. A. Klink, and S. E. Trumbore. 1995. Pasture soils as carbon sink. *Nature* 376:472–473.

The Economist. 1999. Growth in the prairies. April 10th issue, p.34.

Goedert, W. J. 1990. Estratégias de manejo das savanas. In G. Sarmiento, ed., *Las Sabanas Americanas: Aspectos de su Biogeografia, Ecologia y Utilización*, pp. 191–218. Guanare: Acta Científica Venezolana.

Guidon, N. 1991. *Peintures Préhistoriques du Brésil*. Paris: ERC.

Hees, D. R., M. E. P. C. de Sá, and T. C. Aguiar. 1987. A evolução da agricultura na região Centro-Oeste na década de 70. *Rev. Bras. Geogr.* 49:197–257.

Hoffmann, W. A. and R. B. Jackson. 2000. Vegetation-climate feedbacks in the conversion of tropical savanna to grassland. *J. Climate* 13: 1593–1602.

IBGE (Instituto Brasileiro de Geografia e Estatística). 1999. *Censo Agropecuário 1995–1996*. Brasília: www.ibge.gov.br.

IBGE (Instituto Brasileiro de Geografia e Estatística). 2001. *Censo Demográfico 2000: Dados Preliminares*. Brasília: www.ibge.gov.br.

INPE (Instituto Nacional de Pesquisas Espaciais). 2000. *Monitoring of the Brazilian Amazonian Forest by Satellite*. São José dos Campos: Ministério da Ciência e Tecnologia, www.inpe.mct.gov.br.

Klink, C. A. 1994. Clipping effects on size and tillering of native and African grasses of the Brazilian savannas (the "cerrado"). *Oikos* 70:365–376.

Klink, C. A., R. H. Macedo, and C. C. Mueller. 1995. *De Grão em Grão o Cerrado Perde Espaço*. Brasília: World Wildlife Fund–Brazil and Pró-Cer.

Klink, C. A. and A. G. Moreira. 2000. Valoração do potencial do cerrado em estocar carbono atmosférico. In A. G. Moreira and S. Schwartzman, eds., *As Mudanças Climáticas Globais e os Ecossistemas Brasileiros*, pp. 85–91. Brasília: Editora Foco.

Klink, C. A., A. G. Moreira, and O. T. Solbrig. 1993. Ecological impacts of agricultural development in the Brazilian cerrados. In M. D. Young and O. T. Solbrig, eds., *The World's Savannas*, pp. 259–282. London: The Parthenon Publishing Group.

Landers, J. N. 1996. O plantio direto na agricultura: O caso do cerrado. In I. V. Lopes, G. S. Bastos Filho, D. Bille, and M. Bale, eds., *Gestão Ambiental no Brasil*, pp. 2–33. Rio de Janeiro: Editora da Fundação Getúlio Vargas.

Leeuwenberg, F. and M. Salimon. 1999. *Os Xavantes na Balança das Civilizações*. Brasília: Pratica Gráfica.

Lucarelli, H. Z., N. R. Innocencio, and O. M. B. L. Fredrich. 1989. Impacto da construção de Brasília na organização do espaço. *Rev. Bras. Geogr.* 51:99–138.

Mahar, D.J. 1989. *Government Policies and Deforestation in Brazil's Amazon Region*. Environment Department Working Paper 7. Washington, D.C.: The World Bank.

Maybury-Lewis, D. 1988. *The Savage and the Innocent.* 2nd edition. Boston: Beacon Press.

Mittermeier, R.A., N. Myers, and C. G. Mittermeier. 1999. *Hotspots: Earth's Biologically Richest and Most Endangered Terrestrial Ecoregions.* Mexico: CEMEX and Conservation International.

Moreira, A. G. 2000. Effects of fire protection on savanna structure in Central Brazil. *J. Biogeogr.* 27:1021–1029.

Mueller, C. C. 1990. Políticas governamentais e a expansão recente da agropecuária no Centro-Oeste. *Planejamento e Políticas Públicas* 3:45–73.

Mueller, C. C. and D. V. L. Pufal. 1999. *Atividades Agropecuárias no Cerrado.* Brasília: Instituto Sociedade, População e Natureza.

Oliveira, R. S. 1999. "Padrões Sazonais de Disponibilidade de Agua nos Solos de um Cerrado Denso e um Campo Sujo e Evapotranspiração." Master's thesis, Universidade de Brasília, Brasília, Brazil.

Paterniani, E. and E. Malavolta. 1999. La conquista del "cerrado" en el Brasil: Victoria de la investigación científica. *Interciencia* 24:173–176.

Ratter, J. A., J. F. Ribeiro, and S. Bridgwater. 1997. The Brazilian cerrado vegetation and threats to its biodiversity. *Ann. Bot.* 80:223–270.

Roosevelt, A. C., M. L. da Costa, C. L. Machado, M. Michab, N. Mercier, H. Valladas, J. Feathers, W. Barnett, M. I. da Silveira, A. Henderson, J. Sliva, B. Chernoff, D. S. Reese, J. A. Holman, N. Toth, and K. Schick. 1996. Paleoindian cave dwellers in the Amazon: The peopling of the Americas. *Science* 272:373–384.

Sano, E. A., A. O. Barcellos, and H. S. Bezerra. 1999. Área e distribuição espacial de pastagens cultivadas no Cerrado brasileiro. *Boletim de Pesquisa da Embrapa—Cerrados* 3:1–21.

Schmitz, P. I. 1992. A história do velho Brasil. *Ciência Hoje*, Special Issue Eco-Brasil: 95–102.

World Bank. 1982. *A Review of Agricultural Policies in Brazil.* Washington, DC: World Bank Publications.

Part II

The Plant Community: Composition, Dynamics, and Life History

6

Vegetation Physiognomies and Woody Flora of the Cerrado Biome

Ary T. Oliveira-Filho and James A. Ratter

THE CERRADO BIOME OF TROPICAL SOUTH AMERICA COVERS about 2 million km², an area approximately the same as that of Western Europe, representing ca. 22% of the land surface of Brazil, plus small areas in eastern Bolivia and northwestern Paraguay (fig. 6.1). It extends from the southern borders of the Amazonian forest to outlying areas in the southern states of São Paulo and Paraná, occupying more than 2° of latitude and an altitudinal range from near sea-level to 1,800 m. The distribution of the Cerrado Biome is highly coincident with the plateaux of central Brazil, which divide three of the largest South American water basins: those of the Amazon, Plate/Paraguay, and São Francisco rivers. The cerrados form part of the so-called diagonal of open formations (Vanzolini 1963) or corridor of "xeric vegetation" (Bucher 1982), which includes the much drier Caatinga in northeastern Brazil and the Chaco in Paraguay-Bolivia-Argentina. This corridor runs between the two main areas of moist forest of tropical South America: the Amazonian forest in the northwest, and the Atlantic forest in the east and southeast.

The Cerrado Biome was named after the vernacular term for its predominant vegetation type, a fairly dense woody savanna of shrubs and small trees. The term *cerrado* (Portuguese for "half-closed," "closed," or "dense") was probably applied to this vegetation originally because of the difficulty of traversing it on horseback.

The typical vegetation landscape of the Cerrado Biome consists of savanna of very variable structure, termed cerrado *sensu lato*, on the

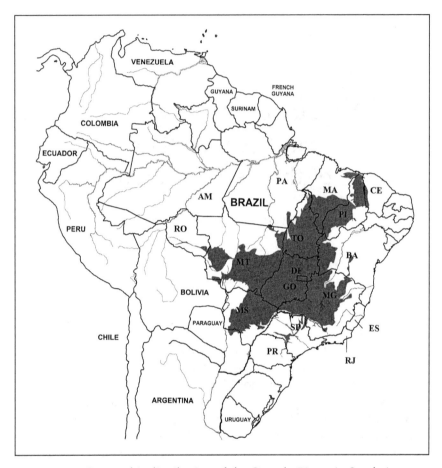

Figure 6.1   Geographic distribution of the Cerrado Biome in South America. The isolated patches of Amazonian savannas are not indicated because they are inserted in a different Biome. Key to state codes: Amazonas (AM), Bahia (BA), Ceará (CE), Distrito Federal (DF), Espírito Santo (ES), Goiás (GO), Maranhão (MA), Minas Gerais (MG), Mato Grosso (MT), Mato Grosso do Sul (MS), Pará (PA), Paraná (PR), Piauí (PI), Rio de Janeiro (RJ), Rondônia (RO), São Paulo (SP), Tocantins (TO). After IBGE (1993), and Ribeiro and Walter (1998).

well-drained interfluves, with gallery forests or other moist vegetation following the watercourses. In addition, areas of richer soils in the biome are clothed in mesophytic forests. In this chapter we consider the main environmental factors associated with the Cerrado Biome, its main vegetation physiognomies, and the diversity and origin of its woody flora. In general, common usage will be followed throughout this chapter, and **cerrado** (**sensu lato**) will be termed **cerrado, cerrados,** or **cerrado vegetation**, while the other vegetation types of the Cerrado Biome will be indicated by clear distinguishing names.

ENVIRONMENTAL VARIABLES DETERMINING THE DISTRIBUTION OF THE CERRADO BIOME

The distribution of the Cerrado Biome shown in figure 6.1 is determined basically by the predominance of cerrado (*sensu lato*) in the landscape. The factors determining the distribution of cerrado vegetation have long been a subject of controversy, but in general the following are considered important: seasonal precipitation, soil fertility and drainage, fire regime, and the climatic fluctuations of the Quaternary (e.g., Eiten 1972; Furley and Ratter 1988; Ratter 1992; Oliveira-Filho and Ratter 1995; Furley 1999). These are, on the whole, the same factors identified as important in maintaining savanna biomes worldwide, although grazing also plays an important role in other continents such as Africa (see references in Werner 1991; Furley et al. 1992). Although the climate of the Cerrado Biome varies considerably, it is mostly typical of the rather moister savanna regions of the world. There is a remarkable variation across the region in both the average annual temperature, ranging from 18° to 28°C, and rainfall, from 800 to 2,000 mm, with a very strong dry season during the southern winter (approx. April–September) (Dias 1992). Nevertheless, as a number of authors point out, rainfall seasonality cannot entirely explain the predominance of cerrado vegetation, as the present climatic conditions would favor the establishment of forests in most of the Cerrado Biome region (e.g., Rizzini and Pinto 1964; Reis 1971; Klein 1975; Coutinho 1978; Van der Hammen 1983).

Soil fertility and moisture are other important factors to be considered in the distribution of cerrado vegetation. Most soils of the Cerrado Biome are dystrophic, with low pH and availability of calcium and magnesium, and high aluminium content (Furley and Ratter 1988; see also chapter 2). Moreover, cerrado only grows on well-drained soils, therefore concentrating on interfluves and avoiding valley bottoms. In fact, within the Cerrado Biome, vegetation physiognomies other than cerrado itself are found on the patches of base-rich soils as well as on sites liable to waterlogging for considerable periods.

Although of fundamental importance, seasonal rainfall and low soil fertility are apparently insufficient to explain the present distribution of cerrado vegetation completely. For instance, vast areas in southeastern Brazil with strongly seasonal rainfall and infertile soils (e.g., eastern Minas Gerais) have a continuous cover of semideciduous forests and show

no trace of cerrado (Oliveira-Filho and Fontes 2000). This leaves disturbance by fire as a possible key determinant of the presence of forest or
cerrado. Most of the flora of cerrado is of fire-adapted species, involving
not only fire tolerance but also fire dependency (Coutinho 1990; Braithwaite 1996; see also chapters 4, 9). Palaeoenvironmental studies have confirmed that forests and cerrado vegetation showed successive expansions
and contractions in the region, following the climatic fluctuations of the
Quaternary (chapter 3). The last expansion of cerrado occurred during
the brief dry episode of the Holocene; however, since then, forests have
not regained their original area, although rainfall apparently returned to
previous levels (Ledru 1993). This failure of forest to expand into the cerrado may be largely determined by human-induced fires, which have been
important elements in the region at least since the mid-Holocene (Ledru
et al. 1998).

Despite past disputes concerning the relative importance of various
environmental factors in determining the distribution of cerrado vegetation (see Eiten 1972; Goodland and Ferri 1979), it is now widely accepted
that climate, soils, and fire are highly interactive in their effect on vegetation within the Cerrado Biome. For example, the seasonal climate favors
the outbreak of fire during the dry season when cerrado plant cover is
highly inflammable; recurrent fires tend to prevent the vegetational succession into forest and also cause soil impoverishment; and increasingly
base-poor soils liable to strong water deficit during the dry season restrict
the establishment of forest species (chapter 4). Therefore one should consider these interactions when trying to understand the distribution of both
the Cerrado Biome and its constituent vegetation types.

VEGETATION PHYSIOGNOMIES OF THE CERRADO BIOME

The vegetation of the Cerrado Biome shows a remarkable physiognomic
variation, and many authors have attempted to produce an efficient classification. A revision is found in Ribeiro and Walter (1998), who also proposed a comprehensive modern treatment of the subject using a pragmatic
but more detailed classification than that adopted here. Most problems
can be explained by the fact that any classification category is actually a
segment of a multidimensional vegetational continuum (particularly
within the cerrado *sensu lato*). Table 6.1 is an attempt to show the major
vegetational categories and the main determining environmental factors.
Traditional Brazilian nomenclature for cerrado vegetation is used here
because it is well accepted, unambiguous, and appropriate.

Table 6.1   Main Vegetation Physiognomies of the Cerrado Biome and Their Association with Soil Fertility Levels and Ground Water Regimes

Groundwater regime	Soil Fertility in Terms of Base Saturation		
	Low	*Intermediate*	*High*
Strongly drained sites with deep water table and seasonal water deficit at topsoil level	Cerrado *s.l.: campo limpo, campo sujo, campo cerrado,* cerrado *sensu stricto,* or dystrophic facies *cerradão*	Mesotrophic facies *cerradão* or mesophytic semideciduous forests on interfluves	Mesophytic deciduous dry forests on interfluves and slopes
High soil moisture for most of the year within catchment areas	Dystrophic facies *cerradão* or valley forests (mesophytic evergreen)	Mesotrophic facies *cerradão* or valley forests (mesophytic semideciduous)	Valley forests (mesophytic semideciduous or deciduous)
Permanently waterlogged to very wet alongside rivercourses	Riverine forests (evergreen): gallery forests (swampy and wet), alluvial forests, or riverside portions of valley forests		
Successive periods of soil water-logging and strong water deficit	Seasonal grasslands: valley-side marshes (*veredas*), floodplain grasslands (*pantanal*), or rocky grasslands (*campo rupestre*). Both valley-side marshes and floodplain grasslands may contain scattered earth-mounds (*campo de murundus*)		

In general, cerrado physiognomies predominate in the landscape because well-drained, low-fertility soils are the predominant substratum. Cerrado vegetation tends to be replaced by forest physiognomies on sites with increased water availability and/or soil fertility, while seasonal grasslands appear where periods of strong water deficit follow periods of waterlogging. It is worth noting that fire, an important element for Cerrado Biome ecosystems, may interfere with the vegetation-environment relationships represented in table 6.1, particularly in forest-cerrado transitions. A brief description of each main vegetation physiognomy and its plant community follows.

CERRADO *SENSU LATO*

Cerrado vegetation is generally characterized by a mixture of plants of two fairly distinct layers. The first, hereafter called the **woody layer**, includes trees and large shrubs; the other, the **ground layer**, is composed of subshrubs and herbs (chapter 7). In cerrado, it is often difficult to distinguish between trees and large shrubs. We define "large shrub" here as a plant bearing a perennial woody stem (i.e., not a hemixyle) and generally attaining a height of at least 1.3 m. The flora of both woody and ground layers has typical features of pyrophytic savanna vegetation. The trees are of low contorted form with thick, corky, fire-resistant bark. Sclerophylly is common: many leaves have thick cuticles, sunken stomata, and greatly lignified and sometimes silicified tissues, and are often of considerable longevity. Xylopodia (swollen, woody underground structures) are well developed in both the woody and ground layers, and the hemixyle growth form, where woody shoots of annual duration are developed from an underground xylopodium, is particularly common (e.g., *Jacaranda decurrens*, *Anacardium humile*, *Andira humilis*). On the other hand, annuals are rare: Warming (1892, 1973) calculated that they constituted less than 6% of the herbaceous species of the cerrados of Lagoa Santa in central Minas Gerais.

The cerrado *sensu lato* encompasses a series of vegetation physiognomies from open grasslands to dense woodlands, and more or less recognizable stages of this continuum are given vernacular names. Dry grassland without shrubs or trees is called **campo limpo** ("clean field"); grassland with a scattering of shrubs and small trees is known as **campo sujo** ("dirty field"). Where there are scattered trees and shrubs and a large proportion of grassland, the vegetation is termed **campo cerrado** ("closed field"); the next stage when the vegetation is obviously, at least visually, dominated by trees and shrubs often 3–8 m tall and giving more than 30% crown cover but with still a fair amount of herbaceous vegetation between them is known as **cerrado** (*sensu stricto*). The last stage is an almost closed woodland with crown cover of 50% to 90%, made up of trees, often of 8–12 m or even taller, casting a considerable shade so that the ground layer is much reduced; this form is called **cerradão** (Portuguese augmentative of *cerrado*). It is unfortunate that, in common usage, the term *cerrado* should refer both to Brazilian savanna vegetation in its generic sense (cerrado *sensu lato*), and to one of its subvariants (cerrado *sensu stricto*). Clearly the dividing line between the five cerrado physiognomies is somewhat arbitrary, but workers in the field usually agree surprisingly well on the classification. Some authors exclude *campo limpo* from cerrado *sensu lato* as it has no woody layer, but we prefer to include it since it is usually composed of the characteristic cerrado ground layer and thus forms the most open part of the continuum. Examples of these physiognomies are given in figure 6.2.

Many factors are probably operative in determining which of these physiognomies of cerrado vegetation occurs in a given site. Goodland and

Figure 6.2   Cerrado physiognomies. (A) Large expanse of cerrado *sensu stricto*, Gilbués, Piauí; (B) *campo limpo*, Chapada dos Veadeiros, Goiás; (C) *campo sujo*, Brasília, Federal District; (D) *campo cerrado*, Alter do Chão, Pará; (E) cerrado *sensu stricto*, Loreto, Maranhão; (F) mesotrophic facies *cerradão*, Doverlândia, Goiás.

Pollard (1973) correlated increased production of woody elements with an increasing soil fertility gradient in the Triângulo Mineiro (western Minas Gerais), while Lopes and Cox (1977), who studied over 500 sites covering much of the central core cerrado area, arrived at the same conclusion. The data of other authors, however, fail to demonstrate this correlation, and sometimes show well-developed *cerradão* on no more fertile soils than sparser forms of cerrado nearby (e.g., Ribeiro 1983; Ribeiro and Haridasan 1990). The explanation of these contradictory results lies at least partly in the occurrence of two floristically different forms of *cerradão*, which have not been distinguished by the majority of authors.

Mesotrophic facies cerradão occurs on soils of intermediate fertility in the cerrado landscape, particularly in terms of calcium and magnesium levels (see fig. 6.2F). This community is readily recognized by the presence of a number of indicator species such as *Magonia pubescens, Callisthene fasciculata, Dilodendron bipinnatum* and *Terminalia argentea* (Ratter 1971, 1992; Ratter et al. 1973, 1977; Oliveira-Filho and Martins 1986, 1991; Furley and Ratter 1988). This facies of *cerradão* is of very widespread occurrence in the Cerrado Biome and is often associated with the transition to mesophytic forest, the climax vegetation of the most base-rich soils in the Cerrado Biome. In fact, it indicates a soil intermediate in fertility between that of the more dystrophic forms of cerrado and that of the mesophytic deciduous dry forest (see table 6.1). Many of the indicator species also occur in deciduous dry forest and the arboreal *caatinga* vegetation (xeric thorn woodland) of NE Brazil, to which the deciduous dry forest is closely related.

A floristically different type of *cerradão*, the dystrophic facies cerradão (Ratter 1971; Ratter et al. 1973, 1977; Oliveira-Filho and Martins 1986, 1991; Furley and Ratter 1988), is found on base-poor soils and has its own indicator species, such as *Hirtella glandulosa, Emmotum nitens, Vochysia haenkeana,* and *Sclerolobium paniculatum*. This facies of *cerradão* is also very widespread in the Cerrado Biome and tends to be associated with the savanna-forest transition on base-poor and often sandy soils (see table 6.1). Therefore, it is commonly found fringing valley forests at the feet of sandstone plateaux and in the transition to Amazonian forests on sandy soils.

Fire is undoubtedly an extremely important factor affecting the density of the woody layer of cerrado vegetation (chapters 4, 9). Although most woody species are strongly fire-adapted, fires at too frequent intervals damage them and favor the ground layer, thus producing more open physiognomies; conversely protection from fire allows the woody vegetation to close, and characteristic *cerradão* tree species (e.g., *Emmotum nitens, Protium heptaphyllum, Virola sebifera*) to establish themselves. It seems probable that, in the past, before the advent of frequent manmade fires, the denser arboreal physiognomies of cerrado occupied a much larger area than they do today. Some authors, such as Warming (1892, 1973) and Coutinho (1978, 1990), believe that the climax vegetation of most of the Cerrado Biome is actually *cerradão* and that the other more open physiognomies exist as successional phases determined mainly by fire regime, all tending to evolve to *cerradão* in the absence of fire. Succession from cerrado to *cerradão* in the Distrito Federal and Mato Grosso is discussed in Ratter (1980, 1986) and Ratter et al. (1973, 1978a), respectively. In both cases there are strong indications that the succession proceeds to forest, probably reflecting the re-expansion of forests after the brief Holocenic dry episode.

Soil moisture is another factor which may affect the physiognomic continuum of cerrado, as this vegetation, mostly restricted to soils which are well drained throughout the year, commonly shows pronounced physiognomic and floristic changes as it approaches seasonally waterlogged grasslands. Where interfluvial cerrados are bordered by *veredas* (valley-side marshy grasslands, see below) it is common to observe a decline in mean tree height and an increasing density of woody plants toward the cerrado margin, where a distinct community of cerrado tree species more resistant to soil saturation occurs (Ratter et al. 1973; Oliveira-Filho et al. 1989). A similar community appears on raised islands of ground which appear in both *veredas* and seasonally flooded alluvial grasslands (Diniz et al. 1986; Oliveira-Filho 1992a). The distribution of these islands frequently produces a landscape known as **campo de murundus**, consisting of an expanse of open grasslands dotted with a regular pattern of raised earthmounds bearing cerrado trees and shrubs, and often termitaria. Landscapes of this type are particularly common in such seasonally inundated floodplains of the Mato Grosso *Pantanal* (Ratter et al. 1988; Dubs 1992), Varjão do Araguaia, and Ilha do Bananal (Ratter 1987). Tree species which tolerate soil saturation and are commonly found in these marginal cerrado communities are *Curatella americana, Byrsonima crassifolia, B. orbignyana, Dipteryx alata, Tabebuia aurea*, and *Andira cuiabensis* (Furley and Ratter 1988; Oliveira-Filho 1992b).

Goodland (1971) focused attention on the role of aluminium in the cerrado, and other authors have since concentrated on the subject. Levels of this element are so high in the dystrophic cerrado soils (chapter 2) as to be extremely toxic to most cultivated plants, but most native species are aluminium-tolerant, as would be expected. The tolerant species include a number of diverse unrelated families that accumulate aluminium in their tissues, particularly in leaves, but also in roots (Haridasan 1982), such as the Vochysiaceae, various Rubiaceae, some Myrtaceae, *Miconia* spp. (Melastomataceae), *Symplocos* (Symplocaceae), *Strychnos pseudoquina* (Loganiaceae), *Myrsine* spp. (Myrsinaceae), and *Vellozia* (Velloziaceae). Some families, including the Vochysiaceae, are obligate aluminium accumulators and cannot grow in its absence.

MESOPHYTIC SEASONAL FORESTS

The occurrence of **mesophytic seasonal forests**—comprising both **deciduous** and **semideciduous forests**—within the Cerrado Biome is very extensive and generally underestimated (Oliveira-Filho and Ratter 1995). Where more fertile soils occur in the region, the climax vegetation is definitely mesophytic forest (see table 6.1). These forests are found, for example, on base-rich alluvial deposits in the Mato Grosso Pantanal (Prance and Schaller 1982; Ratter et al. 1988; Dubs 1992), on calcareous outcrops (Ratter et al. 1973, 1977, 1978b, 1988; Prado et al. 1992), on soils originated from basalt (Oliveira-Filho et al. 1998), and in valleys where the topography has cut into more mineral-rich underlying rocks (e.g., silts and mudstones). One of the largest extensions of these more fertile areas covered by mesophytic forests, the "Mato Grosso de Goiás," is estimated to have had an area of 40,000 km^2 before agriculture destroyed it almost entirely (chapter 5). During a road journey from the Distrito Federal to Estreito in Maranhão, crossing ca. 1,400 km of the states of Goiás and Tocantins, one of us (J.R.) estimated that probably nearly 50% of the vegetation traversed was degraded mesophytic forest or the closely related mesotrophic facies *cerradão*.

The soils of mesophytic forests are particularly good for agriculture; consequently, the vegetation has been devastated to such an extent that in many areas the past role of this forest as an important or even dominant land cover has been obscured. In our experience it is rare to encounter cerrado regions where at least some of these forests do not occur: even the most dystrophic plateaux have small fertile forested valleys. These so-called **valley forests** (see fig. 6.3A) are favored not only by higher soil fertility, but also by higher water availability for most of the year; normally the higher the soil fertility, the higher the forest deciduousness (see table 6.1). In fact, there seems to be no clear-cut ecological and floristic differences between deciduous and semideciduous forests. The level of deciduousness probably depends on the conjunction of soil moisture and chemical properties. Often there are quite local differences in deciduousness in a single valley forest: for instance, at Vale dos Sonhos, Mato Grosso, the same forest community is deciduous on the well-drained valley sides but semideciduous in the moister valley bottom (Ratter et al. 1978b). In the Chapada dos Guimarães, Mato Grosso, the composition and deciduousness of the valley forest changes as the underlying bedrock changes from sandstone to slate, thereby increasing soil fertility (Pinto and Oliveira-Filho 1999). In an area surveyed at Três Marias, Minas Gerais, the forest changes from evergreen to semideciduous and deciduous before meeting the cerrado in a forest strip no more than 150 m wide along the Rio São Francisco (Carvalho et al. 1999).

Deciduous dry forests are particularly common on the base-rich soils of the peripheral areas that connect the Cerrado to the Caatinga Biome in the northeast, and to the Chaco Biome in the western boundaries of the

Figure 6.3 Forest and grassland physiognomies of the Cerrado Biome. (A) Expanse of mesophytic semideciduous (valley) forest, Serra da Petrovina, Mato Grosso; (B) interior of mesophytic deciduous dry forest during the rainy season, Torixoreu, Mato Grosso (note abundant Maranthaceae in ground layer); (C) deciduous dry forest during the dry season, Sagarana, northern Minas Gerais; (D) the same site during the rainy season; (E) swampy gallery forest flanked by *veredas*, Chapada dos Veadeiros, Goiás; (F) *vereda* with buriti palmery, Nova Xavantina, Mato Grosso.

Mato Grosso Pantanal (Ratter et al. 1988; Prado et al. 1992; see figs. 6.3B–D). On the other hand, large extensions of semideciduous forests predominate in the complex transition between the Cerrado and the Atlantic Rainforest Biomes in southeastern Brazil (Oliveira-Filho and Fontes 2000). Deciduous dry forests in Central Brazil are characterized by a species-poor woody plant community dominated by a few indicator species such as *Myracrodruon urundeuva*, *Anadenanthera colubrina*, *Aspidosperma subincanum*, *Tabebuia impetiginosa*, *Dipteryx alata,* and *Dilodendron bipinnatum* (Ratter 1992). Semideciduous forests, however, tend to be considerably richer in species and actually represent a floristic crossroad. The flora is intermediate between those of deciduous dry forests and either rainforests (both Amazonian and Atlantic), on a geographic scale, or evergreen riverine forests, on a local scale. They also share many species with the two types of *cerradão*, thereby representing a connection between the cerrado and rainforest floras (Oliveira-Filho and Fontes 2000).

A particular form of deciduous dry forest is the so-called **mata calcárea** (calcareous forest) which appears on calcareous outcrops throughout the Cerrado Biome. These have already been regarded by Prado and Gibbs (1993) as relics of a once even more extensive deciduous dry forest that during the glacial maxima would have connected the *caatingas*, in northeastern Brazil to the semideciduous forests in southeastern Brazil and southern Paraguay, and to the piedmont forests in central-western Argentina (see chapter 3 for palynological evidence). It is reasonable to think that an intense process of soil leaching and acidification, possibly helped by fire, following the return of more humid climates to the cerrado region, would have favored the establishment of cerrado vegetation in most places and the isolation of decidous dry forests on the present-day islands of mesotrophic and calcareous soils (Ratter et al. 1988).

RIVERINE FORESTS

Riverine forests are ubiquitous throughout the Cerrado Biome; nearly all water bodies of the region are fringed by forests. This forest net is determined by the year-round high soil moisture, which, despite the long dry season of the region, provides a suitable habitat for a large number of typical moist forest species. The nomenclature of riverine forests in Central Brazil is complex, as many names are given to the various forms throughout the region (Mantovani 1989; Ribeiro and Walter 1998). Most of the striking variation of riverine forests both in physiognomy and floristic composition results from variation in topography and drainage characteristics, together with soil properties (Felfili et al. 1994; Silva Júnior et al. 1996; Haridasan et al. 1997; Silva Júnior 1997). Narrow forest strips, found along streams and flanked by grasslands or cerrados, are often called **gallery forests** because tree crowns form a "gallery" over the watercourse (fig. 6.3E). These may be swampy galleries, where slow water flow increases soil anoxia, or wet (nonswampy) galleries, where flow is faster and the soil better drained (Oliveira-Filho et al. 1990; Felfili and Silva Júnior 1992; Walter and Ribeiro 1997). When they border wider rivers, riverine forests are often called **matas ciliares** (literally, eyelash forests) because they fringe both margins like eyelashes (Ribeiro and Walter 1998). Where these forests are on alluvial beds under the strong influence of the river flooding regime, they may alternatively be called **alluvial forests** (Oliveira-Filho and Ratter 1995). Such forests are often characterized by the presence of a raised levee running along their riverside margin. On some steep valleys, where the riverine forests are flanked by wider areas of mesophytic forest instead of cerrado or seasonal grasslands, they are part of **valley forests** (see previous section).

These many forms of riverine forest are not always well defined in the field, as they may replace each other either very gradually over large areas or through short local transitions. Furthermore, the separation of mesophytic seasonal forests and riverine forests sometimes breaks down, as mixed associations often occur in areas of valley forest. In fact, many forests in the Cerrado Biome are formed by narrow strips of evergreen riverine forest stretching alongside the river courses, sided by more or less wide tracts of mesophytic seasonal forests on the adjacent slopes.

A number of species are good indicators of the groundwater regime (Ratter 1986; Oliveira-Filho et al. 1990; Walter and Ribeiro 1997; Schiavini 1997; Felfili 1998). For example, *Xylopia emarginata, Talauma ovata, Calophyllum brasiliense, Hedyosmum brasiliense*, and *Richeria grandis* are typical species of swampy conditions; *Protium spruceanum, Endlicheria paniculata, Pseudolmedia laevigata*, and *Hieronyma alchorneoides* are characteristic of wet but better drained soils; while *Inga vera, Salix humboldtiana*, and *Ficus obtusiuscula* are common in seasonally flooded forests. The abundant light at the sharp transition to cerrado or grassland favors the occurrence of typical forest edge species such as *Piptocarpha macropoda, Lamanonia ternata, Vochysia tucanorum,* and *Callisthene major*. The shady interior favors species such as *Cheiloclinium cognatum* and *Siparuna guinensis*. However, many of riverine forest species are habitat-generalists (e.g., *Schefflera morototoni, Casearia sylvestris, Protium heptaphyllum, Tapirira guianensis, T. obtusa, Virola sebifera, Copaifera langsdorffii*, and *Hymenaea courbaril*), many of which are abundant in the interface with the semideciduous forest and/or the cerrado vegetation.

The key factor for the occurrence of riverine forest within the Cerrado Biome is high soil water availability throughout the year, making up for the overall water deficit during the dry season. This has led a number of authors to suggest that the central Brazilian riverine forests represent floristic intrusions of the Amazonian and/or Atlantic forests into the cerrado domain (Cabrera and Willink 1973; Rizzini 1979; Pires 1984). In fact, a considerable number of species shared by the two great South American forest provinces do cross the Cerrado Biome via the riverine forests (e.g., *Ecclinusa ramiflora*, *Protium spruceanum*, *Cheiloclinium cognatum*, and *Margaritaria nobilis*). Others extend their range into the Cerrado Biome along the riverine forests but do not complete the crossing (e.g., from the Amazonian forest, *Tapura amazonica*, *Elaeoluma glabrescens*, *Oenocarpus distichus*, and even species of rubber-tree [*Hevea* spp.], and from the Atlantic side, *Euterpe edulis*, *Hedyosmum brasiliense*, *Geonoma schottiana*, and *Vitex polygama*). However, there are also a few species that are exclusive to these forests, such as *Unonopsis lindmannii*, *Vochysia pyramidalis*, and *Hirtella hoehnei*.

SEASONAL GRASSLANDS

The alternation of periods of water excess and deficit normally favors the occurrence of seasonal grasslands within the Cerrado Biome. There are three main vegetation physiognomies of the type in the region. **Veredas** are valley-side marshes where the water table reaches or almost reaches the surface during the rainy season; they are commonly found in the middle of topographic sequences, between gallery forests and cerrado. *Veredas* are very widespread in the Cerrado Biome, particularly near headwaters, and may include palm groves of *Mauritia flexuosa* (buriti-palm) (see fig. 6.3E–F). **Floodplain grasslands** are found on areas of even topography liable to more or less long periods of inundation; they are usually restricted to the vicinity of large rivers, such as the Paraguay (Mato Grosso Pantanal) and Araguaia (Varjão and Bananal island). **Rocky grasslands** (*campo rupestre*, *campo de altitude*) are mostly restricted in the Cerrado Biome to the tops of plateaus and mountain ridges, where the soils are shallow or confined to cracks between rocks. As they have very limited water storage capacity, these soils are often soaked during the rainy season but extremely dry during the dry season. As these physiognomies are poor or totally lacking in woody vegetation, they are better treated in chapter 7.

THE BIODIVERSITY OF THE CERRADO BIOME

The combination of the great age of the Cerrado Biome and the relatively recent (Quaternary) dynamic changes in vegetation distribution patterns has probably led to its rich overall biodiversity, estimated at 160,000 species of plants, animals, and fungi by Dias (1992). The figure for vascular plants is still very approximate, but Mendonça et al. (1998) list 6,429 native species from all communities of the biome. Future investigations will certainly add many species to the list. For instance, recent surveys over a large part of the cerrado area show many more woody species in the cerrado *sensu lato*. The eventual total may indeed reach the 10,500 estimate given by Dias (1992). In a recent publication, Myers et al. (2000) recognize the cerrado among 25 global biodiversity "hotspots" and estimate that it contains 4,400 endemic higher plant species, representing no less than 1.5% of the world's total vascular plant species.

An important aspect of biodiversity of the Cerrado Biome, and one of profound ecological importance, is the loss of large mammalian fauna, as Ratter et al. (1997) explain:

> unlike the African savannas, it [the cerrado] has lost the fauna of large mammals with which it must have co-evolved throughout the Tertiary. The large herbivores (grazers and browsers) must have been eliminated as a result of competition with North American fauna which migrated across the Panama Land-Bridge in the Great American Interchange 3 million years ago in the late Pliocene, or later in Man's Pleistocene and Holocene Overkill. The only remnants of the ancient neotropical mammalian fauna now occurring in the Cerrado Biome are some Edentates (such as the tamanduá anteaters and armadillos), marsupials (such as opossums), platyrrhine monkeys (such as marmosets, howlers and capuchins), and various rodents (such as agoutis, pacas, capybaras, and many mouse-sized species). Many larger-fruited plants species probably lost their natural mode of dispersal as a result of the extinction of their native mammalian vectors (see Janzen and Martin 1982). The reintroduction of grazers in the form of cattle and horses into the natural cerrado vegetation in the last few hundred years probably partially restored the balance of the vegetation to the situation prior to the Great American Interchange.

The levels of information on the diversity of the various communities of the Cerrado Biome are very unequal and are considered separately below.

CERRADO *SENSU LATO*

The characteristic arboreal flora of the savanna elements of the Cerrado Biome is relatively well known. A useful base list was provided by Rizzini (1963) and added to by Heringer et al. (1977). In all, these authors record

774 woody species belonging to 261 genera, of which 336 species (43%) are regarded as endemic to cerrado *sensu lato*. Since 1977 much research has been carried out on the floristics and phytosociology of the cerrados, and the number of species recorded has increased. A recent compilation by Castro et al. (1999) gives 973 species and 337 genera identified "with confidence" and in addition mentions a large number of records of undetermined, or partially determined, taxonomic entities. These authors suggest that the total arboreal and large shrub flora of the cerrado *sensu lato* may be 1,000–2,000 species. The latter figure, however, must be approached with a great degree of caution, as stressed by Castro et al. (1999) in an extremely succint discussion. They consider that a reasonably secure minimum estimate for the arboreal-shrubby cerrado flora is around 1,000 species, 370 genera, and 90 families. However, they point out that "three main objections might be raised: (1) the list includes a large number of species that certainly would not be regarded as typical cerrado species (e.g. *Talauma ovata*); (2) a number of species that are not typically woody in most sites are also included (e.g., *Oxalis*); (3) some unrecorded rarer species are likely to be 'hidden,' having been misidentified as common cerrado species" (Castro et al. 1999). We consider (1) and (2) to be particularly potent factors in inflating estimates of cerrado woody species, although fully accepting the arguments of Castro et al. on the difficulties of separating (a) "characteristic" and "accessory" species, and (b) "ground" from "arboreal or large shrub" species when the same taxon may show contrasting growth forms in different localities. Cerrado research is in a very dynamic phase, and more accurate estimates will be available in the next few years. In the meantime it is interesting to note that the present data base of the ongoing Anglo-Brazilian collaborative *Conservation and Management of the Biodiversity of the Cerrado Biome* (BBC) project records approximately 800 species for 300 surveys throughout the cerrado region (Ratter et al. 2000).

The most important families in terms of species numbers, using the fairly conservative figures of Heringer et al. (1977), are Leguminosae (153 spp., all three subfamilies), Malpighiaceae (46 spp.), Myrtaceae (43 spp.), Melastomataceae (32 spp.), and Rubiaceae (30 spp.). However, in many areas the vegetation is dominated by Vochysiaceae (with 23 arboreal species in the cerrado) because of the abundance of the three species of *pau-terra* (*Qualea grandiflora*, *Q. parviflora* and *Q. multiflora*). The largest genera are *Byrsonima* (Malpighiaceae, 22 spp.), *Myrcia* (Myrtaceae, 18 spp.), *Kielmeyera* (Guttiferae, 16 spp.), *Miconia* (Melastomataceae, 15 spp.) and *Annona* (Annonaceae, 11 spp.).

Heringer et al. (1977) analyzed the geographic affinity of the 261 genera they listed and found that 205 had species in common with the Brazilian Atlantic Forest, 200 with the Amazonian forest, 30 with the mesophytic forests, and 51 with the cerrado ground layer, while seven (three of which are monotypic) did not occur in any other vegetation type.

Recent work by Ratter and Dargie (1992), Ratter et al. (1996), Castro (1994), and Castro et al. (1999) has been directed toward discovering patterns of geographic distribution of cerrado vegetation by comparison of large numbers of floristic surveys using multivariate techniques. The studies have covered almost the entire cerrado area and have also included some isolated Amazonian savannas. In all, Ratter et al. (1996) compared 98 sites, while Castro studied 78 areas and 145 species lists. The results of the two groups seem to be very much in accord. Ratter et al. (1996) demonstrated a strong geographic pattern in the distribution of the flora, which allowed the provisional recognition of southern (São Paulo and south Minas Gerais), southeastern (largely Minas Gerais), central (Federal District, Goiás and parts of Minas Gerais), central-western (largely Mato Grosso, Goiás and Mato Grosso do Sul), and northern regions (principally Maranhão, Tocantins and Pará), as well as a disjunct group of Amazonian savannas (see fig. 6.4). This work is continuing as a part of the *Biodiversity of the Cerrado Biome* (BBC) project, and results based on the comparison of more than 300 sites will soon be available.

Diversity of trees and large shrubs occurring at a single site (alpha diversity) may reach 150 species per hectare, but is generally much lower than this, while at the other extreme it can be less than 10 species in isolated Amazonian savannas. Diversity tends to be lower on the more mesotrophic sites where dominance of characteristic "indicator" species such as *Callisthene fasciculata, Magonia pubescens, Terminalia argentea,* and *Luehea paniculata* occurs. The comparison of 98 sites by Ratter et al. (1996) revealed a remarkable intersite heterogeneity (beta diversity). In total 534 tree and large shrub species were recorded at these sites. Of these, 158 (30%) occurred at a single site only; no species occurred at all sites; and only 28 (5%) were present at 50% or more of the sites. The most widespread species was *Qualea grandiflora,* which occurred at 82% of sites. High levels of instersite heterogeneity have also been demonstrated in surveys of the same land unit, the Chapada Pratinha, in the Federal District, Goiás, and Minas Gerais (Felfili and Silva Júnior 1993; Filgueiras et al. 1998). The extreme floristic heterogeneity (beta diversity) of cerrado vegetation has important consequences for conservation planning, since many protected areas will need to be established in order to represent biodiversity adequately. The sites we have recorded with the highest species numbers ("biodiversity hotspots") are in Mato Grosso,

Ocotea, 31 *Protium*, 30 *Inga*, or 19 *Eschweilera* species found in the 10,000 hectares of Amazonian rainforest of the Reserva Ducke near Manaus (Ribeiro et al. 1999), but the number of individuals of each species found in these cerrado mixed populations is probably significantly greater. In addition, the diversity of growth form of the cerrado congeners is much greater than in the forest.

The diversity of plants of the ground layer (the so-called *vegetação rasteira*, consisting of herbs, subshrubs, and smaller shrubs) is much richer than for trees and large shrubs, and species numbers are so high that detailed floristic lists are only available for comparatively few localities (chapter 7). Rizzini (1963) gives the figure of more than 500 genera for smaller plants against less than 200 for trees and large shrubs; more detailed information can be extracted from a number of works. In an exhaustive survey of the IBGE Ecological Reserve in the Federal District conducted over many years, Pereira et al. (1993) record 636 species in the *vegetação rasteira* against 84 arboreal species (a ratio of 7.6:1), while Ratter (1986) lists 400 ground species against 110 trees (3.6:1) in the nearby Fazenda Água Limpa. The difference between these two ratios is probably partly explained by the length of time spent in observations at the two localities: at Fazenda Água Limpa the bulk of the work was done in one year's intensive study, while at the IBGE Reserve detailed observation was extended over 15 years, thus allowing many rare smaller species to be recorded. Figures for São Paulo state are somewhat lower, with ratios of 3:1 and 2:1 recorded by Mantovani and Martins (1993). Ratter and Ribeiro (1996) suggested that extrapolation of the Federal District figures taking 1,000 as the number of tree/large shrub species would give an estimate of over 5,000 ground species, while Castro et al. (1999) estimate 2,919–6,836, working on the same basis from their data (see chapter 7).

MESOPHYTIC FORESTS

As already mentioned, mesophytic deciduous and semideciduous forests occur throughout the cerrado landscape where richer (mesotrophic to eutrophic) soils occur (table 6.1). Their flora belongs to a floristic province which was probably continuous, or almost continuous, during drier and cooler periods in the Pleistocene and is now represented by three main nuclei: the arboreal *caatingas* of NE Brazil; the "Misiones" forests of Corumbá–Puerto Suarez, extending into Paraguay, Argentinian Misiones, and Brazilian Santa Catarina; and the "Piedmont" forests of Bolivia and northen Argentina (Ratter et al. 1988; Prado 1991; Prado and Gibbs

1993; Pennington et al. 2000). The islands of mesophytic forests crossing the region of the Cerrado Biome now form a very discontinous bridge between the *caatinga* vegetation and the "Misiones forests."

The diversity of arboreal species of deciduous dry forests is much lower than that of the cerrado or the riverine forests of the Cerrado Biome. Semideciduous forests also tend to be considerably richer than deciduous dry forests, as they share many species with riverine forests. The total arboreal floristic list is approximately 100 species for deciduous dry forests occurring in the Planalto and central-western areas of the biome, but normally the number of species found in any one locality is very much less than this. Most of the species require higher levels of soil calcium, but a few of the taller trees such as the leguminous *Apuleia leiocarpa, Copaifera langsdorffii*, and *Hymenaea courbaril* var. *stilbocarpa*, and the anacardiaceous *Tapirira guianensis*, are more tolerant of dystrophic conditions and thus can be found in other forest communities (dystrophic galleries, Amazonian forest, etc.). About 20% of the species occurring in central Brazilian deciduous dry forests are also found in mesotrophic *cerradão* (e.g., *Astronium fraxinifolium, Dilodendron bipinnatum, Dipteryx alata*, and *Platypodium elegans*). The floristic relationship between deciduous dry forest and mesotrophic *cerradão* has been noted by various authors (Rizzini and Heringer 1962; Magalhães 1964; Ratter et al. 1977; Furley and Ratter 1988) and is to be expected considering the pedological similarities of the two communities. Deciduous dry forests usually show a high degree of dominance of a few species, which in our experience is a characteristic of communities inhabiting richer soils. Extreme examples of this are shown in surveys of deciduous dry forests in the Triângulo Mineiro. For example, Araújo et al. (1997) found 65 tree species in a forest near Uberlândia, Minas Gerais, where *Anadenanthera colubrina* var. *cebil* (under the synonym *A. macrocarpa*) comprised 60.51% of the relative dominance and had an IVI of 73.41. In Santa Vitória, Oliveira-Filho et al. (1998) registered 60 species, with the top five, including *A. colubrina*, accounting for 61% of all individuals and 70% of the relative dominance.

There is a considerable variation in the floristics and other characteristics of mesophytic forests across the vast area of the Cerrado Biome. Thus forests such as the remnants of the "Mato Grosso de Goiás" on deep soils in the core area of the biome are much more mesic than those on calcareous outcrops and/or near the drier areas lying close to the Caatinga Biome margin. As would be expected, the latter show a greater abundance of more extreme *caatinga* species (associated with more xeric conditions) such as tall *Cereus jamacaru* cacti growing among the trees, *Commiphora leptophloeos, Spondias tuberosa, Zizyphus joazeiro, Schinopsis brasiliensis, Cavanillesia arborea*, etc. Conversely, in forests closer to the "Misiones" floristic nucleus, particularly those south of the Mato Grosso Pantanal, species characteristic of that element, such as *Calycophyllum multiflorum, Pterogyne nitens, Chrysophyllum marginatum, Terminalia triflora, Prosopis* spp., and, once again, *Commiphora leptophloeos* are found.

RIVERINE FORESTS

The riverine forests following the drainage throughout the vast Cerrado Biome cover probably less than 10% of its total area but harbor an enormous floristic and faunal diversity. A number of recent surveys have shown a much greater diversity of the tree/large shrub species in the gallery forests than in the cerrado vegetation itself; for instance, Ramos (1995) found 260 species in the galleries of the Brasília National Park but only 109 species in the cerrado, while Pereira et al. (1993) list 183 species from the galleries and 84 from the cerrado of the IBGE Ecological Reserve, also in the Federal District. In a wider area on the Planalto Central (Federal District, Goiás, and Minas Gerais), Silva Júnior, Nogueira, and Felfili (1998) recorded 446 woody species in 22 gallery forests. Figures for total floristic diversity of riverine forests throughout the complete Cerrado Biome are not yet available. The database of Oliveira-Filho and Ratter (1994) contains 627 species of trees for 17 riverine forests of the Cerrado Biome, while a total of 771 species of trees and shrubs of riverine forests can be extracted from the list prepared by Mendonça et al. (1998) for all communities of the Cerrado Biome. However, as there are still relatively few surveys of these forests, it is certain that they contain many more species than suggested by those figures.

The reason for so much diversity can be ascribed to two main factors: (1) the environmental heterogeneity occurring both within and between riverine forests and (2) the diverse floristic elements from which the communities of riverine forests are derived in different parts of the region.

A number of authors have carried out detailed studies of particular galleries and demonstrated how the habitat, generally defined according to topography and drainage, can be broken into environmental subdivisions and characterized by floristic differences (Felfili et al. 1994; Felfili 1998). Oliveira-Filho et al. (1990) studied the headwaters of the Córrego da Paciência, near Cuiabá, Mato Grosso, and recognized four distinct communities: dry *cerradão* (dystrophic), wet *cerradão* (mesotrophic), wet forest (semideciduous), and swampy forest (evergreen). An extremely detailed study was carried out in the Reserva do Roncador, Federal District (Silva Júnior 1995; Silva Júnior et al. 1996) to investigate the association of particular floristic communities with differing environmental conditions. Dry upslope, intermediate, and wet downslope communities were defined, with all species showing strong habitat preferences and only one species, *Tapirira guianensis* (a geographically widespread generalist) occurring in all three communities. Similarly, Schiavini (1997) and Van den Berg and Oliveira-Filho (1999) have demonstrated the distinct association of woody species with three bands of differing humidity and exposure to light (edge-effect) in two gallery forests near Uberlândia and Itutinga, respectively, both situated in Minas Gerais.

A recent study by Silva Júnior, Felfili, Nogueira, and Rezende (1998) demonstrates the heterogeneity of gallery forests in the Federal District. In all, 15 forests containing a total of 446 arboreal species were compared. These were reduced for analysis to the 226 species with more than five individuals, of which only two species (*Copaifera langsdorfii* and *Tapirira guianensis*) were present at all sites. The percentage breakdown was 27.4% of species occurring only at a single site, 38.9% at two to four sites, 17.2% at five to eight, 10.6 at nine to 12, and 1.8% at 13–15. The results showed the presence of a few species with a wide distribution and many with a very restricted occurrence. The forests with the lowest diversity were, following the normal pattern, the mesophytic examples associated with better soils. Sørensen similarity indices were calculated using the full species lists, including rare species. They were as high as 73.8% for two nearby sites on the IBGE Ecological Reserve and as low as 11.0% for two more distant sites. The majority, however, lie between 30.1% and 47.0%, indicating fairly low levels of similarity between the galleries of the Federal District. All the galleries studied were in the drainage of the Plate/Paraguay system. Had the observations included those galleries also occurring in the Federal District but on the Araguaia-Tocantins and São Francisco drainages, the results would probably have shown greater diversity and many low-similarity indices.

A study of the origins of Central Brazilian forests by the analysis of plant species distribution patterns (Oliveira-Filho and Ratter 1995) involved the comparison of 106 forest surveys containing a total of 3,118 species. The result of multivariate analysis revealed that the riverine forests of the north and west of the Cerrado Biome have a strong relationship with the Amazonian rainforests, while those of the center and south show stronger affinity with the montane semideciduous forests of southeastern Brazil. Introduction of species from adjacent forest nuclei into the riverine forests crossing the cerrado region has probably been very active during postglacial forest expansion, although a great number of the characteristic gallery forest species are widespread generalists. An important aspect of gallery forests is that they have interfaces with many other types of vegetation, including rainforests, mesophytic forests and the cerrado itself. They are thus subjected to very different floristic influences resulting in great heterogeneity.

CONCLUDING REMARKS ON CONSERVATION

The Cerrado Biome is one of the richest savanna biomes in the world and harbors an immense floral and faunal diversity. Much of this is a consequence of its great antiquity, possibly going back to a prototypic cerrado in the Cretaceous, followed by long evolution over the Tertiary and a dynamic phase leading to much speciation over the glacial and interglacial periods of the Quaternary. Endemicity is very high, estimated at as high as 4,400 species of higher plants by Myers et al. (2000), who rank the Cerrado among 25 global biodiversity "hotspots" of absolute importance for conservation. Already about 40% of the original Cerrado Biome area has been converted to "anthropic landscape," principally as "improved" pastures or intense arable cultivation, and the need for implementation of conservation plans is urgent (Klink et al. 1995; Ratter et al. 1997; Cavalcanti 1999; see chapters 5, 18). The contents of this book demonstrate the multifaceted importance of the biome, and we hope that we are now seeing an eleventh-hour awakening to the needs of conservation of cerrado and dry forest.

ACKNOWLEDGMENTS

This chapter was prepared under the sponsorship of the Royal Society of London (International Exchange Award to A. T. Oliveira-Filho) and the Royal Botanic Garden Edinburgh. We acknowledge their support with gratitude. We also thank Samuel Bridgewater, Toby Pennington, and Luciana Botezelli for their invaluable help during the preparation of the manuscript.

REFERENCES

Araújo, G. M., L. A. Rodrigues, and L. Ivizi. 1997. Estrutura fitossociológica e fenologia de espécies lenhosas de mata decídua em Uberlândia, MG.

In L. L. Leite and C. H. Saito, eds., *Contribuição ao Conhecimento Ecológico do Cerrado: Trabalhos Selecionados do 3° Congresso de Ecologia do Brasil (Brasília, 6–11/10/96)*, pp. 22–28. Brasília: Editora da Universidade de Brasília.

Braithwaite, R. W. 1996. Biodiversity and fire in the savanna landscape. In O. T. Solbrig, E. Medina, and J. F. Silva, eds., *Biodiversity and Savanna Ecosystem Processes*, pp. 121–142. Berlin: Springer-Verlag.

Bucher, E. H. 1982. Chaco and Caatinga: South American arid savannas, woodlands and thickets. In B. J. Huntley and B. H. Walker, eds., *Ecology of Tropical Savannas*, pp. 48–79. Berlin: Springer-Verlag.

Cabrera, A. L. and A. Willink. 1973. *Biogeografia de America Latina*. Washington: Secretaria General de la Organización de los Estados Americanos.

Carvalho, D. A., A. T. Oliveira-Filho, E. A. Vilela, N. Curi, E. Van den Berg, and M. A. L. Fontes. 1999. *Estudos Florísticos e Fitossociológicos em Remanescentes de Florestas Ripárias do Alto São Francisco e Bacia do Rio Doce–MG*. Boletim técnico, CDD 33.7, Belo Horizonte: Companhia Energética de Minas Gerais.

Castro, A. A. J. F. 1994. "Comparação Florístico-Geográfica (Brasil) e Fitossociológica (Piauí–São Paulo) de Amostras de Cerrado." Ph.D. thesis, Universidade Estadual de Campinas, Campinas, Brazil.

Castro, A. A. J. F., F. R. Martins, J. Y. Tamashiro, and G. J. Shepherd. 1999. How rich is the flora of Brazilian cerrados? *Ann. Miss. Bot. Gard.* 86:192–224.

Cavalcanti, R. B. (scientific coordinator). 1999. *Ações Prioritárias para Conservação da Biodiversidade do Cerrado e Pantanal*. Brasília: Ministério do Meio Ambiente, Funatura, Conservation International, Fundação Biodiversitas, Universidade de Brasília.

Coutinho, L. M. 1978. O conceito de Cerrado. *Rev. Bras. Bot.* 1:17–23.

Coutinho, L. M. 1990. Fire in the ecology of the Brazilian cerrado. In J. G. Goldammer, ed., *Fire in Tropical Biota*, pp. 82–105. Berlin: Springer-Verlag.

Dias, B. F. S. 1992. Cerrados: Uma caracterização. In B. F. S. Dias, ed., *Alternativas de Desenvolvimento dos Cerrados: Manejo e Conservação dos Recursos Naturais Renováveis*, pp. 11–25. Brasília: Fundação Pró-Natureza.

Diniz, M. A. N., P. A. Furley, C. Johnson, and M. Haridasan. 1986. The murundus of central Brazil. *J. Trop. Ecol.* 2:17–35.

Dubs, B. 1992. Observations on the differentiation of woodland and wet savanna habitats in the Pantanal of Mato Grosso, Brazil. In P. A. Furley, J. Proctor, and J. A. Ratter, eds., *Nature and Dynamics of Forest-Savanna Boundaries*, pp. 431–449. London: Chapman and Hall.

Eiten, G. 1972. The cerrado vegetation of Brazil. *Bot. Rev.* 38:201–341.

Felfili, J. M. 1998. Determinação de padrões de distribuição de espécies em uma mata de galeria no Brasil Central com a utilização de técnicas de análise multivariada. *Bol. Herb. E. P. Heringer* 2:35–48.

Felfili, J. M., T. S. Filgueiras, M. Haridasan, R. Mendonça and A. V. Rezende. 1994. Projeto biogeografia do bioma cerrado: Vegetação e solos. *Cad. Geoc. IBGE* 12:75–166.

Felfili, J. M. and M. C. Silva Júnior. 1992. Floristic composition, phytosociology and comparison of cerrado and gallery forests at Fazenda Água Limpa, Federal District, Brazil. In P. A. Furley, J. Proctor, and J. A. Ratter, eds., *Nature and Dynamics of Forest-Savanna Boundaries*, pp. 393–416. London: Chapman and Hall.

Felfili, J. M. and M. C. Silva Júnior. 1993. A comparative study of cerrado (*sensu stricto*) vegetation in Central Brazil. *J. Trop. Ecol.* 9:277–289.

Filgueiras, T. S., J. M. Felfili, M. C. Silva Júnior, and P. E. Nogueira. 1998. Floristic and structural comparison of cerrado (*sensu stricto*) vegetation in Central Brazil. In F. Dallmeier and J. A. Comisky, eds., *Forest Biodiversity in North, Central, and South America, and the Caribbean*, pp. 633–648. Man and the Biosphere Series, vol. 21. Paris: UNESCO and Parthenon Publishing Group.

Furley, P. A. 1999. The nature and diversity of neotropical savanna vegetation with particular reference to the Brazilian cerrados. *Glob. Ecol. Biogeogr.* 8:223–241.

Furley, P. A., J. Proctor, and J. A. Ratter. 1992. *Nature and Dynamics of Forest-Savanna Boundaries*. London: Chapman and Hall.

Furley, P. A. and Ratter, J. A. 1988. Soil resources and plant communities of the Central Brazilian cerrado and their development. *J. Biogeogr.* 15:97–108.

Goodland, R. A. 1971. A physiognomic analysis of the "cerrado" vegetation of Central Brazil. *J. Ecol.* 59:411–419.

Goodland, R. A. and M. G. Ferri. 1979. *Ecologia do cerrado*. Belo Horizonte and São Paulo: Livraria Itatiaia Editora and Editora da Universidade de São Paulo.

Goodland, R. A. and R. Pollard. 1973. The Brazilian cerrado vegetation: A fertility gradient. *J. Ecol.* 6:219–224.

Haridasan, M. 1982. Aluminium accumulation by some cerrado native species in Central Brazil. *Plant and Soil* 65:265–273.

Haridasan, M., M. C. Silva Júnior, J. M. Felfili, A. V. Rezende, and P. E. N. Silva. 1997. Gradient analysis of soils properties and phytosociological parameters of some gallery forests on the Chapada dos Veadeiros in the cerrado region of Central Brazil. In *Proceedings of the International Symposium on Assessment and Monitoring of Forests in Tropical Dry Regions with Special Reference to Gallery Forests*, pp. 259–275. Brasília: Editora da Universidade de Brasília.

Heringer, E. P., G. M. Barroso, J. A. Rizzo, and C. T. Rizzini. 1977. A flora do cerrado. In *Anais do IV Simpósio sobre o Cerrado*, pp. 211–232. Belo Horizonte: Livraria Itatiaia Editora and São Paulo: Editora da Universidade de São Paulo.

IBGE. 1993. *Mapa de vegetação do Brasil*. Rio de Janeiro: Fundação Instituto Brasileiro de Geografia e Estatística.

Janzen, D. H. and P. Martin. 1982. Neotropical anachronisms: What the gomphotheres ate. *Science* 215:19–27

Klein, R. M. 1975. Southern Brazilian phytogeographic features and the probable influence of upper Quaternary climatic changes in the floristic distribution. *Bol. Paran. Geoc.* 33:67–88.

Klink, C.A., Macedo, R., and Mueller, C. 1995. Bit by bit the Cerrado loses space. In Alho, C. J. R. and E. de S. Martins, eds., p. 66. Brasilia: World Wildlife Fund and Pró-Cer.

Ledru, M. P. 1993. Late Quaternary environmental and climatic changes in central Brazil. *Quat. Res.* 39:90–98.

Ledru, M. P., M. L. Salgado-Labouriau, and M. L. Lorscheitter. 1998. Vegetation dynamics in southern and central Brazil during the last 10,000 yr BP. *Rev. Palaeobot. Palynol.* 99:131–142.

Lopes, A. S. and E. R. Cox. 1977. Cerrado vegetation in Brazil: An edaphic gradient. *Agronom. J.* 69:828–831.

Magalhães, G. M. 1964. Fitogeografia do Estado de Minas Gerais. In Anon. ed., *Recuperação do* Cerrado, pp. 69–82. Rio de Janeiro: Ministério da Agricultura, Edições SIA, Estados Brasileiros n° 21.

Mantovani, W. 1989. Conceituação e fatores condicionantes. In *Anais do I Simpósio sobre Mata Ciliar*, pp. 2–10. Campinas: Fundação Cargill.

Mantovani, W. and F. R. Martins. 1993. Florística do cerrado na Reserva Biológica de Moji Guaçu, SP. *Acta Bot. Bras.* 7:33–60.

Mendonça, R. C., J. M. Felfili, B. M. T. Walter, M. C. Silva Júnior., A. V. Rezende, T. S. Filgueiras, and P. E. Nogueira. 1998. Flora vascular do cerrado. In S. M. Sano and S. P. Almeida, eds., *Cerrado: Ambiente e Flora*, pp. 288–556. Planaltina, DF, Brazil: Empresa Brasileira de Pesquisa Agropecuária.

Myers, N., R. A. Mittermeir, C. G. Mittermeir, G. A. B. Fonseca, and J. Kent. 2000. Biodiversity hotspots for conservation priorities. *Nature* 403: 853–858.

Oliveira-Filho, A. T. 1992a. Floodplain "murundus" of Central Brazil: Evidence for the termite-origin hypothesis. *J. Trop. Ecol.* 8:1–19.

Oliveira-Filho, A. T. 1992b. The vegetation of Brazilian "murundus": The island-effect on the plant community. *J. Trop. Ecol.* 8:465–486.

Oliveira-Filho, A. T., N. Curi, E. A. Vilela, and D. A. Carvalho. 1998. Effects of canopy gaps, topography and soils on the distribution of woody species in a central Brazilian deciduous dry forest. *Biotropica* 30:362–375.

Oliveira-Filho, A. T. and M. A. L. Fontes. 2000. Patterns of floristic differentiation among Atlantic forests in south-eastern Brazil, and the influence of climate. *Biotropica* 32:793–810.

Oliveira-Filho, A. T. and F. R Martins. 1986. Distribuição, caracterização e composição florística das formações vegetais da região da Salgadeira, na Chapada dos Guimarães (MT). *Rev. Brasil. Bot.* 9:207–223.

Oliveira-Filho, A. T. and F. R. Martins. 1991. A comparative study of five cerrado areas in southern Mato Grosso, Brazil. *Edinb. J. Bot.* 48:307–332.

Oliveira-Filho, A. T. and J. A. Ratter. 1994. *Database: Woody Flora of 106 Forest Areas of Eastern Tropical South America.* Edinburgh: Royal Botanic Garden Edinburgh.

Oliveira-Filho, A. T. and J. A. Ratter. 1995. A study of the origin of central Brazilian forests by the analysis of plant species distribution patterns. *Edinb. J. Bot.* 52:141–194.

Oliveira-Filho, A. T., J. A. Ratter, and G. J. Shepherd. 1990. Floristic composition and community structure of a central Brazilian gallery forest. *Flora* 184:103–117.

Oliveira-Filho, A. T., G. J. Shepherd, F. R. Martins, and W. H. Stubblebine. 1989. Environmental factors affecting physiognomical and floristic variations in a cerrado of central Brazil. *J. Trop. Ecol.* 5:413–431.

Pennington, R. T., D. E. Prado, and C. A. Pendry. 2000. Neotropical seasonally dry forests and Quaternary vegetation changes. *J. Biogeogr.* 27 (in press).

Pereira, B. A. S., M. A. Silva, and R. C. Mendonça. 1993. *Reserva Ecológica do IBGE, Brasília–DF: Lista das plantas vasculares.* Rio de Janeiro: Fundação Instituto Brasileiro de Geografia e Estatística.

Pinto, J. R. R. and A. T. Oliveira-Filho. 1999. Perfil florístico e estrutura da comunidade arbóreo-arbustiva de uma floresta de vale no Parque Nacional da Chapada dos Guimarães. *Rev. Bras. Bot.* 22:53–67.

Pires, J. M. 1984. The Amazonian forest. In H. Sioli, ed., *The Amazon: Limnology and Landscape Ecology of a Mighty Tropical River and its Basin,* pp. 581–602. Dordrecht: Junk Pub.

Prado, D. E. 1991. "A Critical Evaluation of the Floristic Links Between Chaco and Caatingas Vegetation in South America." Ph.D. thesis. University of Saint Andrews, Saint Andrews, Scotland.

Prado, D. E. and P. E. Gibbs. 1993. Patterns of species distribution in the dry seasonal forests of South America. *Ann. Miss. Bot. Gard.* 80:902–927.

Prado, D. E., P. E. Gibbs, A. Pott, and V. J. Pott. 1992. The Chaco-Pantanal transition in southern Mato Grosso, Brazil. In P. A. Furley, J. Proctor, and J. A. Ratter, eds., *Nature and Dynamics of Forest-Savanna Boundaries,* pp. 431–470. London: Chapman and Hall.

Prance, G. T. and G. B. Schaller. 1982. Preliminary study of some vegetation types of the Pantanal, Mato Grosso, Brazil. *Brittonia* 34:228–251.

Ramos, P. C. M. 1995. "Vegetation Communities and Soils of Brasília National Park." Ph.D. thesis, University of Edinburgh Edinburgh, Scotland.

Ratter, J. A. 1971. Some notes on two types of cerradão occurring in northeastern Mato Grosso. In M. G. Ferri, ed., *III Simpósio sobre o cerrado,* pp. 100–102. São Paulo: Editora da Universidade de São Paulo.

Ratter, J. A. 1980. *Notes on the Vegetation of Fazenda Água Limpa (Brasília, DF, Brazil).* Edinburgh: Royal Botanic Garden Edinburgh.

Ratter, J. A. 1986. *Notas sobre a vegetação da Fazenda Água Limpa (Brasília, DF, Brasil).* Textos Universitários no. 3. Brasília: Editora da Universidade de Brasília.

Ratter, J. A. 1987. Notes on the vegetation of the Parque Nacional do Araguaia (Brazil). *Notes Roy. Bot. Gard. Edinb.* 44:311–342.

Ratter, J. A. 1992. Transitions between cerrado and forest vegetation in Brazil. In P. A. Furley, J. Proctor, and J. A. Ratter, eds, *Nature and Dynamics of Forest-Savanna Boundaries*, pp. 417–429. London: Chapman and Hall.

Ratter, J. A., G. P. Askew, R. F. Montgomery, and D. R. Gifford, 1977. Observações adicionais sobre o Cerradão de solo mesotrófico no Brasil Central. In M. G. Ferri, ed., *IV Simpósio sobre o Cerrado*, pp. 303–316. São Paulo: Editora da Universidade de São Paulo.

Ratter, J. A., G. P. Askew, R. F. Montgomery, and D. R. Gifford. 1978a. Observations on the vegetation of northeastern Mato Grosso: II. Forests and soils of the Rio Suiá-Missu area. *Proc. Roy. Soc. London B.* 203:191–208.

Ratter, J. A., G. P. Askew, R. F. Montgomery, and D. R. Gifford. 1978b. Observations on forests of some mesotrophic soils in central Brazil. *Rev. Bras. Bot.* 1:47–58.

Ratter, J. A., S. Bridgewater, R. Atkinson, and J. F. Ribeiro. 1996. Analysis of the floristic composition of the Brazilian Cerrado vegetation II: Comparison of the woody vegetation of 98 areas. *Edinb. J. Bot.* 53:153–180.

Ratter, J. A. and T. C. D. Dargie. 1992. An analysis of the floristic composition of 26 cerrado areas in Brazil. *Edinb. J. Bot.* 49:235–250.

Ratter, J. A., A. Pott, V. J. Pott, C. N. Cunha, and M. Haridasan. 1988. Observations on woody vegetation types in the pantanal and at Corumbá, Brazil. *Notes Roy. Bot. Gard. Edinb.* 45:503–525.

Ratter, J. A. and J. F. Ribeiro. 1996. Biodiversity of the flora of cerrado. In R. C. Pereira and L. C. B. Nasser, eds., *Anais do VIII Simpósio sobre o Cerrado and Proceedings of the I International Symposium on Tropical Savannas*, pp. 3–6. Planaltina: Empresa Brasileira de Pesquisa Agropecuaria.

Ratter, J. A., J. F. Ribeiro, and S. Bridgewater. 1997. The Brazilian cerrado vegetation and threats to its biodiversity. *Ann. Bot.* 80:223–230.

Ratter, J. A., J. F. Ribeiro, and S. Bridgewater. 2000. Distribuição das espécies lenhosas da fitofisionomia cerrado sentido restrito nos estados compreendidos no bioma cerrado. *Bol. Herb. E. P. Heringer 5* (in press).

Ratter, J. A., P. W. Richards, G. Argent, and D. R. Gifford. 1973. Observations on the vegetation of northeastern Mato Grosso: I. The woody vegetation types of the Xavantina-Cachimbo Expedition Area. *Phil. Trans. Roy. Soc. London B* 226:449–492.

Reis, A. C. S. 1971. Climatologia dos cerrados. In M. G. Ferri, ed., *III Simpósio sobre o Cerrado*, pp. 15–26. São Paulo: Editora da Universidade de São Paulo.

Ribeiro, J. E. L. S., M. J. G. Hopkins, A. Vicentini, C. A. Sothers, M. A. S. Costa, J. M. Brito, M. A. D. Souza, L. H. P. Martins, L. G. Lohman, P. A. C. L. Assunção, E. C. Pereira, C. F. Silva, M. R. Mesquita, and L. C.

Procópio. 1999. *Flora da Reserva Ducke: Guia de identificação das plantas vasculares de uma floresta de terra-firme na Amazônia central*. Manaus: INPA-DFID.

Ribeiro, J. F. 1983. "Comparação da concentração de nutrientes na vegetação arbórea e nos solos de um cerrado e um cerradão no Distrito Federal." Master's thesis, Universidade de Brasília, Brasília, Brazil.

Ribeiro, J. F. and M. Haridasan. 1990. Comparação fitossociológica de um cerrado denso e um cerradão em solos distróficos no Distrito Federal. In *Anais do XXXV Congresso Nacional de Botânica*, pp. 342–353. Brasília: Sociedade Botânica do Brasil.

Ribeiro, J. F. and B. M. T. Walter. 1998. Fitofisionomias do Bioma Cerrado. In S. M. Sano and S. P. Almeida, eds., *Cerrado: Ambiente e Flora*, pp. 87–166. Planaltina: Empresa Brasileira de Pesquisa Agropecuária.

Rizzini, C. T. 1963. A flora do cerrado: Análise florística das savanas centrais. In M. G. Ferri, ed., *Simpósio sobre o Cerrado*, pp. 127–177. São Paulo: Editora da Universidade de São Paulo.

Rizzini, C. T. 1979. *Tratado de Fitogeografia do Brasil: Vol. 2, Aspectos Sociológicos e Florísticos*. São Paulo: Hucitec-Editora da Universidade de São Paulo.

Rizzini, C. T. and E. P. Heringer. 1962. *Preliminares Acêrca das Formações Vegetais e do Reflorestamento no Brasil Central*. Rio de Janeiro: Edições SIA, Ministério da Agricultura.

Rizzini, C. T. and M. M. Pinto. 1964. Areas climático-vegetacionais do Brasil segundo os métodos de Thornthwaite e Mohr. *Rev. Bras. Geogr.* 26: 523–547.

Schiavini, I. 1997. Environmental characterization and groups of species in gallery forests. In *Proceedings of the International Symposium on Assessment and Monitoring of Forests in Tropical Dry Regions with Special Reference to Gallery Forests*, pp. 107–113. Brasília: Editora da Universidade de Brasília.

Silva Júnior, M. C. 1995. "Tree Communities of Gallery Forests of the IBGE Ecological Reserve, Distrito Federal, Brazil." Ph.D. thesis, University of Edinburgh, Edinburgh, Scotland.

Silva Júnior, M. C. 1997. Relationships between the tree communities of the Pitoco, Monjolo and Taquara gallery forests and environmental factors. In *Proceedings of the International Symposium on Assessment and Monitoring of Forests in Tropical Dry Regions with Special Reference to Gallery Forests*, pp. 287–298. Brasília: Editora da Universidade de Brasília.

Silva Júnior, M. C., J. M. Felfili, P. E. Nogueira, and A. V. Rezende. 1998. Análise florística das matas de galeria no Distrito Federal. In J. F. Ribeiro, ed., *Cerrado: Matas de Galeria*, pp. 51–84. Planaltina: Empresa Brasileira de Pesquisa Agropecuaria.

Silva Júnior, M. C., P. A. Furley, and J. A. Ratter. 1996. Variation in the tree communities and soils with slope in gallery forest, Federal District,

Brazil. In M. G. Anderson and S. M. Brooks, eds., *Advances in Hillslope Processes*, vol. 1, pp. 451–469. London: John Wiley and Sons.

Silva Júnior, M. C., P. E. Nogueira, and J. M. Felfili. 1998. Flora lenhosa das matas de galeria no Brasil Central. *Bol. Herb. E. P. Heringer* 2:57–76.

Van den Berg, E. and A. T. Oliveira-Filho. 1999. Spatial partitioning among tree species within an area of tropical montane gallery forest in southeastern Brazil. *Flora* 194:249–266.

Van der Hammen, T. 1983. The palaeoecology and palaeogeography of savannas. In F. Bourliere, ed., *Tropical Savannas*, pp. 19–35. Amsterdam: Elsevier.

Vanzolini, P. E. 1963. Problemas faunísticos do cerrado. In M. G. Ferri, ed., *Simpósio Sobre o Cerrado*, pp. 307–320. São Paulo: Editora da Universidade de São Paulo.

Walter, B. M. T. and J. F. Ribeiro. 1997. Spatial floristic patterns in gallery forests in the cerrado region. In *Proceedings of the International Symposium on Assessment and Monitoring of Forests in Tropical Dry Regions with Special Reference to Gallery Forests*, pp. 339–349. Brasília: Editora da Universidade de Brasília.

Warming, E. 1892. *Lagoa Santa: Et Bidrag til den Biologiske Plantegeographi*. Copenhagen: K. danske vidensk Selsk., 6.

Warming, E. 1973. Lagoa Santa: Contribuição para a Geographia Phytobiologica (reprint of the 1908 Portuguese translation of the work). In M. G. Ferri, ed., *Lagoa Santa e a Vegetação de Cerrados Brasileiros*, pp. 1–284. Belo Horizonte: Livraria Itatiaia Editora and Editora da Universidade de São Paulo.

Werner, P. A. 1991. *Savanna Ecology and Management, Australian Perspectives and Intercontinental Comparisons*. London: Blackwell.

7

Herbaceous Plant Communities

Tarciso S. Filgueiras

THE HERBACEOUS PLANT COMMUNITIES, CONSISTING OF PLANT life forms not considered trees, can be found as the ground layer of forest habitats such as gallery and semideciduous forests and woodland *cerradão*. But they reach their highest diversity in open habitats such as *campo limpo, campo sujo, campo rupestre*, cerrado *sensu stricto*, and *campo de murundus* (see chapter 6). In such habitats a surprising number of life forms, taxonomic diversity, and adaptations can be found. These features make these communities very attractive both to biologists and to the general public. A *campo limpo* or an area of *campo rupestre* in full bloom during December and January is arguably one of the prettiest natural sights in central Brazil. These rich, challenging communities are described here.

FLORISTICS AND PHYTOSOCIOLOGY

Herbaceous communities are dominated by chamaephytes, hemicryptophytes, geophytes, therophytes, lianas, and epiphytes (Ellenberg and Mueller- Dombois 1967). These diverse life forms are collectively designated by an array of names such as the herbaceous layer (Medina 1987), ground layer (Eiten 1990, 1991), nontrees (Gentry and Dodson 1987), field layer (Munhoz and Proença 1998), herbaceous category (Filgueiras et al. 1998), nonarboreal category (Felfili et al. 1998), and nonwoody vegetation (Castro et al. 1999). These plants are less conspicuous in forests but are the most important elements in the open habitats, where they frequently cover the entire ground and dominate the scene in number of both species and individuals.

The number of species in the herbaceous communities is high, and current data indicate at least 4,700 species (Mendonça et al. 1998) for the cerrado region. The number of species in the nonarboreal versus arboreal category is estimated to vary from 3:1 (Felfili et al 1994) to 4.5:1 (Mendonça et al. 1998). Thus, for each tree species in the cerrados there are between 3 and 4.5 nonarboreal species. The corollary to this statement is that any botanical survey that does not take into account the herbaceous members of the flora is grossly underestimating the total species diversity of the area. Therefore, the available data unequivocally disqualify earlier statements that the herbaceous flora in the Cerrado region is poorer than the woody component. Estimates around 2.5:1 by Castro et al. (1999) should also be disregarded because these authors overlooked recent data such as those of Mendonça et al. (1998), Felfili et al. (1998) and Filgueiras et al. (1998).

The taxonomic composition of the plants in the nonarboreal category is varied. The dominant families are Leguminosae (ca. 780 spp.), Compositae (ca. 560 spp.), Gramineae (ca. 500 spp.), and Orchidaceae (ca. 495 spp.). The genera with the highest number of species are *Chamaecrista* (ca. 120 spp.), *Paspalum* (ca. 117 spp.), *Mimosa* (ca. 113 spp.) *Vernonia* (ca. 100 spp.), *Habenaria* (ca. 70 spp.), and *Panicum* (ca. 63 spp.). On the other hand, a large number of genera are represented by a single species, such as *Langsdorffia, Ottonia, Paragonia, Sanderella, Soaresia,* and *Tatianyx* (Mendonça et al. 1998).

The diversity and abundance of species vary greatly in the different vegetation types (Ratter 1987; Mantovani 1990; Mantovani and Martins 1993) to the extent that the cerrado has aptly been described as a mosaic of resources under any resolution scale (Alho 1982). For example, while surveying the herbaceous communities of Chapada Pratinha (states of Goiás and Minas Gerais; 15°–20° S and 46°–49° W), Felfili et al. (1994) found that the gallery forests showed the lowest number of species (47) against 210 in the cerrado *sensu stricto*. The *cerradão* showed an intermediate number (63 spp.). In the various herbaceous communities at this locality, five families (Gramineae, Leguminosae, Euphorbiaceae, Compositae, and Rubiaceae) out of a total of 64 comprised 53% of all species reported. The remaining 59 families were represented by one to seven species. The grasses alone represented 25% of all species, and the legumes 13%. About 51% of all individuals surveyed were grasses.

In forest formations the more frequent genera and species are as follows: *Coccocypselum* and *Psychotria,* the species *Olyra ciliatifolia, Oplismenus hirtellus, Serjania lethalis,* and the ground orchid *Craniches candida.* Cerrado *sensu stricto* and *campo sujo* (chapter 6) are dominated by species of the genera *Axonopus, Chamaecrista, Croton, Hyptis, Mimosa,* and *Oxalis*; and the species *Echinolaena inflexa, Cissampelos ovalifolia, Trachypogon spicatus, Galactia glaucescens,* and *Andira humilis* (Felfili et al. 1994; Mendonça et al.1998).

Campo limpo is dominated by the Gramineae, and common grasses include *Echinolaena inflexa, Leptocoryphium lanatum, Trachypogon spicatus, Paspalum* spp., *Axonopus* spp., *Mesosetum loliiforme, Schizachyrium tenerum, Tristachya leiostachya,* and *Aristida* spp. Other commonly encountered taxa are *Pradosia brevipes* (= *Chrysophyllum soboliferum), Parinari obtusifolia, Smilax* spp., *Banisteriopsis campestris, Campomanesia* spp., *Cambessedesia espora, Myrcia linearifolia, Spiranthera odoratissima, Senna rugosa, Centrosema bracteosum, Anemopaegma* spp., *Byrsonima* spp., *Calea* spp., *Vernonia* spp., and *Mimosa* spp. (Mendonça et al. 1998).

Campos rupestres are predominantly found in the states of Minas Gerais, Bahia, and Goiás (chapter 6). The rich flora of these habitats is estimated at around 4,000 species (Giulietti et al. 1997). Serra do Cipó, a significant part of the Espinhaço Range (10°–20° S and 40°–44° W) in Minas Gerais, has been carefully surveyed by Giulietti et al. (1987). These authors have documented 1,590 species, the great majority in the nonarboreal category. The best-represented families are Compositae (169 spp.), Gramineae (130 spp.), Leguminosae (107 spp.), Melastomataceae (90 spp.), Eriocaulaceae (84 spp.), and Orchidaceae (80 spp.). Characteristic genera are *Paepalanthus, Leiothrix, Syngonanthus, Vellozia, Barbacenia, Xyris, Marcetia, Lychnophora, Declieuxia, Cambessedesia, Mimosa,* and *Microlicia.*

The flora of the *campo rupestres* is largely endemic, especially in groups such as Velloziaceae (ca. 70%), Eriocaulaceae (ca. 68%), and Xyridaceae (ca. 30%). Some species are narrow endemics, known only from a small area. Examples are found in the genera *Barbacenia, Paepalanthus, Syngonanthus,* and *Xyris.* Other families with endemic species are Iridaceae (*Pseudotrimezia*), Cactaceae (*Cipocereus, Uebelmannia*), Sterculiaceae (*Raylea*), and Compositae (*Bishopella, Morithamnus*). *Aulonemia effusa,* a shrubby bambusoid grass, is known only from the rocky outcrops of the Serra do Espinhaço, Minas Gerais.

The Chapada dos Veadeiros (13°46' S and 47°30' W) is the best-known *campo rupestre* in the state of Goiás. Several vegetation types occur there, and herbaceous communities abound everywhere. In a recent account of the local flora (both woody and herbaceous), Munhoz and Proença (1998) listed 1,310 species distributed in 498 genera and 120 families. The five richest families were Leguminosae (144 spp.), Compositae (125 spp.), Gramineae (115 spp.), Melastomataceae (54 spp.), and Orchidaceae (47 spp.). There, too, the flora is largely herbaceous, and there are many endemics. The flora of this Chapada is only partially known, and further species of trees and ground vegetation are still being discovered. An unexpected biogeographic connection between the Chapada dos Veadeiros, Africa, and Australia recently came to light with the description of a new grass species of the genus *Triraphis*. The genus was previously known only from Africa (6 spp.) and Australia (1 sp.). The discovery of *T. devia* in a *campo rupestre* of the Chapada dos Veadeiros thus presents a puzzling biogeographical enigma (Filgueiras and Zuloaga 1999).

Derived from ultramafic rocks, serpentine soils are characterized by high concentrations of magnesium, iron, nickel, chromium, and cobalt, and abnormally low levels of some essential plant nutrients such as phosphorus, potassium, and calcium (Brooks 1987). Such soils were found by Brooks et al. (1990) in several municipalities in Goiás, along with a serpentine-adapted flora almost exclusively dominated by shrubs, subshrubs, and herbs. In Niquelândia (ca. 14°18' S and 48°23' W) several serpentine endemic grass taxa have been described (Filgueiras et al. 1993; Filgueiras 1995). Because of this high level of endemism, Niquelândia has been included as one of 87 areas selected for special conservation efforts in the cerrado region (chapter 18; locality 33 in fig. 18.1).

The wet habitats (such as the *veredas, campo de murundus*; see chapter 6) tell a different story. In the "Pantanal" of Mato Grosso, a huge seasonally flooded area with a large diversity of landscapes (Allem and Valls 1987; A. Pott 1988; V. J. Pott 1998), the dominant genera are *Nymphaea, Nymphoides, Mayaca, Lycopodium, Pontederia, Eichhornia, Pistia, Ludwigia, Utricularia,* and graminoids, such as grasses, sedges, and Eriocaulaceae.

Outside the Pantanal, wet areas in general are dominated by the grasses *Paspalum hydrophylum, P. pontonale, Leersia hexandra, Sorghastrum setosum, Reimarochloa acuta,* and *Axonopus purpusii*; but *Habenaria*, a genus of ground orchids typical of wet habitats, is represented by at least 70 species (Mendonça 1998). In the Parque Nacional Grande Sertão Veredas (15°00' S and 45°46' W) the aquatic flora is also quite rich (Filgueiras, unpublished data), represented by *Lycopodium, Lycopodiella, Ludwigia, Eleocharis, Bulbostylis, Paspalum morichalense, Panicum pernambucense, Cyperus giganteus, Eriocaulon, Philodice,* and *Syngonanthus*. The aquatic flora of the savannas of Roraima (5°16' N and 1°27' W) is dominated by Alismataceae, Eriocaulaceae, Iridaceae, Mayacaceae, Menyanthaceae, Nymphaeaceae, Pontederiaceae, and Orchidaceae (Miranda 1998).

It is important to stress that various sites differ greatly in species composition. In the Chapada Pratinha (15°–20° S and 46°–49° W), for instance, the five sites surveyed had 52 to 121 species (Filgueiras et al.1998), but only eight species were common to all sites. The total density varied from 595 to 3,278 plants per hectare. *Echinolaena inflexa*, a perennial, rhizomatous, C3 grass, represented 36% of all plants found in that Chapada. In Roraima, a predominantly "savanna" state, 66% of all species surveyed by Miranda (1998) were herbs, including 58 grasses and 40 sedges.

Great taxonomic diversity is commonly found in the same area, throughout the cerrados. For instance, Goldsmith (1974) found that species composition of a *vereda* changed gradually as one goes downslope. Filgueiras (unpublished data) also found in a cerrado *sensu stricto* at Gama/Cabeça de Veado (ca. 15°52' S and 47°58' W) 8 to 28 species per m^2 (mean 11.5 species per m^2). This range is relatively high, considering that the most distant plots were not more than 1,000 meters apart. The consequences of a mosaic habitat to conservation will be addressed later in this chapter.

The general trend for herbaceous communities is to encounter high species densities in the cerrados. Monodominance is rare. A few documented exceptions are (1) at the Parque Nacional das Emas (18°10' S and 52°55' W), where around 70% of the area (ca. 131,800 ha) is dominated by *Tristachya leiostachya*, a caespitose, perennial grass; and (2) huge tracts of land are totally covered by *Actinocladum verticillatum* ("taquari"), resulting in an almost impenetrable thicket (Verdesio and Garra 1987). This thin-culmed bamboo reaches 80–200 cm in height and about 1 cm in diameter. The immense area where it grows is locally called *grameal*. The name denotes a special kind of vegetation and can be loosely translated as "vast quantity of grass."

Essentially herbaceous stemmed vines and lianas are fairly common. The families with the largest number of species with these life forms are Aristolochiaceae (*Aristolochia*), Asclepiadaceae (*Blepharodon, Ditassa* and *Oxypetalum*), Bignoniaceae (*Arrabidaea, Cuspidaria,* and *Paragonia pyramidata*), Convolvulaceae (*Ipomoea),* Cucurbitaceae (*Cayaponia, Melancium*), and Sapindaceae (*Serjania*). *Mikania*, a genus of about 18 vine species, is a member of the Compositae. The parasite *Cassytha filiformis* (Lauraceae) is ubiquitous.

Palms are not very diverse in the Cerrado, but their ecological, economic, and cultural importance greatly surpasses their low taxonomic diversity. They represent valuable resources for wildlife, domesticated animals, and many rural populations. A few acaulescent and shrubby genera and species occur: *Allagoptera* spp., *Astrocaryum campestre, Attalea* spp., *Butia archeri, Syagrus graminea,* and *S. petraea* (Mendonça et al. 1998). There are about 10 bambusoid grasses with a shrubby or herbaceous habit in the Cerrado (e.g., *Actinocladum verticillatum, Apoclada cannavieira, A. arenicola, Aulonemia aristulata, Olyra ciliatifolia, O. taquara, Pharus lappulaceus,* and *Raddiella esenbeckii*) (Judziewicz et al. 1998). The latter, a diminutive plant up to 30 cm tall, resembles a fern and is found on river banks in forests and occasionally in *cerradão* (chapter 6). The bambusoid species of open habitats are fairly common throughout the cerrado and under certain ecological conditions form vast colonies.

A small number of species of cacti occur. A few examples of low-growing taxa are *Cipocereus, Epiphyllum phyllanthus, Melocactus paucipinus, Pilosocereus pachycladus, Rhipsalis* spp., and *Uebelmannia* (Taylor and Zappi 1995). Other genera and species of cacti have been described (Diers and Esteves 1989; Braun 1990). The epiphytic flora is not particularly diverse in the cerrado, as compared with other regions such as Amazonia (Daly and Prance 1989) or the Atlantic forest (Lima and Guedes-Bruni 1994). The families best represented are Orchidaceae (*Bulbophyllum, Cattleya,* and *Oncidium),* Bromeliaceae (*Aechmaea, Bilbergia,* and *Tillandsia*), Araceae (*Philodendron*), Piperaceae (*Peperomia*), and Cactaceae (*Rhipsalis* spp.). Epiphytic ferns are also common (*Adiantum, Anemia,* and *Elaphoglossum*) (Mendonça et al. 1998).

Hemiparasitic taxa are restricted to the Loranthaceae (*Phthirusa, Psittacanthus,* and *Struthanthus*). The most abundant species is *Phthirusa ovata,* whose fruits (red and sticky when ripe) are easily dispersed by birds. Species of *Psittacanthus* and *Struthanthus* are also rather common in *cerradão* and cerrado *sensu stricto.* When in full bloom, their showy yellowish or reddish flowers stand out in the cerrado vegetation.

The commonest true parasites and saprophytes are *Apteria aphylla, Cuscuta* spp., *Helosia brasiliensis, Langsdorffia hypogea, Cassytha* spp., *Pilostyles* spp., and *Voyria flavescens.* Insectivorous droseras are found in swamps, damp places, *veredas,* and wet *campos.* Although they are quite small (usually only 3 to 10 cm tall), they are easily recognizable by their reddish color and conspicuous glandular hairs. A single shrubby gymnosperm (*Zamia bronguiartii*) is found in cerrado *sensu stricto* and on limestone outcrops in the state of Mato Grosso (Guarim Neto 1987).

Invasive species are found practically in all anthropic habitats, be it forest, cerrado *sensu stricto, campo sujo, vereda,* agricultural field, pasture, or land around human dwellings (chapter 5). Mendonça et al. (1998) listed 456 invasive species in the cerrado. Invasive species can pose a serious threat to the cerrado flora by successfully competing with the natives or even eliminating them altogether (Filgueiras 1990). Loss of biodiversity is therefore one of the most serious consequences of the introduction of weeds in cerrado ecosystems. Some introduced African grasses of special concern include *Andropogon gayanus, Brachiaria* spp, *Hyparrhenia rufa, Melinis minutiflora,* and *Panicum maximum.* It is postulated that plants in the nonarboreal category are probably more sensitive to human-related disturbances than trees in the cerrados (Filgueiras et al. 1998). Therefore, plants in that category may be used as a parameter to assess the level of human intervention in cerrado ecosystems.

SPECIAL ADAPTATIONS AND COMPARISONS WITH OTHER ECOSYSTEMS

Many cerrado species display an unusual behavior: their aerial stems die and revert to the underground once a year, during the dry season, even in the absence of fire. When conditions improve, they sprout again from special subterranean structures, the xylopodia. These plants are called recurrent shrubs and semi-shrubs by Eiten (1984), and hemixyles or geoxyles by others (Ellenberg and Mueller-Dombois 1967). An even more peculiar trait of several species in this category is the existence of extensive subterranean woody systems. Apparently Reinhardt (in Warming 1973:58) was the first author to note this peculiarity. While traveling in central Brazil he noted that "les grands arbres souterrains à tige verticale cachée dans le sol, sont une des peculiarités les plus curieuses de la flore de ces régions."

Warming (1973) himself was quite impressed by this singularity of some cerrado species at Lagoa Santa, Minas Gerais. He described and illustrated a plant of *Andira laurifolia* (= *A. humilis*) with a well-developed subterranean system. He emphasized that an area of about 10 m in diameter was occupied by a single individual plant. The investigation of the subterranean growth forms of several cerrado species has been subsequently pursued by researchers such as Rawitscher and Rachid (1946), Rizzini and Heringer (1966), and Paviani (1987).

Some of these subterranean systems are clearly means of vegetative reproduction. The plant sprouts in different directions, but all the shoots are interconnected below ground, forming an extensive, complex system. An example is *Parinari obtusifolia,* in which the aerial stems give the appearance of individual plants, but excavation reveals that all the stems are shoots from the same system of subterranean axes. If considered in its entirety, this classical example of a cerrado subshrub could more accurately be classified as a kind of "subterranean tree." A similar pattern can be seen in *Anemopaegma arvense*, where the aerial stems can be as far as 120 cm apart.

This phenomenon seems to be quite widespread and should be investigated in depth because it has serious implications for the definition of the individual plant in certain cerrado vegetation types, clearly a critical definition in phytosociological studies. In some cases, what has been regarded as several plants is in reality a single individual whose aerial stems are sometimes more than one meter apart but all of whose underground parts are connected. Colonies of *Annona pygmaea*, *Andira humilis*, *Pradosia brevipes*, *Parinari obtusifolia*, and *Smilax goyazana*, among others, can reach several meters in diameter. Each of these colonies may very well derive from a single seed. The same can be said of some grasses with extensive stoloniferous growth habit, such as *Axonopus purpusii*, *Paspalum morichalense*, and *Reimarochloa acuta*.

Another intriguing aspect of the cerrado herbaceous flora is the low occurrence of annual species, representing less than 5%. Thus, perenniality is clearly at an advantage. The success of the perennial versus the annual habit in the cerrado may be due to the limited amount of "topsoil" water during some months of the year (May–September), such that plants need to possess extensive root systems or other adaptations to withstand the dry periods (chapter 10). However, in Gilbués, Piauí (ca. 9°34' S and 44°55' W), a marginal cerrado area (Castro et al. 1999) subjected to serious desertification processes, Filgueiras (1991) found 30 times more annuals than perennials. Under prolonged dry periods, many perennials cannot survive except as seeds. Apparently this is precisely what happens in some areas of the adjacent arid *caatinga* where, among the herbaceous plants, annuality is the rule.

Data on the biomass of plants in the herbaceous layer of cerrados is meager. However, it is well established that they provide the bulk of the fuel for the seasonal fires (Coutinho 1990). In the Distrito Federal, central Brazil, Kauffman et al. (1994) found that the biomass of fuel loads varied greatly from *campo limpo* (7,128 kg/ha) to cerrado *sensu stricto* (10,031 kg/ha) (chapter 4). Grasses comprised 91% to 94% of the total aboveground biomass in *campo limpo* and *campo sujo*, whereas in the *campo cerrado* and cerrado *sensu stricto* communities the graminoids represented only 27%. The remainder of fuel load biomass was composed of deadwood (18%), dicot leaf litter (36%), and dicot and shrub leaves (18%). In Roraima, Miranda and Absy (1997) found that grasses and sedges were the most important families in terms of total biomass. In the 10 areas sampled, the grasses comprised 43% of the biomass; the sedges,

35%. Silva and Klink (1996) found significant variation in biomass at the species level, in a study that compared the biomass of two native grasses, *Schizachyrium tenerum* (C4) and *Echinolaena inflexa* (C3). After 24 weeks of growth, *S. tenerum* was taller, had more tillers and leaves, and allocated more biomass to roots and shoots than *E. inflexa*.

ECOLOGICAL, ECONOMIC, AND ANTHROPOLOGIC IMPORTANCE

It is not surprising that the highly diverse flora of the cerrado is useful to humans and fauna in a variety of ways. The usefulness of these plants falls into several categories, such as food, forage, medicine, ornament, and genetic resources. Because they cover vast tracts of land, these plant communities, especially the *veredas* and the aquatic and semi-aquatic communities, play a key role in watershed protection. Invariably, wherever there is a stream or river source, the presence of herbaceous communities covering the soil minimizes erosion. On the other hand, their absence indicates that erosion processes will begin and much soil will be lost in a short time. To maintain the landscape intact, it is essential to allow a rich, diverse herbaceous community to establish and thrive locally; the important role of nonarboreal plants in natural habitats becomes apparent when they are removed and the soil is exposed. The removal of the herbaceous layer in the cerrado means the disruption of countless ecological processes. Besides the obvious demise of the plant life involved, the local fauna that depends directly or indirectly on the plants will be greatly reduced or, more likely, altogether eliminated. Animals lose their food, vital space, breeding places, and escape routes. Erosion processes may begin quickly. The consequences of erosion and the silting of river channels are well documented, especially in agricultural systems (Goedert et al. 1982; Dedecek et al. 1986).

An important but little discussed aspect of the ecological function of the nonarboreal flora is the effective protection offered to some native bird and small mammal populations (chapters 13, 14). These animals feed, breed, and raise their young in the habitats dominated by these plants. Some small mammals, especially rodents, have very limited home ranges (Alho 1982), and their spatial distribution is closely connected with the distribution of natural resources. Some are generalists, but a few are specialists. For example, the rodent *Bolomys* (= *Zygodontomys*) *lasiurus* feeds mainly on grass "seeds" and is found in a large number of habitats, whereas *Oxymycterus robertii* occurs mostly in open grasslands dominated by *Tristachya leiostachya*. Many birds, mammals, and especially insects are obligatory pollinators and seed dispersal agents of a number of nonarboreal cerrado species. Entire guilds of insects depend exclusively on these plants during their life cycles (chapter 17). The role that all these animal species play in the maintenance of cerrado biodiversity is crucial albeit not readily apparent.

A great extent of cerrado land is used for grazing (chapter 5). About 40% of the Brazilian cattle industry depends on this native grazing land (Filgueiras and Wechsler 1992). A. Pott (1988) presented a list of 145 grasses, 70 legumes, and 60 forb and browse species that feed four million cattle in the *pantanal* of Mato Grosso, western Brazil. The more important taxa in the *pantanal* are *Axonopus purpusii*, *Mesosetum chaseae*, *Hemarthria altissima*, *Leersia hexandra*, *Paratheria prostrata*, *Paspalidium paludivagum*, *Paspalum plicatulum*, and species of the genera *Reimarochloa*, *Aeschynomene*, *Discolobium*, *Galactia*, *Rhynchosia*, *Teramnus*, and *Vigna*. In the Distrito Federal alone, central Brazil, Filgueiras (1992) listed 134 native forage grass species, 13 of which are of very high quality.

Legumes are extremely important in the cerrados. Kirkbride (1984) listed 548 legume species in the region, distributed in 50 genera. Among the genera with fodder species are *Stylosanthes*, *Zornia*, *Desmodium*, *Aeschynomene*, *Eriosema*, and *Arachis*. Because of their high forage value, *Aeschynomene* (ca. 52 spp.; Fernandes 1996) and *Stylosanthes* (ca. 11 spp.; Ferreira and Costa 1979) deserve special attention. Besides increasing the nutritional value of fodder, legumes also increase natural soil fertility by adding nitrogen to the system through the fixation of atmospheric nitrogen by associated *Rhizobium*. An evaluation of the genetic resources of legumes and grasses native to Brazil is presented by Valls and Coradin (1986).

Although grasses and legumes make up the bulk of the grazed plants of the cerrado, they are by no means the only fodder resources there. Macedo et al. (1978) found 83 cerrado species, belonging to 33 families, eaten by cattle in the cerrado. The pattern of consumption of leaves, flowers, and seeds by cattle and other domestic animals depends on the plant species, habitat, and season. Similarly, Pereira (1984) cites 35 species in the cerrado of the Distrito Federal that are frequently eaten by cattle, 26 of which are neither grasses nor legumes.

There is a wealth of medicinal plants amongst the nonarboreal species: roots, bark, stems, leaves, fruits, and seeds of over one hundred species are used by local populations and are also regularly commercialized in local markets and even exported to other areas outside the cerrado region. In many large Brazilian cities it is possible to buy medicinal plants to "cure" almost any kind of malady. The so-called raizeiros (literally, root dealers, or root healers) display their plants in fairs and open markets. The plants are sold in natura or as *garrafadas* (i.e., plant parts bottled in alcohol; Guarim Neto 1987), wine, or *pinga* (i.e., white rum, also known as *cachaça*). The plants supplying this market are harvested directly from the wild. It is noteworthy that there is always a fresh supply of plants to sell. This means that the harvesting is continuous throughout the year and through the years. The impact of continuous harvesting of wild plants from natural populations has not yet been evaluated.

The effectiveness of popular herbal treatments is understandably subject to dispute, but a few species have been scientifically tested with extremely good results (e.g., *Psychotria ipecacuanha* and *Renealmia exaltata*). A growing number of Brazilian botanists are studying the native medicinal flora (Siqueira 1988; Guarim Neto 1987; Barros 1982). Graduate programs that include research on medicinal plants have been established in several Brazilian universities. A small number of selected species has become the prime target for detailed investigation (e.g., *Heteropterys aphrodisiaca*, *Zamia brogniartii*, *Gomphrena officinalis*, *Pfaffia jubata* and *Macrosiphonia velame*). Other medicinal plants commonly used are *Centrosema bracteosum*, *Lychnophora ericoides*, *Palicourea coriacea*, *P. marcgravii*, *P. rigida*, *Anemopaegma arvense*, and *Galactia glaucescens*. Aromatic plants are also found. The roots of *Croton adenodontus* ("arcassu") are used to flavor milk (the taste and odor are described as better than chocolate!), and the fleshy roots of *Escobedia grandiflora* are used as a local substitute for expensive saffron (Pereira 1992).

Toxic plants are found in several taxonomic groups (Ferreira 1971), but they are particularly abundant in the 35 species of *Psychotria. Palicourea marcgravii* ("erva-de-rato") is a major cattle poisoner in certain areas. It is generally held that frequently even toxic plants can be of medicinal importance, depending on the specific dosage of the active substance (Hoehne 1939).

Exploration of the phytochemistry and pharmacology of cerrado herbaceous plants can be rewarding (Gomes et al. 1981, Gottlieb et al. 1996). The potential of extracting lauric acid from *Cuphea* is being explored (Arndt 1985; Hirsinger 1985), and there are 43 native species of *Cuphea* in the cerrado (Mendonça et al.1998). The possibility of extracting essential oils from the 68 native species of *Hyptis* or even from the lemon grasses (*Elionurus* spp.) should also be explored. The cultivation (domestication) of the more promising species is a possible solution for the exploitation of natural stocks. The biosynthesis of new products and drugs is an area where close cooperation between governmental agencies and the private initiative is likely to make a difference. The search for new drugs and new food items has been strongly defended by prominent biologists such Raven (1990), Heywood (1992), and Wilson (1992).

Sustainability in the exploitation of medicinal plants must be a priority. The most-used species are suffering great reduction on their natural stocks due to decades of relentless exploitation. Such is the case of *Lychnophora ericoides* and of *Pseudobrickellia pinifolia* (both called "arnica"), whose stems and leaves are collected by the tons and sold everywhere in central Brazil as a kind of natural antibiotic. Xylopodia and roots of *Anemopaegma arvense* ("verga-tesa") and of *Centrosema bractesoum* ("rabo-de-tatu") are also harvested and sold everywhere as an aphrodisiac and against stomach maladies, respectively. The worst case is that of *Psychotria ipecacuanha* ("poaia," "ipeca," or "ipecacuanha"). At Serra do Tapirapuã, state of Mato Grosso, there used to be a unique vegetation type (known as "poaia forest") where the undergrowth was dominated by this species (Ferreira 1999). After many decades of exploitation of the roots of "poaia," the species is now scarce in the area and is doomed to local extinction. One might solve this problem through cultivation of promising wild species using simple but sound agronomic techniques such as those described by Mattos (1996).

Honey-producing herbaceous plants are also abundant and exploited commercially. Pereira (1990) found 220 species (of which 50 are Leguminosae and 40 Compositae) regularly visited by honeybees in the cerrado of the Distrito Federal in central Brazil.

The flourishing business of dried wildflower arrangements is a source of income for many families. It is the so-called sempre-viva industry (Giulietti et al. 1997). In Goiás and Distrito Federal the trade is called "Flores do Planalto" (Ferreira 1974; Filgueiras 1997). In the state of Minas Gerais this type of business has been quite active for more than 30 years. The plants are dried, tied in bundles, and sold as long-lasting bouquets. Eriocaulaceae, Gramineae and Compositae supply the bulk of the plants collected for this purpose. Many families in the states of Minas Gerais and Goiás set up "firms" and live exclusively on the "sempre-viva" trade. During past decades this was a thriving business. Since then, the rate of exportation has declined considerably.

A large number of native cerrado herbaceous species have obvious potential as ornamentals. Orchids, bromeliads, and cacti need not be discussed here, because there is already an active market that thrives on the legal and illegal commercialization of these plants, a large percentage of which are harvested in the wild. But there is a great potential to establish legal enterprises to exploit the trade (through cultivation) of ornamental cerrado herbaceous plants. All the known species of *Paepalanthus* are very ornamental, but *P. speciosus* and *P. hillairei* are outstanding. The same can be said of many other species, including all the previously mentioned species of bamboos and palms. The genera *Hippeastrum, Allamanda, Mandevilla, Anthurium, Asterostygma, Philodendron, Spathiphyllum, Begonia, Lobelia, Cochlospermum, Calea, Lagenocarpus, Calliandra, Chamaecrista, Collaea, Mimosa, Hibiscus, Pavonia, Maranta, Cambessedesia, Lavoisiera,* and *Microlicia* are examples from an extensive list of plants whose ornamental value merits close investigation. The perfume industry might have potential in exploring the unique fragrance of certain cerrado flowers, such as *Spiranthera odoratissima, Protium ovatum,* and *Mimosa* spp.

The conservation of crop genetic resources is vital for humankind. It is well established that genetic uniformity can make crops vulnerable to epidemics of pests and diseases (Myers 1983). The nonarboreal flora of the cerrados encompasses a large number of species related to crops that represent invaluable resources in crop improvement programs. Despite their obvious potential as sources of wild genes, they have been largely neglected by agronomists and plant breeders. The following list will provide a small sample of the potential in this field: in the cerrados there exist at least 42 species of *Manihot,* 37 *Dioscorea,* 33 *Arachis,* 30 *Ipomoea,* 28 *Solanum,* 26 *Psidium,* 25 *Piper,* 22 *Passiflora,* 12 *Annona,* 11 *Cissus* (related to *Vitis*), 8 *Vanilla,* 7 *Vigna,* 5 *Hibiscus,* 4 *Anacardium,* 3 *Oryza,* 2 *Ananas,* and 1 *Phaseolus* (*P. uleanus*) (Mendonça et al. 1998).

Hoyt (1988) argues that most protected areas are established to preserve a unique landscape or a rare animal species, but seldom to preserve a wild relative of an important crop. It is time for Brazil to protect the native cerrado genetic resources effectively, especially those related to crops and medicinal gene pools. A significant step in this direction has been the founding of the Center for Genetic Resource and Biotechnology (CENARGEN) of the Brazilian Research Enterprise (EMBRAPA). CENARGEN maintains a germplasm bank and an active cerrado plant collecting program (Schultze-Kraft 1980) considered as a model for developing countries. Of special interest are the collections of *Arachis* spp. and forage grass species (especially *Paspalum* spp.).

RESEARCH NEEDS AND CONCLUSIONS

A remarkable paper by Constanza et al. (1997) demonstrates, in economic terms, the value of the services that natural ecosystems provide to humanity. The various ecosystems are viewed as natural capital. The idea that natural ecosystems are worth money is not new; nonetheless it is astonishing to see that the actual estimated mean value of the ecological services rendered by the world's ecosystems and the natural capital that generates them amounts to U.S.$33 trillion.

In a recent workshop held in Brasília, and sponsored by the Ministry of the Environment and three nongovernmental organizations to address this issue, 87 areas were selected for special conservation efforts (see chapter 18). The document resulting from this workshop further states that the conservation of cerrado biodiversity can be achieved by establishing conservation units, doing inventories, supporting herbaria, and monitoring natural populations. The nonarboreal flora of the cerrado is so rich and vital to many human endeavors that its study should be greatly encouraged. Nonetheless, most species in this category are of notorious taxonomic difficulty (such as orchids, grasses, sedges, Compositae, legumes, etc.). A way to overcome this difficulty would be to publish inexpensive field guides, such as those published for the herbs (Correa et al. 1998a) and for ferns and epiphytes of Panama (Correa et al. 1998b). In these publications the specimens representing each species are simply photocopied and bound. This simple initiative would most certainly widen interest in the study of these valuable components of the cerrado flora.

Wilson (1992) offered two important suggestions regarding the future preservation and conservation of biodiversity in general: a survey of the world's total flora, and the search for new biological products such as foods and drugs. A cerrado checklist such as that of Mendonça et al. (1998) is extremely useful and a promising beginning, but a "Cerrado Flora" is now desirable. If a cerrado flora is to be produced, inventories have to be made, herbaria have to be equipped and supported, and human resources must be trained and subsequently hired to work on their specialties.

Special attention should be given to the education of a new generation of plant taxonomists. They should be motivated and financially supported. Once properly trained, they should document the immense floristic wealth of the Cerrado Biome while opening new avenues for the rational uses of cerrado natural resources. The data presented here indicate that efforts and investments toward this most timely goal will be highly rewarding.

ACKNOWLEDGMENTS

I gratefully thank Drs. J. H. Kirkbride, Jr. and Jeanine M. Felfili for helpful suggestions on the manuscript. Two anonymous reviewers greatly improved the text. Dr. Taciana Cavalcanti provided valuable literature assistance, and Maria S.S. Amorim checked the bibliography for errors and inconsistencies. I also thank the Conselho Nacional de Desenvolvimento Científico e Tecnológico (CNPq) for a productivity research grant (Proc. No 301190/86-0).

REFERENCES

Alho, C. J. R. 1982. Brazilian rodents: Their habitats and habits. *The Pymatuning Symposia in Ecology. Special Publication Series. University of Pittsburgh*, 6:143–166.

Allem, A . C. and J. F. M. Valls. 1987. *Recursos Forrageiros Nativos do Pantanal Mato-grossense*. Brasília: Empresa Brasileira de Pesquisa Agropecuária.

Arndt, S. 1985. *Cuphea*: Diverse fatty acid composition may yield oleochemical feedstock. *J. Am. Oil Chem. Soc.* 62:6–12.

Barros, M. A. G. 1982. Flora medicinal do Distrito Federal. *Brasil Florestal* 12:35–45.

Braun, P. J. 1990. *Melocactus estevesii* P.J.Braun: Eine neue Kakteenart aus Nord Brasilien. *Kakteen und andere Sukkulenten* 41:6–10.

Brooks, R. R. 1987. *Serpentine and its Vegetation*. Portland: Dioscorides Press.

Brooks, R. R., R. D. Reeves, A. J. Baker, J. A. Rizzo, and H. D. Ferreira. 1990. The Brazilian serpentine expedition (Braspex), 1988. *Nat. Geogr. Res.* 6:205–219.

Castro, A. A. J. F., F. R. Martins, J. Y. Tamashiro, and G. J. Shepherd. 1999. How rich is the flora of Brazilian cerrados? *Ann. Miss. Bot. Gard.* 86:192–224.

Constanza, R., R. d'Arge, R. de Groot et al. 1997. The value of the world's ecosystem services and the natural capital. *Nature* 387:1–13.

Correa, M. D. A. 1998a. *Guía Preliminar de Campo:* Vol. 3, *Flora del Parque Nacional Altos de Campana, Panamá: Hierbas*. Chicago: Universidad de Panamá and Smithsonian Tropical Research Institute.

Correa, M. D. A. 1998b. *Guía preliminar de campo:* Vol. 4, *Flora del Parque Nacional Altos de Campana, Panamá: Bejucos y Epífitas*. Chicago: Universidad de Panamá and Smithsonian Tropical Research Institute.

Coutinho, L. M. 1990. Fire in the ecology of the Brazilian cerrado. In J. G. Goldammer, ed., *Fire in Tropical Biota*, pp. 82–105. Berlin: Springer-Verlag.

Daly, D. C. and G. T. Prance. 1989. Brazilian Amazon. In D. G. Campbell and H. D. Hammond, eds., *Floristic Inventory of Tropical Countries: The Status of Plant Systematics, Collections, and Vegetation, plus Recommendations for the Future*, pp. 401–426. New York: The New York Botanical Garden.

Dedecek, R. A., D. V. S. Resck, and E. Freitas Jr. 1986. Perdas de solo, água e nutrientes por erosão em latossolo vermelho-escuro nos Cerrados em diferentes cultivos sob chuva natural. *Rev. Bras. Ciênc. Solo* 10:265–272.

Diers, L. and E. Esteves. 1989. *Arrajadoa beatae* Braun and Esteves, eine neue Art (Cactaceae) aus Minas Gerais/Brasilien. *Kakteen und andere Sukkulenten* 40:250–256.

Eiten, G. 1984. Vegetation of Brasília. *Phytocoenologia* 12:271–292.

Eiten, G. 1990. A vegetação do Cerrado. In M. N. Pinto, ed., *Cerrado: Caracterização, ocupação e perspectivas*, pp.17–66. Brasília: Editora Universidade de Brasília.

Eiten, G. 1991. What's a herb? *Veröff. des Geobotanishchen Institutes der Eidg. Techn. Hochschule Stiftung Rübel* 106:288–304.

Ellenberg, H. and D. Mueller-Dombois. 1967. A key to Raunkiaer plant life forms with revised subdivisions. *Berichte des Geobotanishchen Institutes der Eidg. Techn. Hochschule Stiftung Rübel* 37:56–73.

Felfili, J. M., T. S. Filgueiras, M. Haridasan, M. C. da Silva Jr., Mendonça, R. C., and A. V. Rezende. 1994. Projeto biogeografia do bioma Cerrado: Vegetação e solos. *Cad. Geoc. IBGE* 12:75–166.

Felfili, J. M., M. C. da Silva Jr., T. S. Filgueiras and P. E. Nogueira. 1998. Comparison of Cerrado (s.s.) vegetation in central Brazil. *Ciênc. Cult.* 50:237–243.

Fernandes, A.. 1996. *O Táxon Aeschynomene no Brasil.* Fortaleza: Edições Editora da Universidade Federal do Ceará.

Ferreira, M. B. 1971. As plantas tóxicas no Distrito Federal. *Cerrado* 4:26–30.

Ferreira, M. B. 1974. "Flores do Planalto," divisas para Brasília. *Cerrado* 6:4–7.

Ferreira, M. B. and N. M. S. Costa. 1979. O Gênero *Stylosanthes* no Brasil. Belo Horizonte: Empresa de Pesquisa Agropecuária de Minas Gerais.

Ferreira, M. S. F. D. 1999. *Mato Grosso: Um Pouco de seus Ecossistemas.* Cuiabá: Instituto de Biociências, Universidade Federal de Mato Grosso.

Filgueiras, T. S. 1990. Africanas no Brasil: Gramíneas introduzidas da África. *Cad. Geoc.* 5:57–63.

Filgueiras, T. S. 1991. Desertificação em Gilblués, Piauí: Uma análise agrostológica. *Cad. Geoc.* 7:23–27.

Filgueiras, T. S. 1992. Gramíneas forrageiras nativas no Distrito Federal. *Pesquisa Agropecuária Brasileira* 27:1103–1111.

Filgueiras, T. S. 1997. Distrito Federal: Brazil. In S. D. Davis, V.H. Heywood, O. Herrera-MacBryde, J. Villa-Lobos. and A . C . Hamilton, eds., *Centres of Plant Diversity*, pp. 405–410. Oxford: World Widlife Fund and International Union for Conservation of Nature and Natural Resource.

Filgueiras, T. S. 1995. *Paspalum niquelandiae* (Poaceae: Paniceae), a new species from the serpentine outcrops of central Brazil. *Novon* 5:30–33.

Filgueiras, T. S., G. Davidse, and F. O. Zuloaga. 1993. *Ophiochloa*: A new endemic serpentine grass genus (Poaceae:Paniceae) from the Brazilian Cerrado vegetation. *Novon* 3:310–317.

Filgueiras, T. S., J. M. Felfili, M. C. da Silva Jr., and P. E. Nogueira. 1998.

Floristic and structural comparison of cerrado (*sensu stricto*) vegetation in central Brazil. In F. Dallmeier and J. A. Comiskey, eds., *Forest Biodiversity in North, Central and South America, and the Caribbean: Research and Monitoring*, pp. 633–647. Paris: The Parthenon Publishing Group.

Filgueiras, T. S. and F. S. Wechsler. 1992. Pastagens nativas. In B. F. S. Dias, ed., *Alternativas de Desenvolvimento dos Cerrados: Manejo dos Recursos Naturais Renováveis*, pp. 47–49. Brasília: Fundação Pró-Natureza.

Filgueiras, T. S. and F. O. Zuloaga. 1999. A new *Triraphis* (Poaceae: Eragrostideae) from Brazil: The first record of a native species in the New World. *Novon* 9:36–41.

Gentry, A. and Dodson, C. 1987. Contribution of nontrees to the species richness of a tropical rain forest. *Biotropica* 19:149–156.

Giulietti, A. M., N. L. Menezes, J. R. Pirani, M. Meguro, and M. G. L. Wanderley. 1987. Flora da Serra do Cipó, Minas Gerais: Caracterização e lista das espécies. *Boletim de Botânica* 9:1–151.

Giulietti, A. M., J. R. Pirani, R. M. Harley. 1997. Espinhaço range region: Eastern Brazil. In S. D. Davis, V.H. Heywood, O. Herrera-MacBryde, J. Villa-Lobos, and A . C . Hamilton, eds., *Centres of Plant Diversity*, pp. 397–404. Oxford: World Widlife Fund and International Union for Conservation of Nature and Natural Resource.

Goedert, W. J., E. Lobato, and M. Resende. 1982. Management of tropical soils and World food prospects. *Proceedings of the 12th International Congress of Soil Science*, pp. 317–325. New Delhi, India.

Goldsmith, F. B. 1974. Multivariate analysis of of tropical grassland communities in Mato Grosso, Brazil. *J. Biogeogr.* 1:111–122.

Gomes, C. M. R., O. R. Gottlieb, G.-B. Marini-Bettòlo, F. Delle Monache, and R. Polhill. 1981. Systematic significance of flavonoids in *Derris* and *Lonchocarpus Bioch. Syst. Ecol.* 9:129–147.

Gottlieb, O. R., M. A. C. Kaplan, and M. R. de M. R. Borin. 1996. *Biodiversidade: Um Enfoque Químico-Biológico*. Rio de Janeiro: Editora Universidade Federal do Rio de Janeiro.

Guarim Neto, G. 1987. *Plantas Utilizadas na Medicina Popular do Estado de Mato Grosso*. Brasília: Conselho Nacional de Desenvolvimento Científico e Tecnológico.

Heywood, V. H. 1992. Conservation of germplasm of wild species. In O. T. Sandlund, K. Hindar and A. H. D., eds., *Conservation of Biodiversity for Sustainable Development*, pp. 189–203. Oslo: Scandinavian University Press.

Hirsinger, H. 1985. Agronomic potential and seed composition of *Cuphea*, an annual crop for lauric and capric seed oils. *J. Am. Oil Chem. Soc.* 62:76–80.

Hoehne, F. C. 1939. *Plantas e Susbtâncias Tóxicas e Medicinais*. Rio de Janeiro: Graphicars.

Hoyt, E. 1988. *Conserving the wild relatives of crops*. Gland, Switzerland: World Widlife Fund and International Union for Conservation of Nature and Natural Resource.

Judziewicz, E. J., L. G. Clark, X. Londoño, and M. J. Stern. 1998. *American Bamboos*. Washington, DC: Smithsonian Institution Press.

Kauffman, J. B., D. L. Cummings, and D. E. Ward. 1994. Relationships of fire, biomass and nutrient dynamics along a vegetation gradient in the Brazilian cerrado. *J. Ecol.* 82:519–531.

Kirkbride, J. H. 1984. Legumes of the Cerrado. *Pesquisa Agropecuária Brasileira* 19:23–46.

Lima, M. P. de and R. R. Guedes-Bruni.1994. *Reserva Ecológica de Macaé de Cima, Nova Friburgo, RJ: Aspectos Florísticos das Espécies Vasculares*, vol. 1. Rio de Janeiro: Jardim Botânico do Rio de Janeiro.

Macedo, G. A. R., M. B. Ferreira, and C. J. Escuder. 1978. *Dieta de Novilhos em Pastagem Nativa de Cerrado*. Belo Horizonte: Empresa de Pesquisa Agropecuária de Minas Gerais.

Mantovani, W. 1990. Variação da flora arbustivo-herbácea de diversas fisionomias do cerrado de Itirapina, estado de São Paulo. *Anais do 36° congresso nacional de botânica*, pp. 125–135. Brasília: Sociedade Botânica do Brasil and IBAMA.

Mantovani, W. and F. R. Martins. 1993. Florística do cerrado na Reserva biológica de Moji Guaçu, SP. *Acta Bot. Brasil.* 7:33–60.

Mattos, J. K. de A. 1996. *Plantas Medicinais: Aspectos Agronômicos*. Brasília: Edição do Autor.

Medina, E. 1987. Requirements, conservation and cycles of nutrients in the herbaceous layer. In B. H. Walter, ed., *Determinants of Tropical Savannas*, pp. 39–65. Paris: IUBS Special Issue no. 3.

Mendonça, R. C., Felfili, J. M.., B. M. T. Walter, M. C. da Silva Jr., A . V. Rezende, T. S. Filgueiras and P. E. Nogueira. 1998. Flora vascular do Cerrado. In Sano, S. M. and S. P. de Almeira, eds., *Cerrado: Ambiente e Flora*, pp. 289–556. Planaltina: Empresa Brasileira de Pesquisa Agropecuária.

Miranda, I. S. 1998. "Flora, fisionomia e estrutura das savanas de Roraima, Brasil." Ph.D. thesis, Instituto Nacional de Pesquisas da Amazônia, Manaus, Brasil.

Miranda, I. S. and M. L. Absy. 1997. A flora fanerogâmica das savanas de Roraima. In R. I.. Barbosa, E. J. G. Ferreira and E. G. Castellón, eds., *Homem, Ambiente e Ecologia no Estado de Roraima*, pp. 445–462. Manaus: Instituto Nacional de Pesquisas da Amazônia.

Munhoz, C. B. and C. E. B. Proença. 1998. Composição florística do município de Alto Paraíso na Chapada dos Veadeiros. *Bol. Herb. Ezechias Paulo Heringer* 3:102–150.

Myers, N. 1983. *A Wealth of Wild Species*. Colorado: Westview Press.

Paviani, T. I. 1987. Anatomia do desenvolvimento do xilopódio de *Brasilia sickii* G. M.Barroso. Estágio inicial. *Ciênc. Cult.* 39:399–405.

Pereira, B. A. S. 1984. Plantas nativas do cerrado pastadas por bovinos na região geoeconômica do Distrito Federal. *Rev. Bras. Geogr.* 46:381–388.

Pereira, B. A. S. 1990. Espécies apícolas da flora da Área de Proteção Ambiental (APA) da bacia do Rio São Bartolomeu, Distrito Federal: Estudo preliminar. *Cad. Geoc.* 5:7–19.

Pereira, B. A. S. 1992. A flora nativa. In B. F. S. Dias, ed., *Alternativas de Desenvolvimento dos Cerrados: Manejo dos Recursos Naturais Renováveis*, pp. 53–57. Brasília: Fundação Pró-Natureza.

Pott, A. 1988. *Pastagens no Pantanal*. Corumbá: Empresa Brasileira de Pesquisa Agropecuária.

Pott, V. J. 1998. A família Nymphaeaceae no Pantanal, Mato Grosso e Mato Grosso do Sul. *Acta Bot. Brasil.* 12:183–194.

Ratter, J. A. 1987. Notes on the vegetation of the Parque Nacional do Araguaia (Brazil). *Notes Roy. Bot. Gard. Edinb.* 44:311–342.

Ratter, J. A., J. F. Ribeiro, and S. Bridgewater. 1997. The Brazilian cerrado vegetation and threats to its biodiversity. *Ann. Bot.* 80:223–230.

Raven, P. H. 1990. The politics of preserving biodiversity. *BioScience* 40:769.

Rawitscher, F. and M. Rachid. 1946. Troncos subterrâneos de plantas brasileiras. *An. Acad. Bras. Ciênc.* 18:262–280.

Rizzini, C. T. and E. P. Heringer. 1966. Underground organs of trees and shrubs from some Brazilian savannas. *An. Acad. Bras. Ciênc.* 38:85–112.

Schultze-Kraft, R. 1980. Recolección de plantas nativas con potencial forrajero. In Empresa Brasileira de Pesquisa Agropecuária, ed., *Plantas Forrageiras*, pp. 61–72. Brasília: Empresa Brasileira de Pesquisa Agropecuária.

Silva, D. A. da and C. A. Klink. 1996. Crescimento e alocação de biomassa de duas gramíneas nativas do cerrado. In L. L. Leite and C. H. Saito, eds., *Contribuição ao Conhecimento Ecológico do Cerrado*, pp. 59–63. Brasília: Universidade de Brasília.

Siqueira, J. C. 1988. *Plantas Medicinais: Identificação e Uso das Espécies dos Cerrados*. São Paulo: Edições Loyola.

Taylor, N. P. and D. C. Zappi. 1995. Cactaceae. In B. L. Stannard, ed., *Flora of the Pico das Almas, Chapada Diamantina, Bahia, Brazil,* pp.157–164. Kew: Royal Botanic Gardens.

Valls, J. F. M. and L. Coradin. 1986. Recuros genéticos de plantas forrageiras nativas no Brasil. *Anais do 3° Simpósio Sobre Produção Animal*, pp. 19–34. Campo Grande: Fundação Cargill.

Verdesio, J. J. and F. D. Garra. 1987. *Caracterização Física dos Cerrados do Oeste da Bahia*. Brasília: Instituto Interamericano de Cooperação para a Agricultura.

Warming, E. 1973. Lagoa Santa: Contribuição para a Geographia Phytobiologica (reprint of the 1908 Portuguese translation of the work). In M. G. Ferri, ed., *Lagoa Santa e a Vegetação de Cerrados Brasileiros*, pp. 1–284. Belo Horizonte: Livraria Itatiaia Editora and Editora da Universidade de São Paulo.

Wilson, E. O. 1992. *The Diversity of Life*. Cambridge, MA: Belknap Press, Harvard.

8

Patterns and Dynamics
of Plant Populations

Raimundo P. B. Henriques and John D. Hay

THE UNDERSTANDING OF THE DYNAMICS OF PLANT COMMUNITIES is based on measurements of rates of mortality and recruitment. These population parameters are essential to the study and prediction of responses of vegetation to global changes (Phillips and Gentry 1994; Pimm and Sugden 1994) and short-term climatic change (Condit et al. 1992), as well as patterns in species richness (Phillips et al. 1994). The issue takes on special significance because of its implications for community conservation and management (Primack and Hall 1992).

Long-term monitoring of plant populations has been undertaken in various tropical forests around the world (Swaine, Lieberman, and Putz 1987; Hartshorn 1990; Phillips and Gentry 1994; Phillips et al. 1994). For savannas, the dynamics of plant communities protected from fire have been monitored, and data presented (Brookman-Amissah et al. 1980; San José and Fariñas 1983; Devineau et al. 1984; Bowman and Panton 1995). In contrast, although the Cerrado Biome is the second largest in South America, there are very few long-term studies using permanent plots to observe recruitment, growth, and mortality of woody plants in this vegetation type. Silberbauer-Gottsberger and Eiten (1987) recorded three years of cerrado change in ten quadrats of 100 m^2 each.

Currently the Cerrado Biome is suffering strong anthropogenic change associated with deforestation, high fire incidence, and invasion by alien species (chapter 5). Even protected reserves in central Brazil have only a few dozen hectares that have been free from disturbance of this kind for more than 20 years. Studies of parameters such as annual mortality and recruitment rates in areas protected from human impact are needed to determine their natural dynamics.

THE CERRADO BIOME IN THE CONTEXT
OF SAVANNA DYNAMICS

The pattern of dynamic processes that emerges in savannas appears to depend on four site factors: water, nutrients, herbivory, and fire. Using water and nutrients and, to some extent, herbivory, savannas were classified in four types (Frost et al. 1986): (1) low water availability and high nutrient supply; (2) high water availability with extremely nutrient poor soil; (3) limited in both water and nutrients; and (4) no limitations of water and nutrients. Most cerrado vegetation can be included in type (2), however some are of type (3).

Fire is a widespread disturbance factor in savannas, particularly in the cerrados (chapter 4). The response of vegetation to fire depends largely on season, fire temperature, fuel biomass, and subsequent events such as rainfall, drought, and herbivory. Several studies indicate that fire reduces the recruitment of seedlings, saplings, and small trees, leading to an interruption or stabilization of vegetation development (Walker 1981; Frost and Robertson 1986; Frost et al. 1987). Hoffmann (1996) showed that, even two years after burning, seedling establishment in cerrado *sensu stricto* was lower in a burned area than in a control area (see chapter 9).

The influence of fire suppression on cerrado vegetation, and subsequent effects on invasion, recruitment, and structural stability, are fundamental in determining whether cerrado is a self-maintaining community. In the 19th century the naturalist Lund (1843) proposed that the *campos* and cerrado *sensu stricto* were created by the frequent burning of an original forest that he recognized as the *cerradão* (see chapter 6). This idea was reinforced by the floristic analysis of the woody flora of the cerrados (Rizzini 1963). Based on several lines of evidence, Rizzini (1979) agreed with Lund and stated that in areas without edaphic limitations the cerrado *sensu stricto* arose from frequent burning of *cerradão*. Several other authors recognize the *cerradão* as the original (or climax) vegetation of the Cerrado Biome (Ab'Saber and Junior 1951; Aubréville 1959; Schnell 1961; Eiten 1972).

A similar pattern of origin for the African savannas was proposed by Aubréville (1949) who worked in Africa prior to studying the Brazilian cerrados. African savannas, once established, are often maintained by frequent fires, both natural and man-made. Several studies showing the origin of savannas by burning of forests in Africa and India are reviewed by Backéus (1992).

The complete protection of savanna vegetation from fire results in an increase in tree density and species richness, particularly of fire-sensitive species, evolving toward a woodland community (Brookman-Amissah et al. 1980, Lacey et al. 1982, San José and Fariñas 1983, Swaine et al. 1992, Devineau et al. 1984, Bowman and Panton 1995).

In fire-protected cerrado areas, observations indicate that the vegetation changes, both in species composition and tree density, toward a closed woodland (Coutinho 1982, 1990; Moreira 1992). Simulations using data for five cerrado plant species (Hoffmann 1999) suggest that if current fire frequencies were lowered, woody plant cover and tree density would increase (chapters 4, 9). Unfortunately there are few existing experiments of fire suppression in savannas, and to our knowledge none for the cerrado, which would permit a rigorous evaluation of this hypothesis.

The nature of savanna structure and dynamics reflects the set of species attributes (Noble and Slayter 1980). Many studies have reported the widespread occurrence of vegetative reproduction in savanna woody species (Medina 1982; Menaut 1983). In the Cerrado Biome many woody species are known to reproduce vegetatively via root suckers and rhizomes (Rawitscher 1948), and some studies attribute great importance to vegetative growth (Rizzini 1971). However, these early studies are limited by a lack of quantitative data with which to assess the prevalence of vegetative reproduction at the community level. Henriques (1993) showed that vegetative reproduction in cerrado *sensu stricto* occurs in approximately 60% of the woody species. The role of vegetative growth as a stabilizing element has been observed in frequently burned savannas (Lacey et al. 1982). Hoffmann (1998) showed that fire also tends to increase the importance of vegetative *versus* sexual reproduction, and that, under current fire regimes, fire may be causing a shift in species composition, favoring fire resistant species.

Questions

This chapter deals with succession and stand stability in a 0.5 ha plot of cerrado *sensu stricto* that has been protected from fire since 1973. First we address the following questions about structure and composition: (1) What are the floristic characteristics of a cerrado stand? (2) What is the size class and abundance distribution of species? Second, we summarize questions concerning floristic changes, mortality, recruitment, growth, and population dynamics: (1) Is species richness balanced by immigration and extinction? (2) Are stand structural components (number of individuals, number of stems and basal area) increasing with time? (3) Does mortality match recruitment for various species? (4) What is the pattern of growth rate? (5) What are the rates of change of populations? Finally, we address questions about cerrado stability and succession: (1) Is the cerrado (*sensu stricto*) in an unstable equilibrium state? (2) Is the cerrado a self-maintained community?

THE CERRADO *SENSU STRICTO* AS A NEOTROPICAL SAVANNA CASE STUDY

Study Area

The sample plot was located in the Ecological Reserve of the IBGE (Instituto Brasileiro de Geografia e Estatística). The reserve (15°57'S, 47°53'W) is located 16 km SW of the center of Brasília and occupies 1,350 ha of protected area. Altitude ranges between 1,045 m and 1,125 m, with an average slope of 6%. Soils are mostly Oxisols (U.S. Soil Taxonomy System). Chemically, the soils are acid (pH ≈ 5.4), with low available P levels (<1 ppm), low concentrations of cations (K, Ca, Mg), high concentration of aluminum, and high clay content (Moreira 1992; for further details see chapter 2).

The average annual rainfall, based on a 17-year record, is 1,469 mm. Monthly precipitation is strongly seasonal, with >90% falling from October to April. The dry season occurs from May to September, when monthly precipitation can drop to 0 mm. The temperature regime is typical of a continental subtropical climate, with an annual mean of 21.9°C over the same 17-year period.

Methods

The plot was established within a cerrado area that has been protected from fire since 1973. The stand was sampled with a transect of 20 m × 250 m (0.5 ha), divided into 10 × 10 m subplots. The first census of the plot was conducted in January 1989. Within each subplot all plants ≥ 15 cm circumference (4.8 cm diameter) at ground level were measured and identified by species. Plants with multiple stems resulting from vegetative growth were included in the census if the sum of their circumferences was ≥ 15 cm. Each plant was labeled with numbered aluminum tag and mapped. Since many cerrado plants exhibit vegetative reproduction, the identification of separate individuals was made by excavations to 10 cm depth, between all conspecific stems separated by less than 1 m. Excavations were made at the mid-point of the shortest distance between the center of neighboring stems, and the exposed soil profile was examined for connections. After inspection the soil was replaced into the hole. Care was taken not to harm the subterranean organs of the plants, and no mortality was observed. This excavation procedure facilitated a separate calculation of the number of individuals and the number of stems. These values should be viewed with caution, however, since they do not include individuals with connections below 10 cm depth or whose connections had become separated.

The species were classified into the following growth forms: tree (≥10 cm diameter, and ≥3 m height); thick shrub (≥10 cm diameter, and < 3 m height); thin shrub (<10 cm diameter, and < 3 m height); and palms. This classification, similar to that of Silberbauer-Gottsberger and Gottsberger (1984), is based on our field experience and on the checklist of plants of the Ecological Reserve of IBGE (Pereira et al. 1993). A second census of all plants in the plot was made in July 1991. We used the same criteria previously described to determine mortality, recruitment, and growth.

CERRADO STRUCTURE AND SPECIES COMPOSITION

In the first census in 1989 a total of 980 individuals, with 1,158 live stems, were measured. We recorded 48 species in 40 genera and 29 families. Five families (Palmae, Leguminosae, Vochysiaceae, Erythroxylaceae, and Compositae) accounted for 49.8% of all species. The most species-rich family was Leguminosae (5 species), followed by Palmae (4 species) and Vochysiaceae (4 species). The most species-rich genera were *Byrsonima* and *Erythroxylum* (3 species each). Twenty-one families were represented by only one species. The most abundant family was Ochnaceae, with 126 individuals (12.9%), all belonging to a single tree species, *Ouratea hexasperma*. The second family in terms of number of individuals was Compositae (120 individuals), represented by three species.

The relative abundance of species is shown in figure 8.1. The number of rare species represented by only one individual was 13 (26.5% of all species). Although the species richness in this type of vegetation is considerably low compared with tropical forest, it is among the highest known for non-forest vegetation (Silberbauer-Gottsberger and Eiten 1987). In two nearby 1-ha plots, Eiten and Sambuichi (1996) recorded 57 and 92 woody species respectively. These results, when compared with areas near the southern limit of the Cerrado Biome (Silberbauer-Gottsberger and Gottsberger 1984), suggest that the number of woody plant species is higher in the cerrados of central Brazil.

The study plot at the IBGE Reserve is dominated by tree species (30 species), followed by thick shrubs (7 species), thin shrubs (6 species), palms (4 species), and one treelike rosette, *Vellozia squamata* (Velloziaceae). The number of stems drops rapidly with increasing diameter class, and the number of trees with a diameter greater than 10 cm is remarkably low when compared with tropical forest (Manokaran and Kochummen 1987; Swaine, Hall, and Alexander 1987; Primack and Hall 1992). The inverted-J shape of the frequency distribution is typical of natural forest regenerating from seeds, with high numbers in the smaller size classes and a logarithmic decline with increasing size.

CERRADO DYNAMICS

Floristic Changes

The number of families, genera, and species in the plot increased from 1989 to 1991. This corresponded to two new families, five new genera, and eight new species. No species disappeared during this period. The immigration rate of new species into the plot was 6.6 spp. ha^{-1} yr^{-1}. In another cerrado area the calculated immigration rate during a 3-year period was 3.3 spp. ha^{-1} yr^{-1}, with no losses of species recorded (Silberbauer-Gottsberger and Eiten 1987). These values are greater than the immigration rate of species for the Llaños savanna protected from fire in central plains of Venezuela (0.23 spp. ha^{-1} yr^{-1}), calculated from 16 years of records (San José and Farina 1983), but less than the range of 15.3–19.3 spp. ha^{-1} yr^{-1} for the Lamto savanna in Africa, also protected 16 years from fire (Devineau et al. 1984). Immigration rate curves for savannas protected from fire show a faster increase initially followed by a diminution (Braithwaite 1996; Devineau et al. 1984). The unbalanced and high immigration rates in our study area and the Lamto savanna indicate that these areas are in the early stages of succession.

Stand Dynamics

There was a large structural change in the cerrado plot over time (see table 8.1). In 1991 the density, number of stems, and basal area rose to 1,253 individuals (annual net increase of +11.6%), 1,534 stems (+13.5%), and 73,068 cm^2 (+17.1%), respectively. This high net increase corresponded to a low turnover of plants in the plot, since in all growth forms the high recruitment was not balanced by mortality. There were 304 new individuals recruited (+31.0%) and only 31 deaths (3.2%) over the period. There

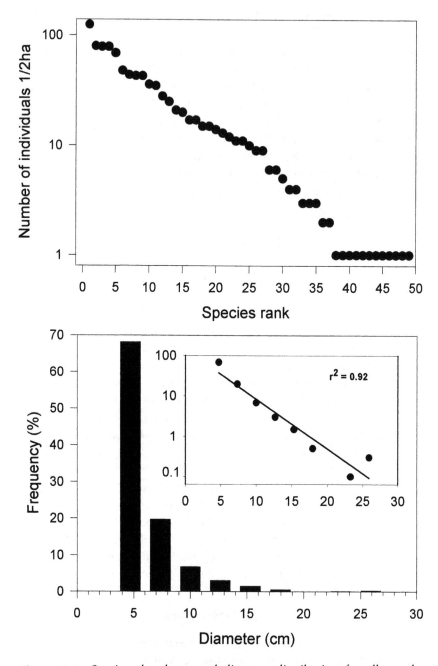

Figure 8.1 Species abundance and diameter distribution for all woody plants ≥ 4.8 cm in diameter in a sample plot of cerrado *sensu stricto* in the Ecological Reserve of IBGE, central Brazil. Top: Species abundance curve; species ranked in descending density. Bottom: Diameter class distribution for all woody plants ≥ 4.8 cm in diameter in a 0.5-ha plot in 1989. Inset: the same data plotted on log scale.

was a total of 420 new stems (36.3%), and the increase in basal area due to recruitment and growth was 22,456 cm^2 (43.3%). Losses due to mortality were 44 stems and 1,194 cm^2 of basal area.

The changes in growth form distribution showed that thick shrubs had the largest annual net increase in density (+18.9%); palms the largest increase in number of stems (+29.3%) and basal area (+63.6%). Since palm species do not have true secondary growth, the high net increase in basal area was due to the high vegetative reproduction of stipes.

The distribution of individuals among growth forms was not significantly different between 1989 and 1991 (χ^2 = 5.98, P > 0.05), but the proportion of number of stems and basal area differed significantly (χ^2 = 15.59, P < 0.005; χ^2 = 6,850.91, P < .0001, respectively). The significantly higher number of stipes and basal area of palms and the treelike rosette plant in 1991 accounted for this difference.

The proportion of stems originating from vegetative reproduction was 18.2% in 1989, and the value increased to 22.4% in 1991 (χ^2 = 28.50, P < .005). This increase was due both to vegetative reproduction of existing individuals and to new recruits. The proportion of vegetatively reproduced stems in other cerrado areas was 9.9% and 11.5% for southeastern and central Brazil, respectively (Silberbauer-Gottsberger and Eiten 1987, Henriques 1993).

Mortality, Recruitment, and Growth

Deaths were recorded in 12 of the 48 species present in 1989. The number of individuals that died over the study period was 31, corresponding to an annual mortality rate of 1.34% yr^{-1}. The losses by deaths of number of stems and basal area were 1.61% yr^{-1} and 0.97% yr^{-1}, respectively (table 8.1). The mortality rate calculated for another cerrado area by Silberbauer-Gottsberger and Eiten (1987) was 5.50% yr^{-1}. These values are in the middle to maximum range of the annual mortality rate recorded in studies of tropical forests (Swaine, Lieberman, and Putz 1987; Hartshorn 1990; Phillips and Gentry 1994; Phillips et al. 1994). Sato and Miranda (1996) showed that after fire the mortality rate of cerrado plants > 5 cm in diameter increases strongly to values ranging between 6.4% to 13.0% yr^{-1}.

The annual mortality rate for the various size classes were: 4.9–7.5 cm (1.61% yr^{-1}); 7.6–10.2 cm (1.19% yr^{-1}); >10.3 cm (0.79% yr^{-1}). These results suggest that the proportion of plants dying diminished with increasing size class, but the sample size is small for a definite conclusion.

The mortality of individuals was highest for thick shrubs and the treelike rosette plant (table 8.1) and lowest for trees. This pattern of mortality rate was observed for both number of stems and basal area. The mortality rate differed markedly among species; the highest rate was shown by *Byrsonima coccolobifolia* (24.9% yr^{-1}). A group of three species, comprising *Kielmeyera coriacea* (Guttiferae), *Connarus suberosus* (Connaraceae), and *Mimosa claussenii* (Leguminosae), showed an intermediate mortality rate, between 1.8% and 7.9% per year. The lowest mortality rate (0.5% to 0.9% per year) was shown by a group of five species, including *Eremanthus goyazensis* (Compositae), *Qualea parviflora* (Vochysiaceae), and *Palicourea rigida* (Rubiaceae). The only other data available for a cerrado tree species, *Vochysia thrysoidea*, showed an average mortality of 1.9% per year, but with temporal and spatial variation (Hay and Barreto 1988).

Of the 48 species marked in 1989, 30 (62.5%) recruited new individuals, and seven (14.6%) produced new stems by vegetative reproduction. By growth form, the proportion of species with recruitment increased in the following order: thin shrubs (50.0%), trees (53.3%), thick shrubs (71.4%), and palms (100%). More than half of the recruits (226, 58.3%) occurred in the 4.8–7.2-cm-diameter class. Another, larger number occurred with less than a 4.8-cm diameter (150, 38.8%), because we included as recruits connected stems with less than a 4.8-cm diameter (see Methods). A few stems (7, 2.9%) were recruited into a larger-diameter class (7.2–9.5 cm).

The annual recruitment rates calculated with the logarithmic model for number of individuals (11.57% per year), number of stems (13.33% per year), and basal area (15.30% per year) are much higher than for any known mature tropical forest (Phillips and Gentry 1994; Phillips et al. 1994). By growth forms the recruitment rates increased in the following order: treelike rosette plant, thin shrubs, palms, trees, and thick shrubs. This order was consistent for number of individuals, number of stems, and basal area.

The most abundant species recruited more than the rare species. All species with more than 42 individuals present in 1989 (nine species) had new recruits, while for those with less than 42 individuals, 46.2% (18 species) failed to recruit. We calculated the per capita recruitment rates for each species as the ratio of new recruits to population size in 1989. Species with less than 42 individuals in 1989 averaged only 53% per capita recruitment, while species with more than 42 individuals averaged 76%.

The mean diameter growth rate, pooled for all individuals and species, but excluding palms and the treelike rosette, was quite low at 1.59 mm per year (SD = 2.43, N = 791). Five percent of all individuals showed apparent negative growth rate in diameter, probably due to natural bark losses or measurement errors. Altogether 41% did not show growth; 51% had a growth rate between 1–5 mm per year; and 3% grew between 5 and 33 mm. Silberbauer-Gottsberger and Eiten (1987) found a mean growth rate of 2.7 mm per year for another cerrado. In their plots 24% of the individuals did not increase in circumference, 52% of individuals increased by 1–3 mm, and 24% above 4 mm per year.

These results for cerrado plants are close to the minimum growth rates reported for tropical forests (Lang and Knight 1983, Primack et al. 1985, Manokaran and Kochummen 1987, Korning and Balslev 1994) and gallery forests in central Brazil (Felfili 1995). The low growth rates observed are consistent with the poor nutrient content of cerrado soils (chapter 2) and a long dry season (4–5 months).

Mean diameter increments, pooled for all individuals and species, increased gradually with diameter. We found a significant linear relationship between individual diameter (x) and mean growth rate (y): $y = -0.97 + 0.30x$ $(r^2 = 0.65, P < .0005)$. Similar relationships were observed in tropical forest communities (Swaine et al. 1987; Condit et al. 1992; Felfili 1995). We hypothesize that plants that achieve larger size grow faster due to deep roots, which provide greater access to water.

Species Population Dynamics

For the whole cerrado community recruitment and mortality were unequal; this was also observed for individuals at the population level (see fig. 8.2). Only two species, *Salacia crassifolia* and *Eremanthus glomerulatus*, showed a net change close to the expected line of zero net change, as would be expected for balanced populations. The other 10 species recruited more individuals than were lost by deaths. *Eremanthus goyazensis* showed the highest population net increase with 33 individuals, followed by *Roupala montana* (21 individuals) and *Kielmeyera coriacea* (15 individuals). Particularly interesting was the persistence of rare species, suggesting that the increase in abundant species does not result in the loss of rare species in this community.

The species studied showed a strong net increase in density (see fig. 8.3). Most species (54%) increased in density by ≤ 2 individuals, 33% increased by 3–15 individuals, and 13% by 16–34 individuals. No species declined during the study period. In terms of growth form, the mean net increase in population density was highest in trees (24.6 ± 6.1 individuals), as compared with thick (11.7 ± 4.2) and thin shrubs (6.9 ± 2.3).

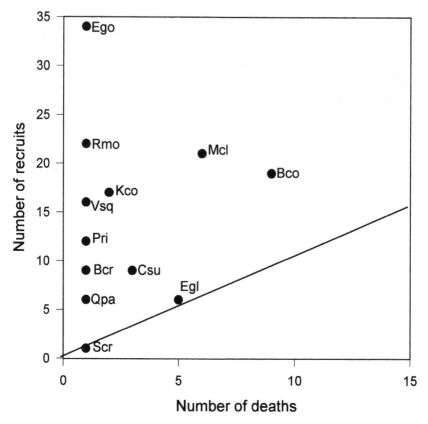

Figure 8.2 Recruitment and mortality for 12 plant species in a cerrado *sensu stricto* at the Ecological Reserve of IBGE, central Brazil, between 1989 and 1991. The diagonal line represents equal mortality and recruitment. The species are: Bco, *Byrsonima coccolobifolia*; Bcr, *Byrsonima crassa*; Csu, *Connarus suberosus*; Egl, *Eremanthus glomerulatus*; Ego, *Eremanthus goyazensis*; Kco, *Kielmeyera coriacea*; Mcl, *Mimosa claussenii*; Pri, *Palicourea rigida*; Qpa, *Qualea parviflora*; Rmo, *Roupala montana*; Scr, *Salacia crassifolia*; and Vsq, *Vellozia squamata*.

STABILITY AND DEVELOPMENT OF CERRADO

Stability and Turnover

Stability could indicate the state of vegetation changes in the absolute number of individuals and basal area over time (Korning and Baslev 1994). Using the terminology of Hallé et al. (1978), after a disturbance, the successional sequence of changing composition and structure begins with a growing phase with a net increase in the number of individuals and

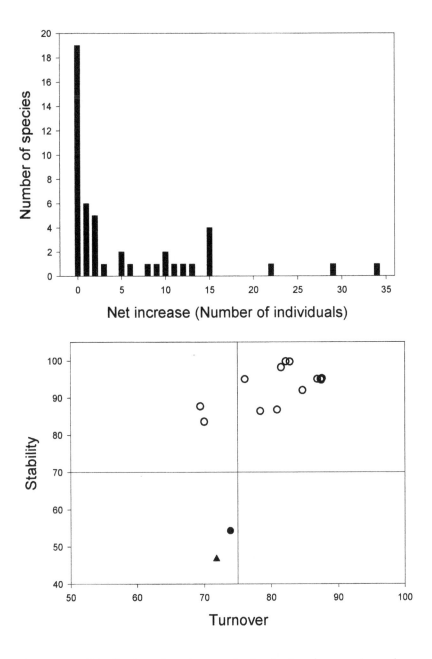

Figure 8.3 Distribution of net increase in density for plant species and stability and turnover for cerrado *sensu stricto* compared with other communities. Top: Distribution of net increase in number of individuals for 48 species. Bottom: Stability and turnover for cerrado (black dot), Lamto savanna (black triangle), and 13 neotropical forests (white dots). Stability is measured as the numerical difference between stand half-life and doubling time. Turnover is measured as the average of doubling time and stand half-life. Source: Phillips and Gentry 1994 (neotropical forests); Devineau et al 1984 (Lamto savanna).

basal area. This phase is followed by a homeostatic phase with accumulation of basal area due to growth, but in which mortality and recruitment are balanced. Another aspect of vegetation dynamics is turnover, which indicates the changes of composition or structure with time (i.e., the time necessary to replace lost stems and basal area). Turnover is measured by averaging rates of recruitment and mortality (Phillips and Gentry 1994: Phillips et al. 1994), calculating the stand half-life, which is the number of years necessary for the initial population to lose 50% (Lieberman et al. 1985; Lieberman and Lieberman 1987; Hartshorn 1990). In this study we measured both stability and turnover using the Korning and Baslev (1994) approach. Stability was estimated as the numerical difference between stand half-life and doubling time. Doubling time is the time needed to double the initial population at the present recruitment rate, using a log model. Turnover was estimated as the average of doubling time and stand half-life. Since low numerical values indicate high stability and turnover, we subtracted the numerical values from 100 to facilitate graphical interpretation.

A comparison of our data with those from neotropical forests (Phillips and Gentry 1994) and a savanna plot protected from fire in the Ivory Coast (Devineau et al. 1984) showed that the cerrado and savanna plots are moderately dynamic but in an unstable growing phase, with a net increase in the number of individuals and basal area (fig. 8.3).

Other results point out the instability of cerrado: (1) the diameter distribution (fig. 8.1) suggests a community with an abrupt decline in size above 10 cm diameter and a decreasing mortality rate with increasing size classes; (2) the species immigration rate was high, and no extinction was recorded for individuals $\geq$ 4.8 cm diameter; (3) the balance between recruitment and mortality over the study period showed a high positive net change (table 8.1).

The Cerrado as a Nonequilibrium Community

Here we present one of the first studies of cerrado dynamics. Data on floristic changes, recruitment, mortality, and growth for 48 plant species were recorded for a 2.4-year period. We are still cautious with these results, because both the period of study and the plot size are in the minimum range for similar studies in tropical forests (Swaine, Lieberman and Putz 1987; Hartshorn 1990; Phillips and Gentry 1994; Phillips et al. 1994). Compared with tropical forests, studies of cerrado vegetation dynamics are still in their infancy. Nevertheless, the data are sufficient to evaluate the state of stand dynamic parameters by growth forms as well as to suggest differences between species. In the light of these results we

Table 8.1   Changes in Density, Number of Stems, and Basal Area (cm²) by Growth Form Over Census Interval of 1989-1991 in 0.5 ha of Cerrado *Sensu Stricto*, in the Ecological Reserve of IBGE, Brasília, Central Brazil

	1989	Deaths 1989–91	Recruits 1989–91	Growth	1991	Annual net change (%)	Annual mortality %/yr	Annual recruitment %/yr
Density								
Trees	598	13	156		741	59.6 (+9.9)	0.92	9.84
Thick shrub	202	17	109		294	38.3 (+18.9)	3.66	19.30
Thin shrub	29	0	6		35	2.5 (+8.6)	0.00	7.84
Palm	72	0	17		89	7.1 (+9.8)	0.00	8.83
Treelike rosette plant	79	1	16		94	6.3 (+7.9)	0.53	7.77
Total	980	31	304		1253	113.8 (+11.6)	1.34	11.57
Stems								
Trees	671	15	167	6	829	65.8 (+9.8)	0.94	9.75
Thick shrub	234	22	113	3	328	39.2 (+16.7)	4.11	18.18
Thin shrub	30	0	8	0	38	3.3 (+11.1)	0.00	9.84
Palm	94	0	62	2	158	26.6 (+28.4)	0.00	21.64
Treelike rosette plant	129	7	37	22	181	21.7 (+16.8)	2.32	16.43
Total	1158	44	387	33	1534	156.7 (+13.5)	1.61	13.33
Basal area								
Trees	34,470	651	3570	6396	43,785	3881 (+11.3)	0.79	10.76
Thick shrub	6292	523	2409	288	8466	906 (+14.4)	3.62	15.98
Thin shrub	682	0	120	122	924	101 (+14.8)	0.00	12.65
Palm	6060	0	604	8649	15313	3855 (+63.6)	0.00	38.62
Treelike rosette plant	4302	20	458	-160	4580	116 (+2.7)	0.19	2.80
Total	51,806	1194	7161	15,295	73,068	8859 (+17.1)	0.97	15.30

Note: Annual net change was computed as follows: [(density, stems or basal area in 1991 – deaths + recruits + growth) – (density, stems or basal area in 1989)]/2.4 years. Annual mortality was estimated as $\log_e$ survivorship vs. time (Swaine and Lieberman 1987). Annual recruitment was estimated using a logarithmic model based on final recruitment and annual mortality (Phillips et al. 1994). See chapter 6 for descriptions of cerrado physiognomies.

evaluate the idea concerning cerrado evolution under fire suppression and reappraise the hypothesis of Lund-Rizzini about the origin of the cerrado physiognomy.

This study found strong evidence that the cerrado *sensu stricto* (chapter 6) has a clear successional nature, as suggested in earlier works (Coutinho 1982, 1990), corroborating the expected increase in cover and density if fire frequency is diminished (Hoffmann 1999; chapter 9). Our results showed that the cerrado, when protected from fire, increases in richness with a net increase in the number of individuals, number of stems, and basal area. The high immigration rate, recruitment greater than mortality, and high positive net growth indicate that the cerrado is in a very early stage of succession, after 16 years of fire suppression.

With fire suppression we expected an increase in fire sensitive species. Particularly noticeable in this respect is the immigration of the thick shrub *Myrsine guianensis* and the high population growth of the tree *Roupala montana*, both indicated by Hoffmann (1996, 1998) as showing positive population growth rates for intervals between fire of 5 and 9 years, respectively.

In an unburned cerrado there was a negative relationship between plant height and population growth rate (Hoffmann 1996). The high net increase in number of individuals and number of stems of trees and thick shrubs in the present study confirm this finding. The same trend was observed in the mean net population density, with values decreasing in the following order: trees, thick shrubs, and thin shrubs.

The great similarity in the results of the present study with those observed in other savannas deserves future investigation, since there are so few comparative studies. Nevertheless, all the present evidence strongly suggests that the savannas have a successional nature and are not equilibrium communities, and that the present structure and physiognomy are maintained by fire frequency below the minimum interval to permit establishment of tree species populations.

Based on the present study, we can only make an educated guess about the origin of the cerrado physiognomy. The extent of impact of fire in the cerrados is yet to be determined. Although there is some evidence that fire has long been present in the Cerrado Biome, the past fire frequency was well below that of the present (Vincentini 1999). The high mortality caused by fire in cerrado plants (Sato and Miranda 1996) and the high sensitivity of *cerradão* species to fire (Moreira 1992; chapter 6), as well as the clearly early successional state of the area studied, is consistent with the Lund-Rizzini hypothesis. The absence of immigration of *cerradão* species into the study plot should not be used to challenge the Lund-Rizzini hypothesis, because the *cerradão* areas that could serve as seed sources are small in total area (< 1%) and distant from the study site. Nevertheless there is strong evidence showing that *cerradão* species are capable of establishment in cerrado *sensu stricto* physiognomy, and that their survival is higher in areas which have been protected from fire for 20 years compared to more recently burned areas (Hoffmann 1996).

Long-term studies are needed to corroborate this result and evaluate whether the savanna physiognomy results from repeated burning of the original forest cover.

REFERENCES

Ab'Saber, A.N. and M.C. Junior 1951. Contribuição ao estudo do Sudoeste Goiano. *Bol. Geográfico* 9:123–138.

Aubréville, A. 1949. Ancienneté de la destruction de la couverture forestière primitive de l'Afrique tropicale. *Bull. Agric. Congo Belg.* 40:1347–1352.

Aubréville, A. 1959. As florestas do Brasil. *An. Brasil. Econ. Flor.* 11:201–243.

Backéus, I. 1992. Distribution and vegetation dynamics of humid savannas in Africa and Asia. *J. Veg. Sci.* 3:345–356.

Bowman, D. M. J. and W. J. Panton. 1995. Munmalary revisited: Response of a north Australian *Eucalyptus tetrodonta* savanna protected from fire for 20 yr. *Aust. J. Ecol.* 20:526–531.

Braithwaite, R.W. 1996. Biodiversity and fire in the savanna landscape. In O. T. Solbrig, E. Medina, and J. F. Silva, eds., *Biodiversity and Savanna Ecosystem Processes*, pp. 121–142. Berlin: Springer-Verlag.

Brookman-Amissah, J., J. N. Hall, M. D. Swaine, and J. Y. Attakorah. 1980. A reassessment of a fire protection experiment in north-eastern Ghana savanna *J. Appl. Ecol.* 17:85–99.

Condit, R., S. P. Hubbell, and R. B. Foster. 1992. Short-term dynamics of a neotropical forest. *BioScience* 42:822–828.

Coutinho, L. M. 1982. Ecological effect of fire in Brazilian cerrado. In B. J. Huntley and B. H. Walker, eds., *Ecology of Tropical Savannas*, pp. 273–291. Berlin: Springer-Verlag.

Coutinho, L. M. 1990. Fire in the ecology of Brazilian cerrado. In J. G. Goldammer, ed., *Fire in the Tropical Biota: Ecosystem Processes and Global Changes*, pp. 82–105. Berlin: Springer-Verlag.

Devineau, J. L., C. Lecordier, and R. Vuattoux. 1984. Evolution de la diversité esécifique du peuplement ligneux dans une succession préforestière de colonisation d'une savane protégée des feux (Lamto, Côte d'Ivoire). *Candollea* 39:103–134.

Eiten, G. 1972. The cerrado vegetation of Brazil. *Bot. Rev.* 38:201–341.

Eiten, G. and R. H. R. Sambuichi. 1996. Effect of long-term periodic fire on plant diversity in a cerrado region. In R. C. Pereira and L. C. B. Nasser, eds., *Biodiversity and Sustainable Production of Food and Fibers in the Tropical Savannas*, pp. 46–55. Planaltina: Empresa Brasileira de Pesquisa Agropecuaria.

Felfili, J. M. 1995. Growth, recruitment and mortality in the Gama gallery forest in Central Brazil over a six-year period (1985–1991). *J. Trop. Ecol.* 11:67–83.

Frost, P., E. Medina, J. C. Menaut, O. Solbrig, M. Swift, and B. Walker. 1986. *Responses of Savannas to Stress and Disturbance.* IUBS Special Issue no. 10:1–82.

Frost, P. G. H. and F. Robertson. 1987. The ecological effects of fire in savannas. In B. Walker, ed., *Determinants of Tropical Savannas*, pp. 93–140. Paris: IUBS Special Issue no. 3.

Hallé, F., R. A. A. Oldeman, and P. B. Tomlinson. 1978. *Tropical Trees and Forests.* Berlin: Springer-Verlag.

Hartshorn, G. 1990. An overview of neotropical forest dynamics. In A. H. Gentry, ed., *Four Neotropical Rainforests*, pp. 585–199. New Haven: Yale University Press.

Hay, J. D. and E. M. J. Barreto. 1988. Natural mortality of *Vochysia thrysoidea* in an unburnt cerrado ecosystem near Brasília. *Biotropica* 20:274–279.

Henriques, R. P. B. 1993. "Organização e Estrutura das Comunidades Vegetais de Cerrado em um Gradiente Topográfico no Brasil Central." Ph.D. thesis, Universidade Estadual de Campinas, Campinas, Brazil.

Hoffmann, W. A. 1996. The effects of fire and cover on seedling establishment in a neotropical savanna. *J. Ecol.* 84:383–393.

Hoffmann, W. A. 1998. Post-burn reproduction of woody plants in a neotropical savanna: The relative importance of sexual and vegetative reproduction. *J. Appl. Ecol.* 35:422–433.

Hoffmann, W. A. 1999. Fire and population dynamics of woody plants in a neotropical savanna: Matrix model projections. *Ecology* 80:1354–1369.

Huntley, B. J. 1982. Southern African savannas. In B. J. Huntley and B. H. Walker, eds., *Ecology of Tropical Savannas*, pp. 101–119. Berlin: Springer-Verlag.

Korning, J. and H. Baslev. 1994. Growth and mortality of trees in Amazonian tropical rain forest in Ecuador. *J. Veg. Sci.* 4:77–86.

Lacey, C. J., J. Walker, and I. R. Noble. 1982. Fire in Australian tropical savannas. In B. J. Huntley and B. H. Walker, eds., *Ecology of Tropical Savannas*, pp. 246–272. Berlin: Springer-Verlag.

Lang, G. E. and D. H. Knight. 1983. Tree growth, mortality, recruitment, and canopy gap formation during a 10-year period in a tropical moist forest. *Ecology* 64:1069–1074.

Lieberman, D. and M. Lieberman. 1987. Forest tree growth and dynamics at La Selva, Costa Rica (1969–1982). *J. Trop. Ecol.* 3:347–358.

Lieberman, D., M. Lieberman, R. Peralta, and G. Hartshorn. 1985. Mortality patterns and stand turnover rates in a wet tropical forest in Costa Rica. *J. Ecol.* 73:915–924.

Lund, P. W. 1843. *Blik paa Brasilien dryeverden. Selsk Skrifter* 11:1–82.

Manokaran, N. and K. M. Kochummen. 1987. Recruitment, growth and mortality of tree species in a lowland dipterocarp forest in Peninsular Malaysian. *J. Trop. Ecol.* 3:315–330.

Medina, E. 1982. Physiological ecology of neotropical savanna plants. In B.

J. Huntley and B. H. Walker, eds., *Ecology of Tropical Savannas*, pp. 308–335. Berlin: Springer-Verlag.

Menaut, J. C. 1983. The vegetation of African savannas. In F. Bourlière, ed., *Tropical Savannas*, pp. 109–150. Amsterdam: Elsevier Scientific Publishing Company.

Moreira, A. G. 1992. "Fire Protection and Vegetation Dynamics in the Brazilian Cerrado." Ph.D. thesis, Harvard University, Cambridge, MA, U.S.A.

Noble, I. R. and R. O. Slayter. 1980. The use of vital attributes to predicts successional changes in plant communities subject to recurrent disturbance. *Vegetatio* 43:5–21.

Pereira, B. A. da S., M. A. da Silva, and R. C. de Mendonça. 1993. *Reserva ecológica do IBGE, Brasília, (DF): Lista das plantas vasculares*. Rio de Janeiro: Instituto Brasileiro de Geografia e Estatística.

Phillips, O. L. and A. H. Gentry. 1994. Increasing turnover through time in tropical forests. *Science* 263:954–958.

Phillips, O. L., A. H. Gentry., S. A. Sawyer, and R. Vásquez. 1994. Dynamics and species richness of tropical rain forest. *Proc. Natl. Acad. Sci. USA* 9:2805–2809.

Pimm, L. S. and A. M. Sugden. 1994. Tropical diversity and global change. *Science* 263:933–934.

Primack, R. B., P. S. Ashton., P. Chai, and H. S. Lee. 1985. Growth rates and population structure of Moraceae trees in Sarawak, East Malaysia. *Ecology* 66:577–588.

Primack, R. B. and P. Hall. 1992. Biodiversity and forest change in Malaysian Borneo. *BioScience* 42:829–837.

Rawitscher, F. K. 1948. The water economy of the vegetation of campos cerrados in southern Brazil. *J. Ecol.* 36:237–268.

Rizzini, C. T. 1963. A flora do cerrado. In M. G. Ferri, ed., *Simpósio sobre o Cerrado*, pp. 126–177. São Paulo: Editora da Universidade de São Paulo.

Rizzini, C. T. 1971. Aspéctos ecológicos da regeneração em algumas plantas do cerrado. In M. G. Ferri, ed., *III Simpósio sobre o Cerrado*, pp. 61–64. São Paulo: Editora Edgard Blücher.

Rizzini, C. T. 1979. *Tratado de Fitogeografia do Brasil: Aspectos Sociológicos e Florísticos*. São Paulo: Hucitec Ltda.

San José, J. J. and M. R. Fariñas. 1983. Changes in tree density and species composition in a protected *Trachypogon* savanna, Venezuela. *Ecology* 64:447–453.

Sato, M. N. and H. S. Miranda. 1996. Mortalidade de plantas lenhosas do cerrado após duas queimadas prescritas. In R. C. Pereira and L. C. B. Nasser, eds., *Biodiversidade e Produção Sustentável de Alimentos e Fibras nos Cerrados*, pp. 204–207. Planaltina: Empresa Brasileira de Pesquisa Agropecuária.

Schnell, R. 1961. Le problème des homologies phytogéographiques entre l'Afrique et l'Amérique tropicales. *Mem. Mus. Hist. Nat., Bot.* 11:137–241.

Silberbauer-Gottsberger, I. and G. Eiten. 1987. A hectare of cerrado: I. General aspects of the trees and thick-stemmed shrubs. *Phyton* 27:55–91.

Silberbauer-Gottsberger, I. and G. Gottsberger. 1984. Cerrado-Cerradão: A comparison with respect to number of species and growth forms. *Phytocoenologia* 12:293–303.

Swaine, M. D., J. B. Hall, and I. J. Alexander. 1987. Tree population dynamics at Kade, Ghana (1968–1982). *J. Trop. Ecol.* 3:331–346.

Swaine, M. D., W. D. Hawthorne, and T. K. Orgle. 1992. The effects of fire exclusion on savanna vegetation at Kpong, Ghana. *Biotropica* 24: 106–172.

Swaine, M. D. and D. Lieberman. 1987. Note on the calculation of mortality rates. *J. Trop. Ecol.* 3:ii–iii.

Swaine, M. D., D. Lieberman, and F. E. Putz. 1987. The dynamics of tree populations in tropical forest: A review. *J. Trop. Ecol.* 3:359–363.

Vincentini, K. R. F. 1999. "História do Fogo no Cerrado: Uma Análise Palinológica." Ph.D. thesis, Universidade de Brasília, Brasília, Brazil.

Walker, B. H. 1981. Is succession a viable concept in African savanna ecosystems? In D. C. West, H. H. Schugart, and D. B. Botkin, eds., *Forest Succession: Concepts and Application*, pp. 431–447. New York: Springer-Verlag.

9

The Role of Fire in Population Dynamics of Woody Plants

William A. Hoffmann and Adriana G. Moreira

UNDERSTANDING THE FACTORS RESPONSIBLE FOR THE GREAT VAR-
iation in woody plant density has been a challenge for ecologists in the
cerrado and other tropical savannas. It is becoming evident that no single
factor determines tree density in the cerrado; rather, nutrient availability,
water stress, and fire interact to determine woody plant cover. Of these
three factors, the role of fire is perhaps most important to understand,
since it alone is largely under human control and is probably the factor
most variable at the interannual to interdecadal scale. Therefore it is likely
responsible for most temporal changes in woody plant density within the
time frame of human observation (see chapter 4).

Evidence of fires in the cerrado has been recorded for before 27,000
years before present (Vicentini 1993) and was probably present long
before then. Thus, woody plants have been exposed to its selective pressure for a considerable part of their evolutionary history. Although the
flora appears very well adapted to normal levels of fire, human activity
has almost certainly increased fire frequency above the natural rate (chapter 5). So while the cerrado plants are in general tolerant of fire, in many
cases they are now subjected to frequencies in excess of the environment
in which they evolved.

The dynamics of plant populations is determined by a suite of vital
rates, including survival, growth, sexual reproduction, vegetative reproduction, and seedling establishment (chapter 8). The impact of fire is
sufficiently severe to affect all of these vital rates, having implications
for population dynamics and thus community dynamics. In this chapter,
we review the known effects of fire on each of these vital rates. Then we

demonstrate the consequences of these changes for population dynamics and finally community dynamics.

EFFECT OF FIRE ON LIFE HISTORIES
OF WOODY PLANTS

Plant Survival

In many fire-prone ecosystems, two principal life history strategies of woody plants can be identified (Keeley and Zedler 1978; van Wilgen and Forsyth 1992; Bond and van Wilgen 1996). The first strategy is that of obligate seeders, species that do not survive fire as adults but have high seedling establishment following burns. The second strategy is that of resprouters, which survive fire by resprouting from stems or roots. In the cerrado, it is this second strategy which predominates, with all or most woody plant species capable of resprouting after fire. Observed community-wide adult mortality rates after fire have ranged from 5% to 19% in the cerrado. The high degree of variability in community-wide estimates may result from differences in fire intensity (Sato 1996), timing, or species composition. Among species, there is considerable variation in adult mortality, with observed mortality rates ranging from 1% for *Piptocarpha rotundifolia* (Hoffmann 1996) to 42%, as in the case of *Eremanthus suberosum* (Sato 1996). As a whole, woody plant mortality rates appear similar to other moist tropical savannas (Rutherford 1981; Lonsdale and Braithwaite 1991) but are much lower than Brazilian tropical forests (see fig. 9.1A) which are not adapted to such high fire frequencies. A number of traits contribute to the high capacity of cerrado species to survive fire, including thick bark (Silva and Miranda 1996), large investment in carbohydrate and nutrient reserves (Miyanishi and Kellman 1986; Hoffmann et al. 2000), and the capacity to resprout from dormant or adventitious buds.

The difference in survival between cerrado and forest species is also observed at the seedling stage. For three forest species studied, no seedlings survived fire when less than 1 year old, whereas 12 of 13 cerrado species were able to survive fire (see fig. 9.1B). Among these 13 species, survival was strongly correlated to seed mass. At a very young age, seedlings develop a deep taproot, which serves not only for carbohydrate and nutrient storage but also to reach moist soil layers during the dry season (Moreira 1992; Oliveira and Silva 1993). Larger seeds should provide the seedling with extra resources for the quick development of root reserves sufficient for postfire resprouting (Hoffmann 2000).

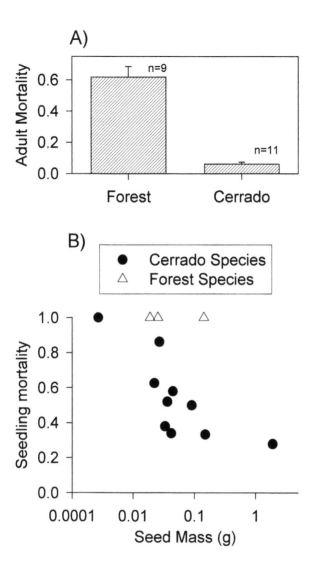

Figure 9.1 Mortality of burned cerrado and forest species. (A) Community-wide estimates of adult mortality in burned stands (mean ± SE). Sample size (n) is the number of burns. Data are from Uhl and Buschbacher (1985); Kauff-man (1991); Sato (1996); Sato and Miranda (1996); Silva, Sato, and Miranda (1996); Holdsworth and Uhl (1997); Silva (1999); Cochrane and Schulze (1999); and Peres (1999). (B) The effect of fire on survival of first-year seedlings. Data are from Franco et al. (1996) and Hoffmann (2000). There is a significant correlation between seed mass and survival for savanna species ($r^2 = 0.59$, $P < .0001$).

Growth

Although fire may not kill an individual, it generally destroys part of the plant. Many larger individuals may lose only leaves and thin branches due to scorching, but smaller individuals may lose most or all of their aerial biomass. In either case, the loss of aerial tissue represents negative growth (i.e., a reduction in the size of an individual). It is well established that size largely determines a plant's rates of growth, survival, and reproduction (Harper 1977), so a reduction in individual size can have negative impacts on future population growth.

The primary trait preventing topkill is the thick bark present in many species of trees and shrubs. The thermal insulation provided by bark protects cambium layers from the high fire temperature (Silva and Miranda 1996; Gignoux et al. 1997). While height is also effective in protecting sensitive tissues from fire, the low stature of cerrado species relative to closely related forest species precludes the possibility that cerrado species have evolved height as an adaptive response to fire (Rizzini 1971).

After the initial reduction in plant size, regrowth of woody plants tends to be vigorous. It is possible that for some species the rate of regrowth is high enough that the plants eventually surpass the size they would have attained had they not been burnt, particularly for subshrubs. However, for larger growth forms, fire normally causes a net reduction in size. As a result, it becomes nearly impossible for small individuals to recruit into large size categories under frequent burning (Gignoux et al. 1997). During two fire cycles of 2 years, Sato (1996) found that in an area initially encompassing 1,212 individuals with stem diameter greater than 5 cm, only 37 new individuals had recruited into this minimum stem diameter. This contrasts with 277 individuals that died within this same interval. Without a prolonged fire-free interval, few individuals are likely to reach a size at which the aerial stem can resist fire. This should be particularly detrimental for tree species that reach sexual maturity at a large size.

Sexual Reproduction

High fire frequency poses a serious constraint to sexual reproduction. If a species has no adaptation to protect seeds from fire, successful reproduction requires that individuals flower and produce seeds and that the resulting seedlings reach a fire-tolerant size in the short period of time between burns. Therefore there should be a selective pressure to accelerate this sequence of events. It was previously shown that many seedlings develop fire tolerance when less than a year old. On the other hand, there is a much greater interspecific variation in capacity for ensuring seedling establishment after burning, whether via fire-protected seeds or rapid postfire flowering.

Seed maturation in the cerrado is concentrated at the end of the dry season and beginning of the wet season (Oliveira 1998). Therefore burns in the dry season destroy many flower buds, flowers, developing fruit, and mature seeds, greatly reducing seed availability (Hoffmann 1998). While some cerrado species produce fruits that protect seeds from burning (Coutinho 1977; Landim and Hay 1996), many do not. And for many species that produce woody fruits that could potentially protect seeds from fire, many seem ineffective because of the phenology of fruit maturation and seed release. For some species, there appears to be a short window of time during which the seeds are mature but have not yet been dispersed (W. A. Hoffmann, personal observation). For many such species, the timing of fire probably determines the success of this strategy; this relation has been observed in the savannas of Northern Australia (Setterfield 1997) but not in the cerrado to date. Similarly, there is a need to understand the contribution of seed banks in postfire regeneration in the cerrado.

For the majority of species, it appears that postfire seedling establishment is dependent upon seeds produced after burning. Among subshrubs and herbs, fire frequently stimulates flowering after fire (Coutinho 1990), but in larger species the opposite is often true. In a study of six species of trees and shrubs, five, *Periandra mediterranea, Rourea induta, Miconia albicans, Myrsine guianensis,* and *Roupala montana,* exhibited reduced seed production after fire. For the first two species, the reduced postburn production was entirely due to the reduction in mean plant size, rather than a reduction in size-specific seed production. For the remaining three species, reduced plant size was compounded by reduced size-specific reproduction. Not all trees exhibit a reduction in postburn seed output, as demonstrated by *Piptocarpha rotundifolia,* which produced nearly twice the number of seeds in the first year after fire (Hoffmann 1998).

Other trees and shrubs reproduce successfully after fire (e.g., Silva, Hays, and Morais 1996; Landim and Hay 1996; Cavalheiro and Miranda 1999), but, in these cases, it is uncertain whether fire has a net positive or negative effect on the number of seeds produced. Since the effect of fire on seed production is often quantitative rather than qualitative, and spatial and temporal variation is high, detailed study is needed to ascertain the response of cerrado species to fire.

Seedling Establishment

In many environments, fire is known to stimulate seedling establishment by removing competing adults and inhibitory litter. Hoffmann (1996) found the opposite to be true in the cerrado. For seven of twelve species, establishment success of experimentally placed seeds was found to be significantly lower in recently burned sites than in sites burned 1 or more years previously, whereas none of the species exhibited enhanced seedling establishment. A possible explanation is that fire reduces litter and canopy cover, which facilitates seedling establishment of several species, probably by ameliorating water stress (Hoffmann 1996).

The poor seedling establishment following fire might explain why many species have not evolved protective fruits that ensure seed availability after fire. If conditions for seedling establishment are poor after burning, there should be little selective pressure to develop such fruits.

Vegetative Reproduction

Rather than produce large cohorts of seedlings following burning, some cerrado species have evolved the strategy of producing large cohorts of suckers (Hoffmann 1998). Following fire, *Rourea induta*, *Roupala montana*, and *Myrsine guianensis* produced large numbers of root suckers, a response documented in species of other regions (Lacey 1974; Farrell and Ashton 1978; Lamont 1988; Lacey and Johnston 1990; Kammesheidt 1999). It is necessary to emphasize that we are defining root suckers as new stems originating from root buds at some distance from the parent individual. When referring to vegetative reproduction, we do not include resprouting from the root crown of the original individual.

Many other cerrado species reproduce vegetatively (Ferri 1962; Rizzini and Heringer 1962; Raw and Hay 1985), but it remains to be confirmed that fire increases vegetative reproduction in these species. The advantages of vegetative reproduction appear to be strong in the cerrado. In general, vegetatively produced offspring tend to be larger than seedlings, making them less prone to stress and disturbance (Abrahamson 1980; Peterson and Jones 1997). This was confirmed in the cerrado for the three species mentioned above. Suckers were much larger than seedlings, and for two of the species, exhibited a greater capacity to survive fire (Hoffmann 1998).

Suckers produced after fire probably benefit from the high light and nutrient availability at this time. However, it is not reasonable to argue that the timing of vegetative reproduction has evolved in response to these selective advantages. There is strong evidence that the timing of sucker production has originated as a physiological constraint on the process of root bud formation. The formation of root buds can be stimulated by the reduction in auxin content resulting from the loss of aerial biomass (Peterson 1975). The influence of auxins on root bud formation has been found in many taxa (Peterson 1975), suggesting that root sucker formation is inextricably linked to loss of aerial biomass. Although this may constrain the timing of vegetative reproduction, natural selection has probably strongly affected not only the quantity of suckers produced, but also the capacity of a species to produce root suckers in the first place.

EFFECT OF FIRE ON POPULATIONS

As shown above, fire simultaneously affects seed production, seedling establishment, vegetative reproduction, survival, and growth. The changes in these vital rates have direct and interactive effects on the population growth rate, so understanding how fire affects only a part of the life cycle yields incomplete information on the response of a species to fire. However, with quantitative information about the effect of fire on the entire plant life history, it becomes possible to synthesize the information into a complete picture using matrix population models. Details on the use of such models for simulating the effects of fire frequency are provided by Silva et al. (1991) and Hoffmann (1999).

Hoffmann (1999) used information on the growth, survival, seed production, seedling establishment, and vegetative reproduction in burned and unburned plots to construct matrix population models for a subshrub (*Periandra mediterranea*), two shrubs (*Miconia albicans* and *Rourea induta*), and two trees (*Myrsine guianensis* and *Roupala montana*). Using these models, four of the five species are predicted to decline under frequent burning but increase under low fire frequencies (see fig. 9.2). The fifth species, *R. induta*, showed little response to burning (fig. 9.2).

The effect of fire on a population depends on its size distribution. Under a constant fire frequency, a population will tend toward a particular stable size distribution. The population growth rates in figure 9.2 represent growth rates after the stable size distribution has been attained. However, for a different size distribution, fire can have a very different effect. Take, for example, the case of *Roupala montana*, as shown in figure 9.3. If a population has not been burned for many years, large adults are common, and the population is predicted to increase at a rate of 5% per year. If this population is then subjected to triennial burning, large individuals produce numerous vegetative offspring, causing a large increase in population density for the first two fire cycles. However, after

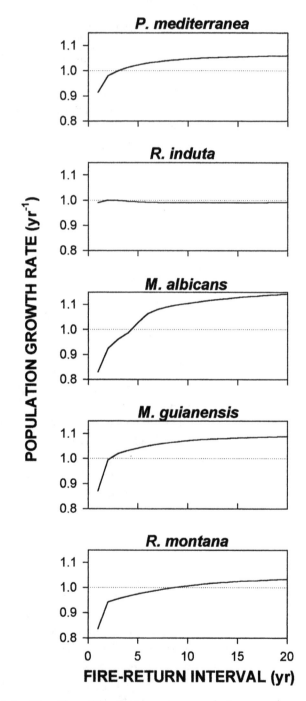

Figure 9.2 The effect of fire frequency on population growth rate (lambda) of five woody plant species, as predicted by matrix population models. Lambda > 1 indicates population growth and lambda < 1 indicates population decline. Growth forms are *P. mediterranea*, subshrub; *R. induta* and *M. albicans*, shrub; *M. guianensis* and *R. montana*, tree. Adapted from Hoffmann (1999).

repeated burns, the population becomes dominated by small individuals, which do not produce sufficient vegetative suckers to balance mortality, so the population experiences decline. Eventually, as the population approaches a new size distribution, the population is expected to decline at a rate of 5% per year (fig. 9.3). This illustrates the importance of long-term studies to shed light on the consequences of fire on population and community dynamics, since short-term studies may yield very different results.

With matrix population models, it is possible to quantify the contributions of sexual and vegetative reproduction to population growth of

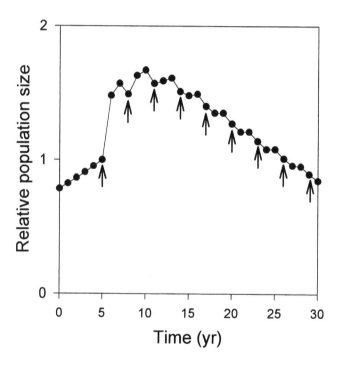

Figure 9.3 The effect of a change in fire frequency for a size-structured population. This graph shows relative population size of *Roupala montana* as simulated by a matrix population model. At the beginning of this simulation, the population is assumed to have been unburnt for many years. The first five years show the population increase under unburned conditions. After the fifth year, the population is subjected to triennial burning, with burns being indicated by arrows. Initially there is an increase in population density, but eventually population decreases as it becomes dominated by small individuals.

these species using elasticity analysis (de Kroon et al. 1986). This analysis revealed that, for the five study species, the contribution of sexual reproduction to population growth was low under frequent burning and increased under less frequent burning (see fig. 9.4). In contrast, for the three suckering species, vegetative reproduction made a large contribution to population growth under frequent burning, whereas it decreased or remained relatively constant under less frequent burning (fig. 9.4). As a result of these contrasting responses to fire frequency, the relative importance of vegetative to sexual reproduction increases under increasing fire frequency (Hoffmann 1999).

For the three species capable of vegetative reproduction, sexual reproduction contributed little to population growth (fig. 9.4), a trend widely observed in clonal plants in other habitats (Abrahamson 1980; Eriksson 1992).

CHANGE IN COMMUNITY STRUCTURE AND PHYSIOGNOMY

As described earlier, woody plants differ widely in their tolerance to fire and in their capacity to recover afterwards. Fire has considerable potential to influence the structure and composition of the vegetation, particularly those considered fire-type vegetation where fires are likely to be recurrent, such as seasonal tropical savannas (Myers 1936; Soares 1990). Savannas are modified by natural fires (Coutinho 1990; Sarmiento 1984), and many tropical savannas are maintained today by frequent anthropogenic fires (Walker 1981). In the cerrado, fire plays a fundamental role for the floristics and physiognomy of savanna stands (Moreira 1996; Meirelles et al. 1997; see also chapter 6).

To understand the role of fire in maintaining vegetation structure in the cerrado, it is useful to examine the consequences of excluding fire from the system. Fire protection in moist savannas induces gradual changes in the density of tree species, leading to denser savannas (Menaut 1977; Brookman-Amissah et al. 1980; Frost and Robertson 1987; San José and Fariñas 1991; Swaine et al. 1992). The same processes occur in the cerrado, where a gradual and progressive increase of woody vegetation after fire exclusion has been reported (see chapter 8). Coutinho (1990) reports that a *campo sujo* became a taller, dense *cerradão* after 43 years of fire and cattle exclusion (chapter 6).

Fire protection increases the frequency of woody plants in both open and closed cerrado physiognomies, particularly in the more open ones,

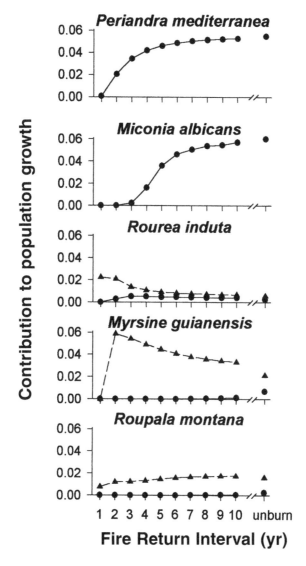

Figure 9.4 The effect of fire frequency on the importance of vegetative and sexual reproduction (data from Hoffmann 1999). The contribution of the two forms of reproduction were calculated from elasticity analysis of the matrix models (de Kroon et al. 1986).

indicating that protection against burning permits woody vegetation regeneration (Moreira 1996). Conversely, reintroduction of fire into previously protected sites causes a reduction in tree and shrub density (chapter 4). Fire reduces the woody plant cover as a whole regardless of physiognomy, but denser formations experience the greatest declines (Moreira 2000).

The reduction in woody plant density in burned areas is compounded by a reduction in mean plant size. In a study comparing the effect of long term fire protection on five cerrado physiognomies, ranging from open *campo sujo* to *cerradão* forest, Moreira (1992, 2000) showed that the number of plants of the smallest size class was higher in burned than in protected sites, while fire protection led to a shift towards taller plants (see fig. 9.5). Contrary to claims that fire affects only individuals below the fire line, usually around 1 to 1.5 m (Brookman-Amissah et al. 1980; San José and Fariñas 1983; Frost and Robertson 1987; Coutinho 1990), fire actually affects woody plants in all size classes, which can lead to striking differences on woody cover (Moreira 1992, 1996).

The negative effect of fire on woody plant density does not occur uniformly across all species. Rather, fire influences species composition, in general driving a shift towards smaller growth forms. It has been widely noted that savanna fire favors herbaceous plants at the expense of woody plants; however, even among woody plants, fire causes a shift toward smaller growth forms (fig. 9.5). The abundance of trees and large shrubs tended to be reduced by fire, whereas subshrubs were favored. It appears that this relationship between growth form and fire tolerance reflects the effect of topkill (Hoffmann 1999). For the same rate of regrowth, larger growth forms require more time to regain their preburn size following topkill. Similarly, topkill makes it unlikely that immature individuals of trees and large shrubs reach a mature size under frequent burning, thus curtailing population growth.

There are many notable exceptions to this relationship between growth form and sensitivity to fire, such as the tree *Bowdichia virgilioides*, which is considered a fire-tolerant species (San José and Fariñas 1983) and was more common in frequently burned sites than in adjacent protected sites (Moreira 2000). The existence of fire-sensitive shrubs has not been previously reported in the cerrado literature, and the shrub flora is always considered to be "typically pyrophytic" (Coutinho 1990). The genus *Miconia*, however, has two fire-sensitive shrub species, *M. albicans* and *M. pohliana*. In fact, *M. albicans* was the most abundant species in a cerrado *sensu stricto* protected for more than 20 years against fire, but was

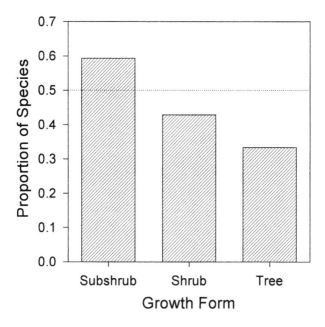

Figure 9.5 Response of three growth forms to frequent fire. The figure shows the proportion of species that were more common in a frequently burned site than in an adjacent unburned site. Smaller growth forms were more likely to be favored by fire (data from Moreira 1992).

completely absent in the adjacent unprotected side (Moreira 1992, 1996). Because this species produces very few seeds in the first two years after burning, and its seedlings are extremely sensitive to fire for at least the first two years of life (Hoffmann 1998), establishment is virtually eliminated under frequent burning.

Fire has a particularly strong effect on species composition of more closed physiognomies such as *cerradão* (chapter 6). Moreira (1996) found that five of the ten most abundant species in a fire-protected *cerradão* were totally absent in sampled areas of adjacent unprotected *cerradão*. *Cerradão* generally includes fire-sensitive species typical of forest, as shown earlier. *Emmotum nitens*, *Ocotea spixiana* and *Alibertia edulis* are forest species typical of *cerradão* that rarely, if ever, are found in more open cerrado (Furley and Ratter 1988). The establishment of *E. nitens* in open cerrado is probably constrained by the absence of a resprouting ability in seedlings of this species. *Emmotum nitens* and *A. edulis* have highly branched roots without any of the enlarged taproot that characterizes many cerrado woody plants (Labouriau et al. 1964; Rizzini 1965; Moreira 1987, 1992; Oliveira and Silva 1993).

Other fire-sensitive species typical of *cerradão* can expand into other physiognomies in the absence of fire. The fire-sensitive *cerradão* species *Blepharocalix salicifolius* and *Sclerolobium paniculatum* were present in fire-protected *campo sujo* and *campo cerrado*. In general fire protection leads to an increase of tree species in these shrub-dominated physiognomies, particularly in *campo sujo*, corroborating the idea that fire protection can lead to the substitution of open *campo sujo* by a tree-dominated denser cerrado physiognomy (Moreira 1992; chapter 6).

In general burning decreases species richness of woody plants in cerrado (Eiten and Sambuichi 1996; Moreira 1996; Sato 1996; Silva 1999). This reduction represents a reduction not only in the number of species per area, as would be expected from an overall reduction in the density of individuals, but also in the total stock of species in an area (Eiten and Sambuichi 1996).

CONCLUSIONS

Fire has a large impact on the population ecology of woody plants in the cerrado. Fire tends to increase plant mortality, reduce plant size, and increase vegetative reproduction. Sexual reproduction is also affected, but the response is highly species specific. The net effect of these changes is a reduction in woody plant cover under the high fire frequecies currently observed. This change is manifested as a reduction in mean plant size within populations, a reduction in the density of individuals, and a shift in species composition towards smaller growth forms. Composed of fire-sensitive species, forest formations within the cerrado region are most susceptible to fire. Because of the large impact of fire on the cerrado, the current high frequency of anthropogenic burning is capable of effecting widespread change in the cerrado ecosystem.

REFERENCES

Abrahamson, W. G. 1980. Demography and Vegetative Reproduction. In O. T. Solbrig, ed., *Demography and Evolution in Plant Populations*, pp. 89–106. Berkeley, CA: University of California Press.

Bond, W. J. and B. W. van Wilgen. 1996. *Fire and Plants*. London: Chapman and Hall.

Brookman-Amissah, J., Hall J. N., Swaine M. D., and J. Y. Attakorah. 1980. A re-assessment of a fire protection experiment in a north-eastern Ghana savanna. *J. Appl. Ecol.* 17:85–99.

Cavalheiro, M. C. and H. S. Miranda. 1999. Estudo da produção de flores em áreas de campo sujo submitidos a diferentes regimes de queima, Brasília, DF. *Rev. Bras. Fisiol. Vegetal, Sup.* 11:82.

Cochrane, M. A. and M. D. Schulze. 1999. Fire as a recurrent event in tropical forests of the eastern Amazon: Effect on forest structure, biomass, and species composition. *Biotropica* 31:2–16.

Coutinho, L. M. 1977. Aspectos ecológicos do fogo no cerrado: II. As queimadas e a dispersão de sementes de algumas espécies anemocóricas do estrato herbáceo-subarbustivo. *Bol. Bot. USP* 5:57–64.

Coutinho, L. M. 1990. Fire ecology of the Brazilian cerrado. In J. G. Goldammer, ed., *Fire in the Tropical Biota*, pp. 82–105. Berlin: Springer-Verlag.

De Kroon, H., A. Plaisier, J. van Groenendael, and H. Caswell. 1986. Elasticity: The relative contribution of demographic parameters to population growth rate. *Ecology* 67:1427–1431.

Eiten, G. and R. H. R. Sambuichi. 1996. Effect of long-term periodic fire on plant diversity in a cerrado area. In R. C. Pereira and L. C. B. Nasser, eds., *Biodiversidade e Produção Sustentável de Alimentos e Fibras nos Cerrados*. Proceedings of the VIII Simpósio Sobre o Cerrado, pp. 46–55. Brasília: Empresa Brasileira de Pesquisa Agropecuária.

Eriksson, O. 1992. Evolution of seed dispersal and recruitment in clonal plants. *Oikos* 63:439–448.

Farrel, T. P. and D. H. Ashton. 1978. Population studies on *Acacia melanoxylon* R.Br.: I. Variation in seed and vegetative characteristics. *Aust. J. Bot.* 26:365–379.

Ferri, M. G. 1962. Histórico dos trabalhos botânicos sobre o cerrado. In M. G. Ferri, ed., *Simpósio Sobre o Cerrado*, pp. 7–35. São Paulo: Editora Edgard Blucher.

Franco, A. C., M. P. Souza, and G. B. Nardoto. 1996. Estabelecimento e crescimento de *Dalbergia miscolobium* Benth. em áreas de campo sujo e cerrado no DF. In H. S. Miranda, C. H. Saito, and B. F. de S. Dias, eds., *Impactos de Queimadas em Áreas de Cerrado e Restinga*, pp. 84–92. Brasília: ECL/Universidade de Brasília.

Frost, P. G. H. and F. Robertson. 1987. The ecological effects of fire in savannas. In B. J. Walker, ed., *Determinants of Tropical Savannas*, pp. 93–140. Paris: IUBS Special Issue no. 3.

Furley, P. A. and J. A. Ratter. 1988. Soil resources and plant communities of Central Brazilian cerrado and their development. *J. Biogeogr.* 15:97–108.

Gignoux, J., J. Clobert, and J.-C. Menaut. 1997. Alternative fire resistence strategies in savanna trees. *Oecologia* 110:576–583.

Harper, J. L. 1977. *Population Biology of Plants*. London: Academic Press.

Hoffmann, W. A. 1996. The effects of cover and fire on seedling establishment in a neotropical savanna. *J. Ecol.* 84:383–393

Hoffmann, W. A. 1998. Post-burn reproduction of woody plants in a neotropical savanna: The relative importance of sexual and vegetative reproduction. *J. Appl. Ecol.* 35:422–433.

Hoffmann, W. A. 1999. Fire and population dynamics of woody plants in a neotropical savanna: Matrix model predictions. *Ecology* 80:1354–1369.

Hoffmann, W. A. 2000. Post-establishment seedling success of savanna and forest species in the Brazilian Cerrado. *Biotropica* 32:62–69.

Hoffmann, W. A., F. A. Bazzaz, N. J. Chatterton, P. Harrison, and R. B. Jackson. 2000. Elevated CO_2 enhances resprouting of a tropical savanna tree. *Oecologia* 123:312–317.

Holdsworth, A. R. and C. Uhl. 1997. Fire in Amazonian selectively logged rain forest and the potential for fire reduction. *Ecol. Appl.* 7:713–725.

Kammesheidt, L. 1999. Forest recovery by root suckers and above-ground sprouts after slash-and-burn agriculture, fire, and logging in Paraguay and Venezuela. *J. Trop. Ecol.* 15:143–158.

Kauffman, J. B. 1991. Survival by sprouting following fire in tropical forests of the Eastern Amazon. *Biotropica* 23:219–224.

Keeley, J. E. and P. H. Zedler. 1978. Reproduction of chaparral shrubs after fire: A comparison of sprouting and seeding strategies. *Am. Nat.* 99:142–161.

Labouriau, L. G., I. F. M. Válio, and E. P. Heringer. 1964. Sôbre o Sistema Reprodutivo de Plantas dos Cerrados. *An. Acad. Bras. Ciên.* 36:449–464.

Lacey, C. J. 1974. Rhizomes in tropical Eucalypts and their role in recovery from fire damage. *Aust. J. Bot.* 22:29–38.

Lacey, C. J. and Johnston, R. D. 1990. Woody clumps and Clumpwoods. *Aust. J. Bot.* 38:299–334.

Lamont, B. B. 1988. Sexual versus vegetative reproduction in *Banksia elegans*. *Bot. Gaz.* 149:370–375.

Landim, M. F. and Hay, J. D. 1996. Impacto do fogo sobre alguns aspectos da biologia reprodutiva de *Kielmeyera coriacea* Mart. *Rev. Bras. Biol.* 56:127–134.

Lonsdale, W. M. and R. W. Braithwaite. 1991. Assessing the effects of fire on vegetation in tropical savannas. *Aust. J. Ecol.* 16:363–374.

Meirelles, M. L., C. A. Klink, and J. C. S. Silva. 1997. Un modelo de estados y transiciones para el Cerrado brasileño. *Ecotrópicos* 10:45–50.

Menaut, J. C. 1977. Evolution of plots protected from fire since 13 years in a Guinea savanna of Ivory Coast. *Actas del IV Simposio Internacional de Ecologia Tropical (Panama)* 2:541–558.

Miyanishi K. and M. Kellman. 1986. The role of root nutrient reserves in regrowth of two savanna shrubs. *Can. J. Bot.* 64:1244–1248.

Moreira, A. G. 1987. "Aspectos Demograficos de Emmotum nitens (Benth.)

Miers em um Cerradão Distrófico do Distrito Federal." Master's thesis, Universidade Estadual de Campinas, Campinas, Brazil.

Moreira, A. G. 1992. "Fire Protection and Vegetation Dynamics in the Brazilian Cerrado." Ph.D. thesis, Harvard University, Cambridge, MA, U.S.A.

Moreira, A. G. 1996. Proteção contra o fogo e seu efeito na distribuição e composição de espécies de cinco fisionomias de cerrado. In H. S. Miranda, C. H. Saito, and B. F. de S. Dias, eds., *Impactos de Queimadas em Areas de Cerrado e Restinga*, pp. 112–121. Brasília: ECL/Universidade de Brasília.

Moreira, A. G. 2000. Effects of fire protection on savanna structure in Central Brazil. *J. Biogeogr.* 27:1021–1029.

Myers, J. G. 1936. Savanna and forest vegetation of the interior Guiana Plateau. *J. Ecol.* 24:162–184.

Oliveira, P. E. 1998. Fenologia e biologia reprodutiva das espécies de cerrado. In S. M. Sana and S. P. de Almeida, eds., *Cerrado: Ambiente e Flora*, pp. 169–192. Planaltina: Empresa Brasileira de Pesquisa Agropecuária.

Oliveira, P. E. and J. C. S. Silva. 1993. Biological reproduction of two species of Kielmeyera (Guttiferae) in the cerrados of central Brazil. *J. Trop. Ecol.* 9:67–80.

Peres, C. A. 1999. Ground fires as agents of mortality is a central Amazonian Forest. *J. Trop. Ecol.* 15:535–541.

Peterson, C. J. and Jones, R. H. 1997. Clonality in woody plants: A review and comparison with clonal herbs. In H. de Kroon and J. van Groenendael, eds., *The Ecology and Evolution of Clonal Plants*, pp. 263–289. Leiden, Netherlands: Backhuys Publishers.

Peterson, R. L. 1975. The initiation and development of root buds. In G. Torrey and D. T. Clarkson, eds., *The Development and Function of Roots*, pp. 125–161. San Diego: Academic Press.

Raw, A. and J. Hay. 1985. Fire and other factors affecting a population of *Simarouba amara* in cerradão near Brasília. *Rev. Bras. Bot.* 8:101–107.

Rizzini, C. T. 1965. Experimental studies on seedling development of cerrado woody plants. *Ann. Miss. Bot. Gard.* 52:410–426.

Rizzini, C. T. 1971. A flora do cerrado. In M. G. Ferri, ed., *Simpósio Sôbre o Cerrado*, pp. 105–154. São Paulo: Editora Edgard Blücher.

Rizzini, C. T. and Heringer, E. P. 1962. Studies on the underground organs of trees and shrubs from some southern Brazilian savannas. *An. Acad. Bras. Ciênc.* 34:235–247.

Rutherford, M. C. 1981. Survival, regeneration and leaf biomass changes in woody plants following spring burns in *Burkea africana–Ochna pulchra* savanna. *Bothalia* 13:531–552.

San José, J. J. and M. R. Fariñas. 1983. Changes in tree density and species composition in a protected Trachypogon savana, Venezuela. *Ecology* 64:447–453.

San José, J. J. and M. R. Fariñas. 1991. Changes in tree density and species composition in a protected Trachypogon savanna protected for 25 years. *Acta Oecol.* 12:237–247.

Sarmiento, G. 1984. *The Ecology of Neotropical Savannas*. Cambridge, MA: Harvard University Press.

Sato, M. N. 1996. "Taxa de Mortalidade da Vegetação Lenhosa do Cerrado Submetida a Diferentes Regimes de Queima." Master's thesis, Universidade de Brasília, Brasília, Brazil.

Sato, M. N. and H. S. Miranda. 1996. Mortalidade de plantas lenhosas do cerrado *sensu stricto* submetidas a diferentes regimes de queima. In H. S. Miranda, C. H. Saito, and B. F. S. Dias, eds., *Impactos de Queimadas em Áreas de Cerrado e Restinga*, pp.102–111. Brasilia: Ecologia, Universidade de Brasília.

Setterfield, S. A. 1997. The impact of experimental fire regimes on seed production in two tropical eucalypt species in northern Australia. *Aust. J. Ecol.* 22:279–287.

Silva, D. M. S., J. D. Hay, and H. C. Morais. 1996. Sucesso reprodutivo de *Byrsonima crassa* (Malpighiaceae) após uma queimada em um cerrado de Brasilia–DF. In H. S. Miranda, C. H. Saito, and B. F. S. Dias, eds., *Impactos de Queimadas em Áreas de Cerrado e Restinga*, pp. 122–127. Brasília: Ecologia, Universidade de Brasília.

Silva, E. P.da R. 1999. "Efeito do Regime de Queima na Taxa de Mortalidade e Estrutura da Vegetação." Master's thesis, Universidade de Brasília, Brasília, Brazil.

Silva, E. P. da R. and H. S. Miranda 1996. Temperatura do câmbio de espécies lenhosas do cerrado durante queimadas prescritas. In R. C. Pereira and L. C. B. Nasser, eds., *Biodiversidade e Produção Sustentável de Alimentos e Fibras nos Cerrados*. Proceedings of the VIII Simpósio Sobre o Cerrado, pp. 253–257. Brasília: Empresa Brasileira de Pesquisa Agropecuária.

Silva, G. T., M. N. Sato, and H. S. Miranda. 1996. Mortalidade de plantas lenhosas em um campo sujo de cerrado submitidos a queimadas prescritas. In H. S. Miranda, C. H. Saito, and B. F. de S. Dias, eds., *Impactos de Queimadas em Áreas de Cerrado e Restinga*, pp. 93–101. Brasília: ECL/Universidade de Brasília.

Silva, J. F., J. Raventos, H. Caswell, and M. C. Trevisan. 1991. Population responses to fire in a tropical savanna grass: A matrix model approach. *J. Ecol.* 79:345–356.

Soares, R. V. 1990. Fire in some tropical and subtropical South American vegetation types: An overview. In J. G. Goldammer, ed., *Fire in the Tropical Biota*, pp. 63–81. Berlin: Springer-Verlag.

Swaine, M. D., W. D. Hawthorne, and T. K. Orgle. 1992. The effects of fire exclusion on savanna vegetation at Kpong, Ghana. *Biotropica* 24:166–172.

Uhl, C. and R. Buschbacher. 1985. A disturbing synergism between cattle ranching burning practices and selective tree harvesting in the eastern Amazon. *Biotropica* 17:265–268.

Van Wilgen, B. W. and G. G. Forsyth. 1992. Regeneration strategies in Fynbos plants and their influence on the stability of community boundaries after fire. In B. W. van Wilgen, D. M. Richardson, F. J. Kruger, and H. J. van Hensbergen, eds. *Fire in South African Mountain Fynbos,*. pp. 54–80. Berlin: Springer-Verlag.

Vicentini, K. R. C. F. 1993. "Análise Palinológica de uma Vereda em Cromínia–GO." Master's thesis, Universidade de Brasília, Brasília, Brazil.

Walker, B. H. 1981. Is succession a viable concept in African savanna ecosystems? In D. C. West, H. H. Shugart, and D. B. Botkin, eds., *Forest Succession: Concepts and Applications*, pp. 431–447. New York: Springer-Verlag.

10

Ecophysiology of Woody Plants

Augusto C. Franco

THE CENTRAL PLAINS OF BRAZIL ARE OCCUPIED BY A COMPLEX OF plant physiognomies such as *cerradão*, cerrado *sensu stricto*, and *campo sujo* (chapter 6). The great spatial variation in woody plant density across the cerrado landscape results in a complex pattern of resource availability, which changes both horizontally across the landscape and vertically within each vegetation type. This is of particular importance for seedlings that colonize the grass matrix typical of a *campo sujo* vegetation or a closed canopy woodland such as a *cerradão*. Like any other neotropical savanna, the cerrados are characterized by a strongly seasonal climate with distinctive wet and dry seasons (see fig. 10.1). Soils are deep and well drained, acidic, extremely low in available nutrients and with high Al content (Goodland and Ferri 1979; Haridasan 1982; Sarmiento 1984; chapter 2). Recurrent fires in the dry season place additional stress on the survival of woody plant seedlings (fire effects are discussed in chapters 4, 9).

Models explaining the structure and function of savanna ecosystems

Figure 10.1 (*opposite page*)   Integrated monthly rainfall, monthly variation in the daily number of hours of sunshine, minimum and maximum air temperatures and relative humidity, and changes in soil water potential for a *campo sujo* (solid squares) and a *cerradão* (open circles). Weather data come from the climate station of Reserva Ecológica do IBGE, Brasília (15°56' S, 47°53' W). Soil water potential is the mean ± SE for 5 soil psychrometers (models PCT-55-15-SF or PST-55-15-SF, Wescor Inc., Logan, Utah, U.S.) at 5, 30 and 85 cm depth, placed in a *campo sujo* and a *cerradão* of Fazenda Água Limpa (15°56' S, 47°55' W), near the Reserva Ecológica do IBGE. The *campo sujo* site had 533 individuals per ha with stem diameter greater than 5 cm at 30 cm from the ground. The *cerradão* site had 2,800 individuals of the same size class per ha. See chapter 6 for descriptions of cerrado physiognomies.

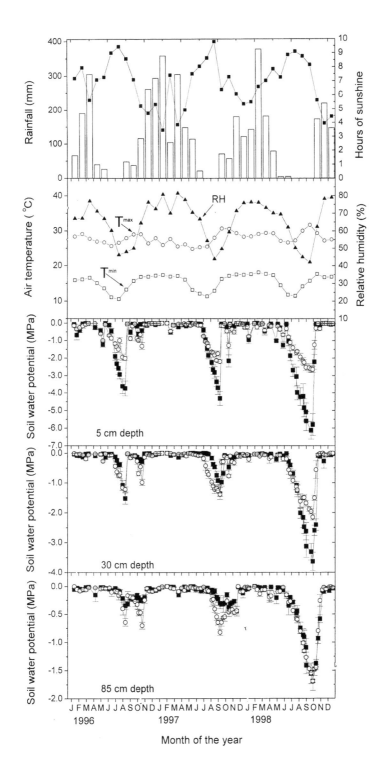

Month of the year

typically involve water and nutrients as limiting resources and a two-layered (grasses versus woody species) soil-water system (Walker and Noy-Meir 1982; Knoop and Walker 1985). According to this model, the shallow roots of grasses make them superior competitors for water in the upper part of the soil profile, whereas deeply rooted woody plants have exclusive access to a deeper, more predictable water source. This two-layered model implies that tree recruitment is dependent on the capacity of seedlings to withstand grass root competition during early growth stages, until their roots reach deeper and more reliable soil water sources to sustain them during dry periods (Medina and Silva 1990).

While the tropical savannas of northern South America are characterized by relatively low woody plant diversity, cerrado communities contain a remarkably complex community structure rich in endemic woody species. More than 500 species of trees and large shrubs are present in the cerrado region, and individual sites may contain up to 70 or more woody species (Felfili et al. 1998). As discussed in chapter 6, the establishment of modern floristic composition is primarily the result of historic and biogeographic events. However, resource availability certainly plays a prominent role in maintaining such high species diversity. Tilman (1982) and Cody (1986) have independently proposed models to explain plant species diversity in resource-limited environments. Their models, based on selection for increased resource partitioning, suggest that seasonality in cerrado rainfall regime and low soil nutrient availability are critical factors allowing the coexistence of many woody species in relatively small areas. The theoretical model of Tilman (1982) is particularly significant in this respect since it explicitly considers the temporal and spatial dynamics of resource availability in selection for divergent strategies of resource capture.

In this chapter I will show that in savannas of high woody species diversity such as the cerrado, plants exhibit an array of physiological and morphological mechanisms to cope with rainfall seasonality. Because these plants are growing on extremely infertile soils, I will also discuss the effects of nutrient deficiency on plant growth and biomass allocation. Finally, I will focus on establishment and growth of seedlings of woody perennials, with an emphasis on the effects of spatial variation in light, an important environmental factor that is generally neglected in savanna studies.

RAINFALL SEASONALITY AND PATTERNS
OF WATER USE

The cerrados are characterized by strong rainfall seasonality (fig. 10.1). The winters are dry and cool. Frost events are uncommon and occur only at the southern limit of the cerrado region. Low relative humidities result in a high evaporative demand during the dry period that generally extends from May to September. During the dry season, the soil begins to dry out from the surface downwards, and soil water potentials (Ψ^s) in the upper soil layers can reach values below −3.0 MPa. Upper soil layers dry much faster in an open *campo sujo* formation than in a closed *cerradão* formation. Deeper soil layers exhibit a much higher degree of water constancy, but they are drier in the *cerradão* (fig. 10.1). Indeed, soil gravimetric water content at a *campo sujo* averaged 4% higher than at a nearby *cerradão* site for the 1–2-m depth range, and 3% higher for the 2–3-m depth range, toward the end of the dry period (Jackson et al. 1999). Thus, increases in woody cover and in tree diversity result in a larger exploitation of soil water resources. However, individuals of three evergreen species showed similar water status and daily water loss in these two contrasting vegetation types (Bucci et al. 2002).

A schematic representation of the most relevant factors affecting the water balance of cerrado woody plants is presented in figure 10.2. Imbalance between water supply and demand results in changes in plant water status, generally assessed by measurements of leaf water potential (Ψ^l). Because nocturnal stomatal closure curtails transpiration and potentially allows for plant rehydration, a water balance between the plant and the soil should be reached by the end of the night. Thus, predawn Ψ^l can be used as a measure of the water potential of the soil adjacent to the root system and therefore of the maximum water status the plant can achieve. Predawn Ψ^l remains between −0.1 and −0.3 MPa in the wet season, and −0.3 to −0.8 MPa in the dry season (Mattos et al. 1997; Mattos 1998; Franco 1998; Meinzer et al. 1999; Bucci et al. 2002). Minimum Ψ^l is reached between midday and early afternoon, when the evaporative demand of the atmosphere is higher. Values are in the range of −1 to −3 MPa in the wet season and −2 to −4 MPa in the dry season (Perez and Moraes 1991; Franco 1998; Meinzer et al. 1999; Bucci et al. 2002). Thus, Ψ^l of adult shrubs and trees decreases in the dry season but is still higher than Ψ^s of upper soil layers. Excavation experiments reported that about 70% of the root biomass was contained in the first 100 cm depth (Abdala et al. 1998). However, a small proportion of root biomass was found to depths of 6 to 8 meters. Thus, these few deeper roots would extract enough water to ensure that rainfall seasonality exerts only a small effect on Ψ^l.

Not all cerrado woody species are deep-rooted. Earlier excavation experiments have already observed a wide range of rooting habits, from shallow-rooted to deep-rooted species (Rawitscher 1948). Measurements of hydrogen isotope differences between plant tissue water and soil water

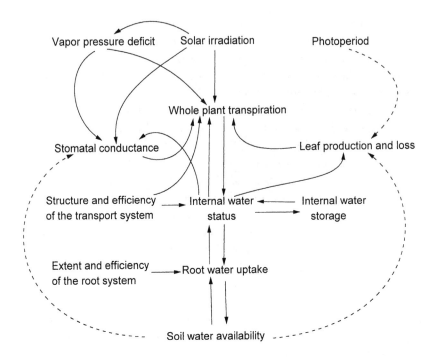

Figure 10.2 A diagrammatic representation of the most relevant factors affecting the water balance of cerrado woody plants. The solid arrows indicate processes that are well established in the literature or that have been reported for cerrado plants. The broken arrows indicate processes that are either controversial (most of them are probably controlled by endogenous growth regulators, including receptor pigments) or poorly known in the cerrado. For the sake of simplicity, the processes affecting soil water availability (e.g., soil water evaporation, precipitation, and soil characteristics) are not detailed.

samples collected at different depths have indicated a complex pattern of water exploitation of the soil profile (Jackson et al. 1999). Deciduous species were extracting water from deep soil layers in the dry season, whereas evergreeen species showed a broader range of water extraction patterns, from shallow-rooted to deep-rooted. These studies suggest that root patterns in cerrado ecosystems are much more complex than predicted by the classical two-layered soil-water model for savannas.

Loss of water vapor by transpiration is driven by the leaf-to-air vapor pressure difference. It occurs mainly through the stomatal pore because of the presence of a highly waterproof cutinized epidermis. The adequate availability of soil water for deep-rooted woody plants would imply that they would not need to rely on regulation of water losses even during the peak of the dry season. This was in agreement with early studies, relying on rapid weighing of detached leaves, that concluded that many cerrado woody species transpired freely throughout the year; some restricted their transpiration only at the end of the dry season; and only a few restricted their transpiration from the beginning of the dry season (e.g., Ferri 1944; Rawitscher 1948). This weak stomatal control of transpiration was also reported in four characteristic woody trees of the Venezuelan savannas (Sarmiento et al. 1985) also common in the Brazilian cerrados. Because of the diurnal and seasonal changes in vapor pressure deficit, conclusions based on measurements of transpiration rates have to be examined with caution. Indeed, studies using gas exchange techniques consistently reported lower maximum photosynthetic rates and stomatal conductances (mainly a function of the degree of stomatal opening), as well as a moderate midday depression of photosynthesis and stomatal conductances in most species during the dry season (Johnson et al. 1983; Moraes et al. 1989; Perez and Moraes 1991; Moraes and Prado 1998). In most species lower stomatal conductances in the dry season resulted in an increase in water use efficiency (the amount of CO_2 assimilated by photosynthesis/water lost by transpiration), which was determined by comparing the carbon isotope ratios of leaves collected in the wet and dry season (Mattos et al. 1997). Medina and Francisco (1994) reported similar results for two common woody species of the Venezuelan savannas. Control of water loss and reductions of photosynthetic rates as a result of partial stomatal closure both in the wet and dry season were also reported (Franco 1998; Moraes and Prado 1998; Maia 1999).

Several studies of regulation of water use at the plant level with sap flow sensors have shown that, despite the potential access to soil water at a depth where availability was relatively constant throughout the year, plants exhibited reduced transpiration due to partial stomatal closure during both the dry and wet seasons (Meinzer et al. 1999; Bucci et al. 2000; Naves-Barbiero et al. 2000). For most species, sap flow increased sharply in the morning, briefly attained a maximum value by about 09:30 to 12:00 h, then decreased sharply, despite steadily increasing solar radiation and atmospheric evaporative demand. This decrease was particularly strong in the dry season, when high values of vapor pressure deficit prevail during most of the daylight hours. In some cases, transpiration rates briefly recovered in the late afternoon (Naves-Barbiero et al. 2000). Thus the reductions in stomatal opening are the result of hydraulic constraints, or a direct response to changes in leaf-to-air vapor pressure deficit.

At the ecosystem level, measurements of water fluxes using eddy covariance methods showed that evapotranspiration increased linearly with solar irradiation in the wet season, but not in the dry season, when this linear increase was maintained only up to solar irradiation values of 200 Wm^{-2} (about one-fourth the energy of full sunlight at this time of year). This was followed by a considerable decline in the slope of the curve, despite steadly increasing solar irradiation (Miranda et al. 1997). These authors also reported that ecosystem surface conductances (an estimate of stomatal opening at the canopy level) were much lower in the dry season and fell gradually as the day progressed both in the wet and dry seasons, probably in response to increases in the evaporative demand of the atmosphere.

In conclusion, stomatal conductance plays a major role in the control of water flow, minimizing the effects of increases in the evaporative demand of the atmosphere (increase in vapor pressure deficit) as a force driving transpiration (fig. 10.2). The values of stomatal conductance which restrict transpiration to levels compatible with the supply of water from the xylem will vary, mainly depending on the magnitude of the vapor pressure deficit and total leaf area, which changes seasonally, but also reflecting some limitation due to soil water availability and stem water storage. Thus, this rapid decrease in flow rate after an early peak may represent the limits of an internal reservoir that is recharged at night, or may be the result of limited capacity of the root system to absorb water in sufficient quantities to sustain high rates of transpiration over a longer period. Little is known about the internal structure of the vascular systems of these species and their effects on the transpiration rates. Morphological studies coupled with physiological measurements of hydraulic conductivity, root distribution, and water extraction patterns at different soil layers are necessary to provide a more complete picture of water relations of cerrado woody plants. Moreover, it is yet to be determined whether leaf abscission and changes in stomatal opening of cerrado plants are biochemically controlled in response to environmental signals, or whether they are a response to physical factors related to changes in cell turgor driven by water availability.

SEASONAL WATER AVAILABILITY, PHOTOSYNTHESIS, LEAF PHENOLOGY, AND GROWTH

The drop in predawn Ψ^l by the end of the dry season indicates that full nocturnal recharge of water reservoirs is not attained and that soil water availability is becoming a critical factor. Plant water reservoirs and nocturnal recharge are still sufficient to accommodate high rates of plant water loss in the dry season, but shoot growth depends on the relief of water stress. Table 10.1 presents leaf characteristics of 13 woody species commonly found throughout the cerrado region. In general, evergreen species tend to rely on shallower water sources and to flush during the wet season. *Vochysia elliptica*, an evergreen with a deep root system, flushes at the end of the dry season, a pattern similar to deep-rooted decid-uous trees. However, the evergreen *Roupala montana* and the deciduous *Pterodon pubescens* apparently have a shallow-rooted system, but both show a leaf flush pattern typical of a deep-rooted tree: that is, they flush at the end of the dry season. Partial (*R. montana*) or total (*P. pubescens*)

Table 10.1 Patterns of Leaf Flush, Maximum CO_2 Assimilation Rates (A_{max}; μmol CO_2 m^{-2} s^{-1}) and Depth of the Root System of 13 Cerrado Woody Species

Species	Family	Leaf flush	A_{max}	Root depth
Evergreen				
Didymopanax macrocarpum	Araliaceae	Throughout the year	18[a]	Shallow
Miconia ferruginata	Melastomataceae	Wet season	14[f]	Shallow
Rapanea guianensis	Myrsinaceae	Wet season	12[e]	Not available
Roupala montana	Proteaceae	End of the dry season	14[b]	Shallow
Sclerolobium paniculatum	Caesalpinoidae	Wet season	20[f]	Shallow
Vochysia elliptica	Vochysiaceae	End of the dry season	14[f]	Deep
Briefly deciduous				
Blepharocalyx salicifolia	Myrtaceae	End of the dry season	12[f]	Not available
Dalbergia miscolobium	Caesalpinoidae	End of the dry season	15[c]	Deep
Pterodon pubescens	Faboideae	End of the dry season	11[c]	Shallow
Qualea grandiflora	Vochysiaceae	End of the dry season	16[f]	Deep
Deciduous				
Caryocar brasiliense	Caryocaraceae	End of the dry season	11[e]	Not available
Kielmeyera coriacea	Guttiferae	End of the dry season	12[d]	Deep
Qualea parviflora	Vochysiaceae	End of the dry season	12[c]	Intermediate

Sources: [a]= Franco 1983; [b]= Franco 1998; [c] = Kozovits 1997; [d] = Cardinot 1998; [e] = Maia 1999; [f] = previously unpublished data.

Note: A_{max} was measured during morning hours under field conditions in the wet season. Depth of the root system was based on comparisons of stable hydrogen isotope composition of stem xylem water, and soil water that was collected at different depths (Jackson et al. 1999). Patterns of leaf flush were compiled from Franco (1998); Jackson et al. (1999); Maia (1999).

leaf fall precedes new leaf production in both species. Stem diameter of several deciduous and evergreen cerrado species increases continuously throughout the entire rainy season, but this stem expansion ceases at the onset of the dry period (Alvin and Silva 1980).

Deciduous and evergreen trees of the tropical dry forest of Guanacaste, Costa Rica, are also able to flush or flower during drought, following stem rehydration (Borchert 1994a,b). Rehydration and bud break during the dry season occur only in trees with access to soil water or to stored water in their trunks, whereas desiccated trees at dry sites rehydrate and flush only after the first heavy rains of the rainy season (Borchert 1994c). In a study of leaf phenology patterns of 49 woody species (tall shrubs and trees) in a tropical eucalypt savanna of North Australia, Williams et al. (1997) reported that leaf fall in all species coincided with the attainment of seasonal minima in predawn Ψ^l, which were about −1.5 to −2.0 MPa in the evergreen and semideciduous (canopy fell below 50% of full canopy in the dry season) species, and about −0.5 to −1.0 MPa in the fully deciduous species. Leaf flushing occurred primarily in the late dry season. However, two evergreen species flushed throughout the dry season with a major peak in the late dry season, while leaf flushing in two fully deciduous species occurred only at the onset of rainy season. Soil moisture at 1m depth did not fall below the permanent wilting point (10% v/v); hence, reserves of soil water were sufficient to support whole-plant rehydration that preceded leaf flushing in the absence of rain. Production of new foliage more or less ceased soon after the beginning of the wet season, and most species were dormant by mid-late wet season. Therefore, we can expect that stem rehydration that precedes leaf flushing in cerrado woody species is the result of adjustments in the tree-water supply and demand, driven by a complex interplay of partial stomatal closure and reductions in leaf area that affects the root:shoot ratio as well as the depth of the root system and the size of the internal water reservoirs (fig. 10.2). Integrative studies of plant water status, cambium activity and leaf phenology of a range of cerrado woody species are critically needed.

The degree of soil water partitioning and variation in the timing of leaf production and loss among cerrado woody species suggests that resource partitioning may play an important role in maintenance of the high diversity of woody species in the cerrado. However, resource partitioning may imply a series of tradeoffs. For example, although a deep rooting pattern may allow for new leaf production during the driest months of the year, it may also impose hydraulic limitations on the amount of water that the plants can extract and transpire daily (Meinzer et al. 1999). Stomatal control of transpiration could be especially critical for evergreen species that maintain a considerable amount of photosynthetically active leaves throughout the whole dry season. Moreover, leaf flushing in the dry season may have important implications in competition for water and nutrients with perennial grasses, which grow only after the first rains.

NUTRIENT DEFICIENCY, LEAF SCLEROMORPHISM, AND PHOTOSYNTHETIC CAPACITY

A well-developed cuticle, epidermal cells with thick cell walls, presence of hypodermis, and abundance of structural elements characterize leaves of most cerrado woody plants. Arens (1958a,b) proposed that the leaf scleromorphism in cerrado woody plants was a consequence of nutrient deficiency, especially nitrogen. According to his hypothesis, high light levels and lack of water stress would result in an abundance of assimilated CO_2, consumed in the production of structural elements in the leaves because of low availability of nutrients for growth. Cerrado soils are generally very deep Oxisols with a high percentage of clay; well drained; strongly acidic; high in Al saturation; and extremely low in available nutrients (Goodland and Ferri 1979; Sarmiento 1984; chapter 2). P levels are generally below 1 ppm. In a study of 40 woody cerrado species, Miranda et al. (1997) reported that N levels (% of dry matter) in mature leaves were between 0.7% and 1.0% for 33% of the species, 1.1% and 1.5% for 50% of the species, and 1.6 and 2.2% for 10% of the species. Only 3 species had N levels above 2.2%. In a study of 8 common cerrado species, Medeiros and Haridasan (1985) reported that leaf nutrient levels varied between 0.05% and 0.07% for P, 0.14% and 0.80% for Ca, 0.28 and 0.87% for K, and 0.07% and 0.28% for Mg. These leaf nutrient levels are within the range of values reported for sclerophyllous leaves with similar specific leaf areas (leaf area: leaf dry mass) on extremely oligotrophic soils of the upper Rio Negro, Venezuela (Medina et al. 1990; Reich et al. 1995). Therefore, low nutrient levels could be a major constraint for plant growth in cerrado ecosystems.

One of the major physiological processes that affects growth is photosynthesis. Low nutrient levels could result in low photosynthetic rates and, as a consequence, low growth rates. On a leaf area basis, maximum photosynthetic rates of cerrado woody plants are moderate, generally in the range of 6 to 20 μmol CO_2 m^{-2} s^{-1} (Prado and Moraes 1997; Moraes and Prado 1998; table 10.1). These values are similar to those of tropical canopy trees (Hogan et al. 1995; Zotz and Winter 1996) and savanna trees of Venezuela (Sarmiento et al. 1985; Medina and Francisco 1994) and Australia (Eamus et al. 1999). On a dry weight basis, values are also moderate, ranging from 39 to 146 μmol kg^{-1} s^{-1} (Prado and Moraes 1997). Factors such as a higher diversion of plant resources to non-photosynthetic tissues, larger leaf construction and maintenance costs, or the degree and duration of crown deciduousness could potentially offset the benefits of maintaining a photosynthetically active crown in the dry season and may explain similar maximum photosynthetic rates in evergreen and deciduous species (table 10.1). Indeed, the combined effects of herbivory, partial leaf loss, and reductions in photosynthetic rates greatly reduced the estimated daily carbon gain of the evergreen *Roupala montana* by end of the dry season (Franco 1998).

Conditions where photosynthetic capacity is high but N availability is low favor carbon storage and biomass partitioning to roots rather than leaves (Fichtner et al. 1995). At the plant level, root:shoot ratios of cerrado woody species are fairly large, and underground structures for storage or vegetative propagation are common (Rizzini and Heringer 1961, 1962). Both evergreen and deciduous species have scleromorphic leaves and a large carbon investment in root biomass. At the ecosystem level, a cerrado *sensu stricto* vegetation (chapter 6) has a root:shoot ratio of about 1, significantly greater than ratios for tropical forests (ca. 0.1 to 0.5), but within the range of values measured for other savanna ecosystems (0.5 to 2.1; Abdala et al. 1998).

Changes in nutrient availability affect growth and biomass partitioning of cerrado woody plants. Growth of seedlings of *Miconia albicans* and *Copaifera langsdorffii* was enhanced in gallery forest soils richer in nutrients as compared with growth on dystrophic cerrado soils (Haridasan 1988; Machado 1990). High nutrient availability had a positive effect on plant biomass and a negative effect on root:shoot ratio and nonstructural carbohydrate concentrations in seedlings of *Dalbergia miscolobium* (Sassaki and Felippe 1998) and *K. coriacea* (Hoffmann et al. 2000). Melo (1999) reported that N fertilization decreased the leaf mass:leaf area ratio of *Eugenia dysenterica* and *Sclerolobium paniculatum* seedlings but did not affect the leaf mass:leaf area ratio of *Dypterix alata* and *Hancornia speciosa*. N fertilization did not enhance growth of these four species, instead decreasing biomass allocation to roots, while P fertilization had a positive effect on growth of all four species.

Mycorrhizal fungi generally play a critical role in soils with low availability in phosphorus. The information about mycorrhizal fungi in cerrado soils is scant. In a survey of mycorrhizal colonization in cerrado species, Thomazini (1974) reported that all species were infected with mycorrhiza.

In a greenhouse experiment with seedlings of four cerrado woody species, Reis (1999) concluded that mycorrhiza effectively colonized and increased growth of these species, but moderate additions of P enhanced the response of these plants to inoculation, increasing their growth.

The presence of nodules could overcome N deficiency in legumes, provided P deficiency was not a constraint for nodule activity. Legumes native to low-phosphorus soils often fix N well on them (Barnet and Catt 1991). The presence of nodules was observed in many cerrado legume trees (Faria et al. 1987, 1994). Total leaf N in cerrado legumes is higher than in other tree and shrub species (Miranda et al. 1997; Kozovits 1997). Based on $\delta^{15}N$ measurements, Sprent et al. (1996) presented some evidence that N fixation by nodules is a significant N source for small nodulated legume shrubs and herbs of the cerrado, but no such studies were performed with cerrado legume trees. Tripartite symbiosis (Rhizobium-mycorrhizal fungi-legume) were reported in hemicryptophyte legumes of *Trachypogon* savannas in Venezuela (Medina and Bilbao 1991). However, no evidence was found that this symbiosis was effective in reducing P deficiency in these small legumes under natural conditions. There is a need for studies evaluating the contribution of mycorrhiza-Rhizobium associations as N and P sources and the cost of such associations for cerrado legume trees and shrubs.

Many species of the cerrado vegetation accumulate aluminum in large quantities in their leaves, but the accumulation of Al does not interfere in the absorption of other cations like K, Ca, and Mg (Haridasan 1982; Medeiros and Haridasan 1985). Aluminum accumulation is particularly common in cerrado species of the families Vochysiaceae, Melastomataceae, and Rubiaceae. High concentrations of Al were found in the leaf phloem of Al-accumulating species and in the walls and contents of the collenchyma of the midrib, epidermal cells, guard cells of the stomata, and spongy parenchyma (Haridasan et al. 1986, 1987). However, the physiological significance of Al accumulation for cerrado plants is still unknown. *Miconia albicans*, an Al-accumulating shrub of the cerrado region, failed to grow in calcareous soils and produced chlorotic leaves, but showed complete recovery when parts of their root systems were grown in an $AlCl_3$ solution containing 10 mg Al/L or transplanted into an acid latosol (Haridasan 1988). Similar results were also found for seedlings of *Vochysia thyrsoidea* (Machado 1985).

In conclusion, the scleromorphic, nutrient-poor leaves of cerrado trees and shrubs do maintain relatively high photosynthetic rates. It appears that most of the assimilated carbon is not used for growth but stored in underground structures or diverged for leaf structural components, although experiments designed specifically to test this hypothesis are lacking. It is, however, somewhat surprising that leaf longevity is not an overwhelming feature in such nutrient-poor soils. Even in most species that are considered evergreen, leaf lifespan is less than a year.

ESTABLISHMENT AND GROWTH
OF SEEDLINGS OF WOODY PERENNIALS

Tree seedling establishment in neotropical savannas is heavily constrained by grass root competition, drought, and fire (Medina and Silva 1990). Survival of tree seedlings during a given rainy season depends on the water availability in the topsoil, where most of the roots of the herbaceous layer are found. Probability of seedling establishment depends on their capability to reach moist soil layers beyond the grass root zone, and on the buildup of underground energy reserves, which allow regrowth of aerial biomass after fire or drought. In this model, water is the basic constraint on seedling establishment and growth.

Grass root competition for soil water should not be a critical factor in the wet season, because the topsoil layers remain wet (high Ψ^s) most of the time (fig. 10.1). However, unpredictable dry spells in the wet season may limit the survival of newly germinated seedlings (Hoffmann 1996). Other factors, such as herbivory and pathogen attack, have to be considered and may play a major role, at least for some species (Nardoto et al. 1998; Braz et al. 2000; chapter 16). On the other hand, seasonal drought was not an important mortality factor for seedlings of three common cerrado trees (Nardoto et al. 1998; Braz et al. 2000; Kanegae et al. 2000).

Seasonal drought may have a major impact not only on seedling survival, but also on seedling growth and carbon metabolism. Information on physioecological characteristics of cerrado tree seedlings such as photosynthetic responses to water stress and carbon budgets is scant. Net CO_2 assimilation rates (A_{CO_2}) of seedlings of cerrado woody plants reach the compensation point ($A_{CO_2} = 0$) at Ψ^l of −2.4 to −3.9 MPa (Prado et al. 1994; Sassaki et al. 1997; Moraes and Prado 1998). Soil water potential of upper soil layers reaches values within this range during the dry season (fig. 10.1). Cerrado woody species allocate a larger proportion of the their biomass to roots than to shoots during the initial growth period (Arasaki and Felippe 1990; Sassaki and Felippe 1992; Paulilo et al. 1993). However, roots of seedlings that germinated in the rainy season would still be exposed to these dry soil layers during the subsequent drought period, and perhaps in the next drought as well (Rizzini 1965; Moreira 1992).

Although generally not considered a limiting factor in savanna environments, canopy shading can restrict seedling growth in the initial phases of plant development. Leaves of cerrado woody species typically reach 90% of the maximum photosynthetic values at photosynthetic photon flux densities (PPFD; the flux of photons between 400 and 700 nm wavelength per unit area) of 600 to 1,200 μmol m^{-2} s^{-1}, which is about 30% to 60% of full sunlight (Prado and Moraes 1997). PPFD compensation point ranges from 10 to 50 μmol m^{-2} s^{-1} at leaf temperatures in the range of 25° to 30°C. Open cerrado vegetation types such as *campo sujo* are covered with a grass layer, typically 40 to 50 cm tall (chapter 6). For instance, PPFD measurements suggested that 5-cm-tall *Kielmeyera coriacea* and *Dalbergia miscolobium* would not receive enough light to reach even 50% of their photosynthetic capacity during the daylight period in a *campo sujo* site (Nardoto et al. 1998; Braz et al. 2000). The effects of canopy shading on CO$_2$ assimilation can become critical for seedling growth and survival in closed canopy vegetation such as *cerradão* physiognomies. Because of shading, species characteristic of open habitats may not be able to grow in closed canopy sites, whereas photoinhibition can be an important stress factor for young plants in fully sun-exposed habitats (Mattos 1998). Thus, reported differences in the range of several species along a gradient from *campo sujo* to *cerradão* (Goodland 1971; Goodland and Ferri 1979; chapter 6) may reflect species differences in shade tolerance.

Cerrado trees grow slowly in natural conditions (table 10.2; Rizzini 1965). This is probably the result of inherently low growth rates, larger

Table 10.2 Size of Cerrado Woody Plants in a *Campo Sujo* Formation Near Brasília, Central Brazil

Species	Plant size (cm)	Plant age (years)
Bowdichia virgilioides[a]	8.3 (0.5; n = 20)	2
Dalbergia miscolobium[b]	23 (2.7; n = 10)	7
Kielmeyera coriacea[c]	8.3 (1.8; n = 7)	5
Qualea grandiflora[a]	5.3 (0.3; n = 38)	1
	19.0 (1.3; n = 4)	5

[a]Established from seeds that were planted into the field site.

[b]Nine-month-old seedlings were transplanted into the field site. This site burned once. At the time of the fire event, plants had an approximate age of 20 months.

[c]Two-month-old seedlings were transplanted into the field site.

Note: The study site (15° 56' S, 47° 55'W) is at the center of the cerrado region. Plant size is the combined length of the main stem and branches, if present. See chapter 6 for description of cerrado physiognomies.

carbon allocation to roots linked to low availability of nutrients, and light limitation by canopy shading. Thus, seedlings of woody plants develop a tree canopy layer in the grass matrix through a slow process. Tree canopy recovery after disturbance in cerrado vegetation is mainly the result of resprouting of existing trees and shrubs. The effects of grass root competition for nutrients need to be evaluated, and the light regimes along the gradient from *campo sujo* to *cerradão* need a better characterization. Research is also needed to characterize shade and high light tolerance and the contribution of light acclimatization to increased carbon gain for plants growing in different cerrado physiognomies.

ACKNOWLEDGMENTS

This research was supported by the Conselho Nacional de Desenvolvimento Científico e Tecnológico (CNPq), the Inter-American Institute for Global Change Research, and the Programa de Apoio a Núcleos de Excelência-PRONEX. I thank Raimundo P. B. Henriques and William Hoffmann for their helpful comments.

REFERENCES

Abdala, G. C., L. S. Caldas, M. Haridasan, and G. Eiten. 1998. Above and belowground organic matter and root: Shoot ratio in a cerrado in central Brazil. *Braz. J. Ecol.* 2:11–23.

Alvin, P. de T. and J. E. da Silva. 1980. Comparação entre os cerrados e a região amazônica em termos agroecológicos. In D. Marchetti and A. D. Machado, eds., *Simpósio sobre o Cerrado: Uso e Manejo*, pp. 143–160. Brasília: Editora Editerra.

Arasaki, F. R. and G. M. Felippe. 1990. Crescimento inicial de *Kielmeyera coriacea*. *Cienc. Cult.* 42:715–720.

Arens, K. 1958a. Considerações sôbre as causas do xeromorfismo foliar. *B. Fac. Fil. Ci.. Letr. Univ. São Paulo, Botânica* 15:23–56.

Arens, K. 1958b. O cerrado como vegetação oligotrófica. *B. Fac. Fil. Ci.. Letr. Univ. São Paulo, Botânica* 15:59–77.

Barnet, Y. M. and P. C. Catt. 1991. Distribution and characteristics of root nodule bacteria isolated from Australian *Acacia* spp. *Plant and Soil* 135:109–120.

Borchert, R. 1994a. Soil and stem water storage determine phenology and distribution of tropical dry forest trees. *Ecology* 75:1437–1449.

Borchert, R. 1994b. Electric resistance as a measure of tree water status during a seasonal drought in a tropical dry forest in Costa Rica. *Tree Physiol.* 14:299–312.

Borchert, R. 1994c. Water status and development of tropical trees during seasonal drought. *Trees* 8:115–125.

Braz, V. S., M. F. Kanegae, and A. C. Franco. 2000. Estabelecimento e desenvolvimento de *Dalbergia miscolobium* Benth. em duas fitofisionomias típicas dos cerrados do Brasil Central. *Acta Bot. Bras.* 14:27–35.

Bucci, S., S. Naves, G. Cardinot, G. Nardoto, M. Bustamante, F. R. Meinzer, G. Goldstein and, A. C. Franco. 2002. Convergence in regulation of water use among Brazilian savanna woody species: The role of plant hydraulic architecture. *Trees* (in press).

Cardinot, G. K. 1998. "Efeitos de Diferentes Regimes de Queimas nos Padrões de Rebrotamento de Kielmeyera Coriacea Mart. e Roupala Montana Aubl., Duas Espécies Típicas do Cerrado." Master's thesis, Universidade de Brasília, Brasília, Brazil.

Cody, M. L. 1986. Structural niches in plant communities. In J. Diamond and T. Case, eds., *Community Ecology*, pp. 381–405. New York: Harper and Row.

Eamus, D., B. Myers, G. Duff, and D. Williams. 1999. Seasonal changes in photosynthesis of eight savanna species. *Tree Physiol.* 10:665–671.

Faria, S. M. de, H. C. de Lima, A. M. Carvalho, V. F. Conçalves, and J. I. Sprent. 1994. Occurrence of nodulation in legume species from Bahia, Minas Gerais and Espírito Santo states of Brazil. In J. I. Sprent and D. McKey, eds., *Advances in Legume Systematics 5: The nitrogen factor*, pp. 17–23. Kew: Royal Botanic Gardens, Kew.

Faria, S. M. de, H. C. de Lima, A. A. Franco, E. S. F. Mucci, and J. I. Sprent. 1987. Nodulation of legume trees from south-east Brazil. *Plant and Soil* 99:347–356

Felfili, J. M., M. C. Silva Junior, T. S. Filgueiras, and P. E. Nogueira. 1998. A comparative study of cerrado (sensu stricto) vegetation in Central Brazil. *Cienc. Cult.* 50:237–243.

Ferri, M. G. 1944. Transpiração de plantas permanentes dos "Cerrados." *B. Fac. Fil. Ci.. Letr. Univ. São Paulo, Botânica* 4:161–224.

Fichtner, K., G. W. Koch, and H. A. Mooney. 1995. Photosynthesis, storage and allocation. In Schulze, E.-D. and M. M. Caldwell, eds., *Ecophysiology of Photosynthesis*, pp. 133–146. Berlin: Springer-Verlag.

Franco, A. C. 1983. "Fotossíntese e Resistência Foliar em Didymopanax macrocarpum." Master's thesis, Universidade de Brasília, Brasília, Brazil.

Franco, A. C. 1998. Seasonal patterns of gas exchange, water relations and growth of *Roupala montana*, an evergreen savanna species. *Plant Ecol.* 136:69–76.

Goodland, R. 1971. A physiognomic analysis of the cerrado vegetation of Central Brazil. *J. Ecol.* 59:411–419.

Goodland, R. and M. G. Ferri. 1979. *Ecologia do Cerrado*. São Paulo: Editora da Universidade de São Paulo.

Haridasan, M. 1982. Aluminium accumulation by some cerrado native species of central Brazil. *Plant and Soil* 65:265–273.

Haridasan, M. 1988. Performance of *Miconia albicans* (SW.) Triana, an aluminium-accumulating species, in acidic and calcareous soils. *Commu. In Soil Sci. Plant Anal.* 19:1091–1103.

Haridasan, M., P. G. Hill, and D. Russel. 1987. Semiquantitative estimates of Al and other cations in the leaf tissues of some Al-accumulating species using probe microanalysis. *Plant and Soil* 104:99–102.

Haridasan, M., T. I. Paviani, and I. Schiavini. 1986. Localization of aluminium in the leaves of some aluminium-accumulating species. *Plant and Soil* 94:435–437.

Hoffmann, W. A. 1996. The effects of cover and fire on seedling establishment in a neotropical savanna. *J. Ecol.* 84:383–393.

Hoffmann, W. A., F. A. Bazzaz, N. J. Chatterton, P. A. Harrison, and R. B. Jackson. 2000. Elevated CO_2 enhances resprouting of a tropical savanna tree. *Oecologia* 123:312–317.

Hogan, K. P., A. P. Smith, and M. Samaniego. 1995. Gas exchange in six tropical semi-deciduous forest canopy tree species during the wet and dry season. *Biotropica* 27:324–333.

Jackson, P. C., F. C. Meinzer, M. Bustamante, G. Goldstein, A. Franco, P. W. Rundel, L. Caldas, E. Igler, and F. Causin. 1999. Partitioning of soil water among tree species in a Brazilian Cerrado ecosystem. *Tree Physiol.* 19:717–724.

Johnson, C., A. C. Franco and L. S. Caldas. 1983. Fotossíntese e resistência foliar em espécies nativas do cerrado: Metodologia e resultados preliminares. *Rev. Bras. Bot.* 6:91–97.

Kanegae, M. F., V. da S. Braz, and A. C. Franco. 2000. Efeitos da disponibilidade sazonal de água e luz na sobrevivência de *Bowdichia virgilioides* em duas fitofisionomias típicas dos cerrados do Brasil Central. *Rev. Bras. Bot.* 23:457–466.

Knoop, W. T. and B. H. Walker. 1985. Interactions of woody and herbaceous vegetation in a Southern African savanna. *J. Ecol.* 73:235–254.

Kozovits, A. R. 1997. "Assimilação de Nitrogênio em Espécies Lenhosas do Cerrado." Master's thesis, Universidade de Brasília, Brasília, Brazil.

Machado, J. W. B. 1985. "Acumulação de Alumínio em Vochysia thyrsoidea Pohl." Master's thesis, Universidade de Brasília, Brasília, Brazil.

Machado, J. W. B. 1990. "Relação Origem/solo e tolerância à saturação hídrica de Copaifera langsdorffii Desf." Ph.D. thesis, Universidade de Campinas, Campinas, Brazil.

Maia, J. M. F. 1999. "Variações Sazonais das Relações Fotossintéticas, Hídricas e Crescimento de Caryocar brasiliense e Rapanea guianensis em um Cerrado *sensu stricto*." Master's thesis, Universidade de Brasília, Brasília, Brazil.

Mattos, E. A. 1998. Perspectives in comparative ecophysiology of some Brazilian vegetation types: leaf CO_2 and H_2O exchange, chlorophyll a fluorescence and carbon isotope discrimination. In Scarano, F. R. and A. C. Franco, eds., *Ecophysiological Strategies of Xerophytic and Amphibi-*

ous Plants in the Neotropics. Series Oecologia Brasiliensis, vol. 4, pp. 1–22. Rio de Janeiro: Universidade Federal do Rio de Janeiro.

Mattos, E. A. de, F. Reinert, and J. A. P. V. de Moraes. 1997. Comparison of carbon isotope discrimination and CO_2 and H_2O exchange between the dry and the wet season in leaves of several cerrado woody species. *Rev. Bras. Fisiol. Vegetal* 9:77–82.

Medeiros, R. A. de and M. Haridasan. 1985. Seasonal variations in the foliar concentrations of nutrients in some aluminium accumulating and non-accumulating species of the cerrado region of central Brazil. *Plant and Soil* 88:433–436.

Medina, E. and B. Bilbao. 1991. Significance of nutrient relations and symbiosis for the competitive interaction between grasses and legumes in tropical savannas. In Esser, G. and G. Overdieck, eds., *Modern Ecology: Basic and Applied Aspects*, pp. 295–319. Amsterdam: Elsevier Science Publications.

Medina, E. and M. Francisco. 1994. Photosynthesis and water relations of savanna tree species differing in leaf phenology. *Tree Physiol.* 14: 1367–1381.

Medina, E., V. Garcia, and E. Cuevas. 1990. Sclerophylly and oligotrophic environments: relationships between leaf structure, mineral nutrient content, and drought resistance in tropical rain forests on the upper Río Negro region. *Biotropica* 22:51–64.

Medina, E. and J. F. Silva. 1990. Savannas of northern South America: A steady state regulated by water-fire interactions on a background of low nutrient availability. *J. Biogeogr.* 17:403–413.

Meinzer, F. C., G. Goldstein, A. C. Franco, M. Bustamante, E. Igler, P. Jackson, L. Caldas, and P. W. Rundel. 1999. Atmospheric and hydraulic limitations on transpiration in Brazilian cerrado woody species. *Funct. Ecol.* 13:273–282.

Melo, J. T. de. 1999. "Resposta de Mudas de Algumas Espécies Arbóreas Nativas do Cerrado aos Nutrientes em Latossolo Vermelho Escuro." Ph.D. thesis, Universidade de Brasília, Brasília, Brazil.

Miranda, A.C., H. S. Miranda, J. Lloyd, J. Grace, R. J. Francey, J. A. Mcintyre, P. Meir, P. Riggan, R. Lockwood, and J. Brass. 1997. Fluxes of carbon, water and energy over Brazilian cerrado: An analysis using eddy covariance and stable isotopes. *Plant Cell Environ.* 20:315–328.

Moraes, J. A. P. V. de, S. C. J. G. de A. Perez, and L. F. Carvalho Jr. 1989. Curso diário do potencial de água e resistência estomática em plantas de cerradão. *Ann. Miss. Bot. Gard.* 27:13–23.

Moraes, J. A. P. V. and C. H. B. A. Prado. 1998. Photosynthesis and water relations in cerrado vegetation. In Scarano, F. R. and A. C. Franco, eds., *Ecophysiological Strategies of Xerophytic and Amphibious Plants in the Neotropics.* Series Oecologia Brasiliensis, vol. 4, pp. 45–63. Rio de Janeiro: Universidade Federal do Rio de Janeiro.

Moreira, A. G. 1992. "Fire Protection and Vegetation Dynamics in the Brazilian Cerrado." Ph.D. thesis, Harvard University, Cambridge, MA, U.S.A.

Nardoto, G. B., M. P. Souza, and A. C. Franco. 1998. Estabelecimento e padrões sazonais de produtividade de *Kielmeyera coriacea* (Spr) Mart. nos cerrados do Planalto Central: Efeitos do estresse hídrico e sombreamento. *Rev. Bras. Bot.* 21:313–319.

Naves-Barbiero, C. C., A. C. Franco, S. J. Bucci, and G. Goldstein. 2000. Fluxo de seiva e condutância estomática de duas espécies lenhosas sempre-verdes no campo sujo e cerradão. *Rev. Bras. Fisiol. Vegetal* 12: 119–134.

Paulilo, M. T. S., G. M. Felippe, and J. E. Dale. 1993. Crescimento e desenvolvimento inicial de *Qualea grandiflora*. *Rev. Bras. Bot.* 16:37–46.

Perez, S. C. J. G. de A. and J. A. P. V. de Moraes. 1991. Determinações de potencial hídrico, condutância estomática e potencial osmótico em espécies dos estratos arbóreo, arbustivo e herbáceo de um cerradão. *Rev. Bras. Fisiol. Vegetal* 3:27–37.

Prado, C. H. B. A. and J. A. P. V. de Moraes. 1997. Photosynthetic capacity and specific leaf mass in twenty woody species of Cerrado vegetation under field conditions. *Photosynthetica* 33:103–112.

Prado, C. H. B. A., J. A. P. V. de Moraes. and E. A. de Mattos. 1994. Gas exchange and leaf water status in potted plants of *Copaifera langsdorffii*: 1. Responses to water stress. *Photosynthetica* 30:207–213.

Rawitscher, F. 1948. The water economy of the vegetation of the "campos cerrados" in southern Brazil. *J. Ecol.* 36:237–268.

Reich, P. B., D. S. Ellsworth, and C. Uhl. 1995. Leaf carbon and nutrient assimilation in species of differing successional status in an oligotrophic Amazonian forest. *Funct. Ecol._9*:65–76.

Reis, M. J. de O. dos. 1999. "Eficiência Micorrízica em Plantas Nativas do Cerrado." Ph.D. thesis, Universidade de Brasília, Brasília, Brazil.

Rizzini, C. T. 1965. Experimental studies on seedling development of cerrado woody plants. *Ann. Miss. Bot. Gard.* 52:410–426.

Rizzini, C. T. and E. P. Heringer. 1961. Underground organs of plants from southern Brazilian savannas, with special reference to the xylopodium. *Phyton* 17:105–124.

Rizzini, C. T. and E. P. Heringer. 1962. Studies on the underground organs of trees and shrubs from some Southern Brazilian savannas. *An. Acad. Bras. Ciênc.* 34:235–251.

Sarmiento, G. 1984. *The Ecology of Neotropical Savannas*. Cambridge, MA: Harvard University Press.

Sarmiento, G., G. Goldstein, and F. Meinzer. 1985. Adaptive strategies of woody species in neotropical savannas. *Biol. Rev.* 60:315–355.

Sassaki, R. M. and G. M. Felippe. 1992. Remoção dos cotilédones e desenvolvimento inicial de *Dalbergia miscolobium*. *Rev. Bras. Bot.* 15:5–16.

Sassaki, R. M. and G. M. Felippe 1998. Response of *Dalbergia miscolobium* Benth. seedlings, a cerrado tree species, to mineral nutrient supply. *Rev. Bras. Bot.* 21:55–72.

Sassaki, R. M., Machado, E. C., Lagôa, A. M. M. A., and G. M. Felippe. 1997. Effect of water deficiency on photosynthesis of *Dalbergia mis-*

colobium Benth., a cerrado tree species. *Rev. Bras. Fisiol. Vegetal* 9: 83–87.

Sprent, J. I., I. E. Geoghegan, and P. W. Whitty. 1996. Natural abundance of ^{15}N and ^{13}C in nodulated legumes and other plants in the cerrado and neighbouring regions of Brazil. *Oecologia* 105:440–446.

Thomazini, L. I. 1974. Mycorrhiza in plants of the "Cerrado". *Plant and Soil* 41:707–711.

Tilman, D. 1982. *Resource Competition and Community Structure*. Princeton, NJ: Princeton University Press.

Walker, B. H. and I. Noy-Meir. 1982. Aspects of the stability and resilience of savanna ecosystems. In B. J. Huntley and B. H. Walker (eds), *Ecology of Tropical Savannas*, pp. 577–590. Berlin: Springer-Verlag.

Williams, R. J., B. A. Myers, W. J. Muller, G. A. Duff, and D. Emaus. 1997. Leaf phenology of woody species in a north Australian tropical savanna. *Ecology* 78:2542–2558.

Zotz, G. and K. Winter. 1996. Diel patterns of CO_2 exchange in rainforest canopy plants. In *Tropical Forest Plant Ecophysiology*, S.E. Mulkey, R. L. Chazdon and A. P. Smith (eds), pp. 89–113. New York: Chapman and Hall.

Part III

The Animal Community: Diversity and Biogeography

11

Lepidoptera in the Cerrado Landscape and the Conservation of Vegetation, Soil, and Topographical Mosaics

Keith S. Brown Jr. and David R. Gifford[1]

MANY CHAPTERS IN THIS BOOK EMPHASIZE THE COMPLEXITY, antiquity, and singularity of the biological systems of the Central Brazil Plateau. The widespread misconception that these mixed-savanna systems are species-poor has been definitively set aside by these chapters, as well as by those in a recent book on gallery forests in the region (Rodrigues and Leitão-Filho 2000). The only poverty now apparent is that of our data and sampling of the many profoundly different biological systems that occur in bewilderingly complex mosaics throughout the Cerrado Biome, often determined by varying soil characteristics (chapter 2) and water availability (chapter 6, see also Oliveira-Filho and Ratter 1995, 2000; Castro et al. 1999; Oliveira-Filho and Fontes 2000).

Many authors have also noted the difficulty in biological inventory, monitoring, and conservation priority-setting in the region, as a result of the great variation of environments in time and space and the resulting unstable ecological mosaics (chapter 18). Such heterogeneous landscapes defy analysis or classification by large mobile mammals like us. Smaller animals and most plants are very sensitive to immediate environmental

---

[1]Dr. David Gifford was working in the Ecology Program of the University of Brasília when he passed away in June 1981, having left parts of this paper as an unpublished manuscript (1979).

factors, however, and thus can be used effectively as indicators of system structure, richness, and history (see Brown 1991, 1997, 2000; Brown and Freitas 2000). How can we "get a handle" on the diversity and importance of a given site in order to describe and characterize its ecology and compare it with others for management and conservation of its biological diversity?

This chapter will examine various groups of specialized plant-feeding Lepidoptera (moths and butterflies) of the Planalto region in central Brazil, sampled throughout this area dominated by various types of cerrado vegetation (chapter 6) and at its peripheries (see fig. 11.1), to evaluate their usefulness as indicators both of the history and biogeographical subdivision of the region and of the variable community structure and species richness at the landscape and local levels. The objectives will be to answer the following three questions:

a. Do Lepidoptera show clear endemism in the cerrado region, with biogeographical divisions or transitions within the region or at its peripheries, similar to those seen in woody plants, eventually referable to broad historical factors acting on the landscape?
b. Which local environmental factors have the greatest effects on the structure and richness of the Lepidopteran community in a given site?
c. Are the broad or local patterns revealed in Lepidoptera coherent enough to qualify as good indicators for landscape evaluation, conservation, and management in the region?

DATABASE: COLLECTION AND ANALYSIS

The data (regional and site lists of species) for answering these questions were compiled from many sources, both published (Brown and Mielke 1967a, 1967b, 1968; Brown 1987a; Mielke and Casagrande 1998; Camargo and Becker 1999) and unpublished (Brown and Mielke 1972; Gifford 1979; Motta 2002; Callaghan and Brown in preparation). Acceptably complete lists (at least 50% of the expected community in smaller groups, or 24 species in larger groups, with two exceptions) were obtained by repeated sampling in 23 to 33 sites (see appendix table) of species in four readily encountered and recognized groups of Lepidoptera: three of Nymphalid butterflies (bait-attracted groups except Satyrinae—126 taxa, Heliconiini—25 taxa, and Ithomiinae—43 taxa) and saturniid moths (169 species; see also Camargo and Becker 1999). Less complete data, useful for regional lists, were obtained simultaneously for sphingid

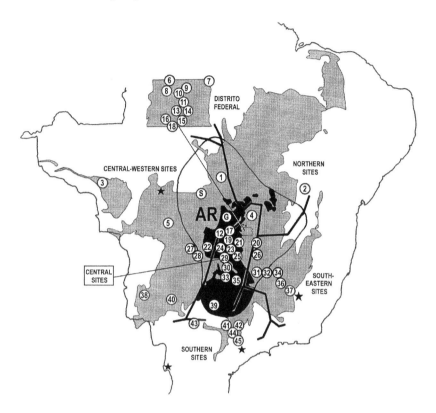

Figure 11.1   Biogeographical division of the Brazilian cerrado region (see fig. 6.4). Sampling sites 1–45 for Lepidoptera are in order of increasing latitude, with 11 Federal District sites in the enlarged rectangle at top (see appendix table). S = Rio Suiá-Missu, G = northern sector of the "Mato Grosso de Goiás." The "Araguaia subspecies-endemic center" (AR, encircled by a heavy line) follows Brown (1979, 1982a, 1982b, 1987b, 1987c). Black areas within AR are regions of rich mesotrophic and eutrophic soils. Stars indicate four outlying limits of the distribution of a suite of cerrado-endemic stenotopic species of *Hypoleria* and *Pseudoscada* (sedentary transparent Ithomiinae confined to superhumid habitats) to the south, southeast, and northwest of the region.

moths and many other butterfly groups. Butterflies, most common, diverse, and easily sampled from February to July, were censused with transect walks and bait-traps (Brown 1972; Brown and Freitas 2000). A permanent notebook was used to record lists of all species observed in a given day and place, as well as observation conditions including observers, effort, maps, and biological data. Moths were attracted to and identified on a light-colored solid surface or screen reflecting near-UV light ("black" 15-W fluorescent tubes or 250-W mixed mercury-vapor lamps); they were best inventoried at the beginning to middle of the rainy season (September to January).

These data were used for quantitative analyses and comparison with geographical and ecological factors, whose purpose was to reveal patterns of species occurrence and distribution in relation to environment, landscape, and the fauna of adjacent regions (see table 11.1, fig. 11.1). Since estimates of the abundance of species were not fully standardized among the years and sites, the records were kept in binary form (recorded/unrecorded in each site).

Lepidoptera in the cerrado region, like most humidity-sensitive insects, tend to be concentrated in gallery forests and other dense vegetation near water (see Brown 2000). There are also many endemic species of more open vegetation, however. The UV light source is invariably placed in an open area for nocturnal census and attracts adults from all habitats within many hundreds of meters. Flowers and larval host plants on borders and in open vegetation, helpful in the attraction and maintenance even of shade-loving butterflies, were also regularly monitored in many sites, along with transects in larger savanna areas.

The high levels of natural disturbance in the cerrado landscape frustrated attempts at objective evaluation of the degree of short- and long-term anthropic disturbance in many sites, especially those censused over many years during various landscape reorganization episodes.

The site lists were analyzed by PC-ORD, STATISTICA (StatSoft 1995), FITOPAC (Shepherd 1995), and CANOCO (Ter Braak 1987–1992), in search of patterns of biogeographical distribution, and for discovery of principal local environmental factors acting on the Lepidoptera communities in each site and in the region (fig. 11.1; appendix table).

BIOGEOGRAPHY OF CERRADO LEPIDOPTERA, ENDEMISM, AND HISTORY

Distribution analysis of a variety of cerrado Lepidoptera showed that the proportions of endemic, widespread, and peripheral-affinity species vary widely among groups (table 11.1). Of the 802 taxa analyzed, over 33% (including 70% of the wide-flying Sphingidae) are widespread in South America, while only 19% are endemic to the Cerrado region. Species showing primary affinities with the dry areas to the northeast (*caatinga*) and southwest (*chaco*) are very few (only 2.6%, two-thirds of these Saturniidae), though a number of species native to these regions have been found in some collections on their borders (see table 3 of Camargo and

Table 11.1   Biogeographical Affinities of Cerrado Species in Various Lepidopteran Groups

REGIONS:	SE/S	NW/N	SW/S	NE/E	Endemic to All Areas	
Widespread	Atlantic	Amazon	Chaco	Caatinga	Cerrado (%)	Total
Saturniidae						
Arsenurinae	5	[1] 7	[6] 2	[3]	—	4 (14)	[10] 18
Ceratocampinae	18	[1] 11	[2] 6	[3] 1	[1] 1	8 (15)	[7] 45
Hemileucinae	12	[3] 20	[9] 17	[1] 3	1	15 (19)	[13] 68
Saturniinae	5	3	—	—	—	—	8
Saturniidae (total)	40	[5] 41	[17] 25	[7] 4	[1] 2	27 (16)	[30] 139
Sphingidae (total)	67	11	2	2	—	9 (10)	91
Papilionidae (total)	4	[9] 10	[9] 2	[2]	[1]	10 (21)	[21] 26
Pieridae (total)	17	10	[2] 1	—	—	3 (9)	[2] 31
Riodininae (total)	40	39	51	—	—	42 (24)	172
Myrmecophilous genera	14	9	8	—	—	21 (40)	52
Nonmyrmecophilous genera	26	30	43	—	—	21 (18)	120
Nymphalidae						
Libytheana-Dana.-Ithomiinae	4	[5] 13	[3]	—	—	15 (38)	[8] 32
Morph.-Brassol.-Satyrinae	40	21	[1] 12	—	—	23 (36)	[1] 96
Apat.-Colob.-Cyrest.-Limen.	10	11	[3] 3	—	—	2 (7)	[3] 26
Charaxinae-Biblidinae	34	[1] 17	5	—	[1]	14 (31)	[2] 70
Nymphalinae-Heliconiinae	10	[1] 21	[9] 3	[1]	—	7 (29)	[11] 41
Nymphalidae (total)	98	[7] 83	[16] 23	[1]	[1]	61 (21)	[25] 265
Grand total:	266	[21] 194	[44] 104	[10] 6	[3] 2	152 (19)	[78] 724
Percent of total:	33.2%	26.8%	18.5%	2%	0.6%	19%	100%

Sources: Brown and Mielke 1972 (all groups); Gifford 1979 (Heliconiini and Ithomiinae); Tyler et al. 1994 (Papilionidae); Camargo and Becker 1999 (Saturniidae); Callaghan and Brown in preparation (Riodininae).

Note: Bracketed numbers include marginal species invading the Cerrado from the indicated adjacent biomes.

Becker 1999). The primary link, as with plants (chapter 6), is with the Atlantic forests to the southeast (26.8%, seen in all groups) (table 11.1), followed by the Amazonian forests to the northwest (18.5% of taxa, almost a third peripheral, found only in central Mato Grosso sites 3 and 5 and the region marked *S* in fig. 11.1).

Progressively higher levels of endemism in the cerrado region are shown by well-marked species and geographical subspecies of Papilionidae (21% endemic, or 47% for the Troidini alone), Riodininae (24%, or 40% for myrmecophilous species), Satyrinae (27%), Biblidinae (= Eurytelinae auctt.) (27%), and Ithomiinae (43%, not considering the four widespread Danainae and one Libytheine) (table 11.1). When the mostly grass-feeding satyrs and hesperiine skippers (not analyzed here) are better studied, they should show still higher endemism, considering the great wealth of potential host species in the cerrado (chapter 7). The large proportion of endemic myrmecophilous Riodininae (40%), with many still awaiting description, may be related to the specific and intense interactions between ants and plants in the cerrado vegetation (chapter 15); endemism is much lower in species with non-myrmecophilous juveniles (18%).

Does this endemism have primarily historical or ecological roots? As with most basic ecological questions, the most likely answer is "yes"— that is, both are important and necessary. Geomorphologic analysis (chapter 2) and paleopollen records (Ledru 1993; chapter 3) show that the region has had a long history of landscape changes (especially in its contacts with the Amazon Basin to the north), but a surprising apparent stability of interlinked forest/savanna matrices during the climatic fluctuations of the late Pleistocene. The very complex geology and sharp relief of the central plateau (chapter 2) probably always included gallery and headwater swamp forests, providing refuge and connectivity (especially southward with the Atlantic forests) to diversified forest biotas even during the least favorable climatic periods. Orographically and edaphically determined deciduous and semideciduous forests (including mesotrophic headwater woods and spring-fed copses at the base of escarpments) are predictably green at least at certain seasons of each year (Ratter et al. 1973, 1978) and were probably also widely distributed at all times, preserving endemic taxa adapted to this mosaic landscape.

An "Araguaia endemic center" for subspecies of Heliconiini and Ithomiinae, recognized by Brown (1979, 1982a, 1982b, 1987b, 1987c; see also Gifford 1979), is outlined in figure 11.1 (AR, one-third of maximum isocline for corrected endemism based on eight taxa, reinforced with the 18 taxa included today, table 11.1). This center covers a large region characterized by cerrados mixed with forests in a complex dystrophic/eutrophic soil matrix (Ratter et al., 1973, 1978). Gifford's observations in the upper Xingu (Suiá-Mussu, Serra do Roncador; marked S in fig. 11.1), along the Araguaia (site 1), and in the northern part of the "Mato Grosso de Goiás" (12, 17, 22, G in fig. 11.1) show many species in these two groups as recent marginal invaders from the Amazonian forests, mixing with the usual endemic fauna found in greater abundance farther

southeast, often on rich soils (black areas in fig. 11.1). The relatively stable topographical, soil, and vegetation mosaics in the Brazilian Planalto
could have conserved the adapted and differentiated "Araguaia"—
endemic Lepidoptera and plants through all the ecological cataclysms that
so greatly affected the lowland sedimentary regions and their biotas in the
Neotropics. As a typical example, five well-differentiated, stenotopic
transparent Ithomiinae species in the genera *Hypoleria* and *Pseudoscada*
are ubiquitous in interconnected humid gallery forests (Brown 2000) with
deep shade throughout the cerrado landscape all the way to its various
peripheries (see fig. 11.1), along with many other similarly restricted and
endemic species of insects, plants, birds, and reptiles (chapters 6, 7, 12–14;
Rodrigues and Leitão-Filho 2000). All of these animals are excellent indicators of complex forest/cerrado mosaic landscapes in the region today,
and suggest their relative continuity in the past, not as climatically induced
and edaphically defined "forest islands" or refuges (as proposed for the
more level Amazon Basin, Brown 1979), but as interlinked and geomor-
phologically stabilized vegetation mosaics. These mosaics could conserve
and select both forest and savanna species over time (Gifford 1979).

LANDSCAPE HETEROGENEITY
IN THE CERRADO REGION

The sites (fig. 11.1) were compared for similarity of their faunal lists for
each of the four censused Lepidoptera groups (appendix table). The
results (see fig. 11.2) support a landscape- and vegetation-based grouping
in the saturniid moths (fig. 11.2A) but a geographic base for community
composition in the butterflies (fig. 11.2B-D). Thus, sites in a given biogeographical subregion (fig. 11.1; Ratter et al. 1997), even those with different vegetation types, tend to cluster together in the three butterfly
analyses (fig. 11.2B-D); note the proximity of nearby site numbers in the
dendrograms—the average difference in numbers between pairs and triads for all three groups is 6.6. In contrast, adjacent sites are scattered over
the dendrogram in the Saturniidae analysis (fig. 11.2A); note the lack of
clusters of proximal site numbers (average difference in pairs and triads
13.3), with, however, a homogeneity of vegetation in two of the three
large clusters. This indicates a strong association of the Planalto saturniid
moths with specific landscapes and vegetation (Ce, FCe, SDF) rather than
with geographical subregions; the light-sampling procedure would bring
them in from their typical habitats within the broad local vegetation
mosaic (see also below, and Camargo 1999; Camargo and Becker 1999).

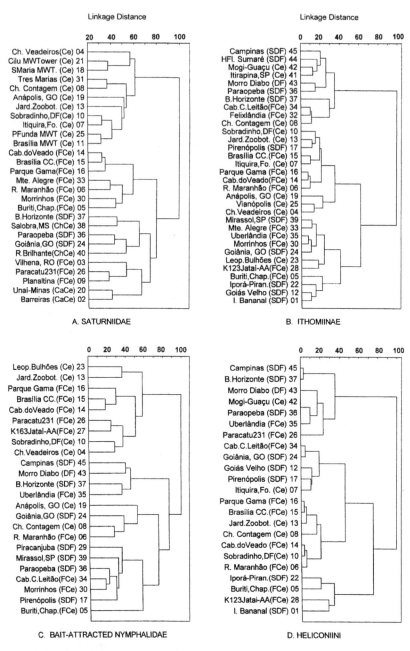

Figure 11.2 Dendrograms showing similarity among Lepidoptera communities in the cerrado region and its peripheries. Linkage (percentage of maximum) is by Ward's minimum-variance clustering, favoring formation of pairs and triads; similarity is as 1-Pearson's *r*. The four groups have different sets of sites. Note the strict vegetation grouping in Saturniidae moths (A), but more biogeographical grouping (proximate site numbers) in butterflies (B-D). See appendix table for vegetation codes, site characteristics, and numbers of species in each group recorded in each site.

The butterflies, more restricted to microhabitats and food plants, are recorded along the specific transects chosen for census. They will thus reflect a biogeographical affinity smaller in scale than the saturniid moths' affinity with general vegetation or landscape (fig. 11.1).

Within each subregion (fig. 11.1), the presence of a species in a site is most likely to be determined by various factors in the local environment, often a mosaic of cerrado and forest physiognomies including galleries along permanent watercourses. This is evident in Principal Components and CANOCO analyses of the sites and their butterfly faunas (see tables 11.2 and 11.3 and fig. 11.3). The PCA (with latitude and longitude removed) grouped the various sites along major axes reflecting not only geography through broad climatic patterns (fig. 11.3A), but also a wide variety of factors of topography, soil, and vegetation, especially their complex mosaics (table 11.2), thereby revealing each site's strong relationship to landscape heterogeneity. Even with the inclusion of many factors (especially edaphic)

Table 11.2   Statistics of the PCA Analysis for 14 Environmental Factors in 45 Sites in the Brazil Planalto

Environmental Factors % of variation explained	Axis 1 25.0%	Axis 2 14.7%	Axis 3 13.8%
Mean altitude	−.360	.270	−.399
Topographical Surface (1–5)	.368	−.119	.173
General topography	.130	.333	−.154
Permanent water availability	−.155	.028	.327
Vegetation category	.368	−.125	−.252
Vegetation mosaic	.169	.089	.240
Bamboos	.338	.081	.049
Soil category	.213	−.222	−.493
Soil mosaic	.144	.512	−.142
Soil bases (fertility)	.404	.075	−.231
Annual rainfall	−.242	.268	.029
Length of the dry season (days)	.189	.430	−.076
Temperature (yearly average)	.305	.099	.452
Temperature variation	−.004	−.434	−.172
"Flavors" of each axis	Soil bases, vegetation, altitude	Soil mosaic, climate (dry, temp.variat.)	Soil type, temperature, altitude

Note: Data taken from the appendix table, not including latitude or longitude. The most important factors are in **bold**. Axis 4 (vegetation mosaic, water) explained a further 10.6%; Axis 5 (topography, bamboos), 8.3% of the variation. See figure 11.3A for site ordination along the first two axes, with maximized scatter of points representing different types of environment. Note significant inclusion of variation in soil, vegetation, and topographical factors in the first three axes. This would suggest a large variation of microhabitats within and between cerrado sites.

Table 11.3　Statistics of the Principal Vectors in the CCA Analysis

Factors	F	P	%
(1) Saturniidae (153 taxa, 28 sites) (Fig. 3B)			
Mean altitude	1.01	0.001	19
Temperature variation	1.72	0.001	16
Annual rainfall	1.49	0.004	14
Vegetation category	1.38	0.022	12
(Soil mosaic —not significant)	1.24	0.097	<u>11</u>
Total explained (four significant vectors)			61
(2) Heliconiini (25 taxa, 23 sites)			
Temperature (annual mean)	4.07	0.001	26
Bamboos	2.59	0.004	16
Vegetation mosaic	2.39	0.006	13
Annual rainfall	1.70	0.065	9
Soil mosaic	1.81	0.044	<u>9</u>
Total explained (five significant vectors)			73
(3) Bait-attracted Nymphalidae (126 taxa, 24 sites) (Fig. 3C)			
Vegetation mosaic	2.19	0.001	15
Temperature variation	1.95	0.002	13
Length of the dry season	1.63	0.051	10
Mean altitude	1.55	0.016	10
Soil mosaic	1.50	0.047	<u>9</u>
Total explained (five significant vectors)			57
(4) Ithomiinae (43 taxa, 33 sites) (Fig. 3D)			
Temperature variation	3.59	0.001	16
Vegetation mosaic	2.51	0.003	10
Mean altitude	2.37	0.001	9
Length of the dry season	1.84	0.014	7
Vegetation category	1.84	0.015	7
Annual rainfall	1.66	0.036	6
Soil mosaic	1.51	0.056	5
General topography	1.54	0.056	<u>5</u>
Total explained (eight significant vectors)			65

Note: The data represent 14 environmental factors (latitude/longitude eliminated) and four Lepidoptera communities in 45 sites in the Brazilian cerrado region.

in the first three PCA axes, together they explained only about half the variations among the sites (table 11.2).

　　When the species of the four Lepidoptera groups were directly compared with various local geoecological variables (table 11.3, fig. 11.3) by Canonical Community Ordination, the most influential ecological factors

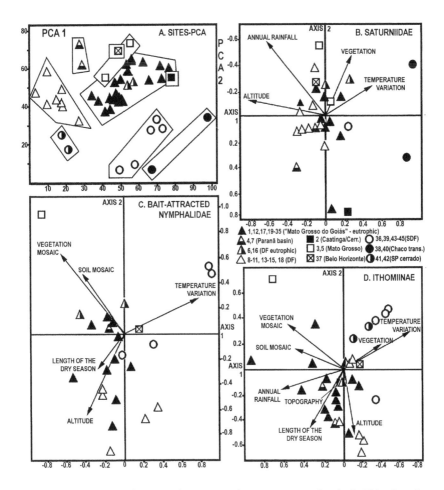

Figure 11.3 Site ordination by Principal Components Analysis (A) of environmental variables and by Canonical Community Ordination of three Lepidoptera groups (B-D, see table 11.3). The most significant relations (other than latitude and longitude, not ecological) are expressed as vectors in B (Saturniidae), C (bait-attracted Nymphalidae), and D (Ithomiinae; Heliconiini are very similar) (see tables 11.2, 11.3). The appendix table gives the 45 site names, environmental characteristics, and species richness of each group. In 15 sites, only one of the four groups was sampled, typically (10) Saturniidae (Camargo and Becker 1999). The second axis in B and the first in D have been inverted, to place the temperature variation vector always to the upper right and site 5 in the upper left quadrant. The variable positions of some vectors in B-D is due to the smaller number of significant factors in site ordination for each lepidopteran group, in accord with their different environmental responses.

on the community in each site (also after removing latitude and longitude) invariably included climate (a regional phenomenon), altitude (a restricted topographical factor, highly correlated with lower temperature), and vegetation (especially its fine mosaic, an intensely local factor; soil mosaic was also very important, see table 11.3). All these factors helped to explain the community structure and composition (table 11.3), while reflecting the geographical subregions and contributing to the principal axes defining the environment in each site (table 11.2, fig. 11.3A). Their combination can be expressed as a single composite term, *environmental heterogeneity*, that effectively determines Lepidoptera community composition and richness throughout the region. This is hardly a surprising result, given the predominance of landscape mosaics, predictability of a marked dry season, continued presence of ever-humid swamp and gallery forests, and strong ecological specializations of the animals in the region. This is true even at a fine scale in the nuclear cerrado region, as seen in the elongation of the cluster of positions of neighboring sites in the Federal District and the Mato Grosso de Goiás (fig. 11.1; open or black triangles in fig. 11.3) along the vegetation mosaic and altitude vectors in the canonical analyses, and in the separation of sites outside this "central" region from this cluster. Much less scattering is seen along the climatic vectors (fig. 11.3).

LEPIDOPTERA AND THE "CONSERVATION LANDSCAPE"

The close relationship of the structure of the Lepidoptera communities with the complex landscape mosaics in the Cerrado region (fig. 11.3, tables 11.2, 11.3), and their typical association with ecotones and gallery forests (Pinheiro and Ortiz 1992; Brown 2000), give a clear direction for effective conservation of the widest range of landscapes and genetic variation in the region. This can start with rigorous protection of the landscape factors that have always been designated as reserves by Brazilian law: water-springs, marshes, riparian forests, and areas of steep or complex topography. In the cerrado, these will include most of the vegetation types and the species endemic to the region. A plan for a "conservation landscape" occupied by humans and their economic activities should also take other humid areas into account (such as depressions on high plateaus, with many endemic monocots and their herbivores), as well as any topographic factors (such as breaks between the geomorphic surfaces; see chapter 2) that create complex mosaics. Fragile but fertile Surface-3 mesotrophic soils and associated semideciduous forest mosaics are likely to be important also (Ratter et al. 1973, 1978, 1997). Indeed, a Cerrado landscape without mesotrophic, headwater, and gallery forests would be much poorer in insect species, like most of the hydrologic savannas in the Amazon or Llanos regions (Ratter et al. 1997).

Lepidoptera are easily monitored in the Cerrado landscape. In any season, butterflies are attracted in the morning to flowers and fruits on borders, and later in the day to resources or baits within the forest (Brown 1972; Brown and Freitas 2000). Moths can be called from afar by near-UV light in the open areas and identified when they sit down on the illuminated light-hued sheet, net, or wall (as in the microwave tower blockhouses, in Santa Maria, Anápolis, Cilu, and Ponte Funda among others), especially on foggy or moonless nights in the spring and summer (see Camargo 1999). The richness, composition, and mosaic structure of the local biota can thus be continuously censused and monitored as landscape conversion or effective use increases. Each site will have its characteristic community (note the scatter of points in figure 11.3), closely tied with local edaphic and vegetation mosaics and water availability (fig. 11.3, tables 11.2, 11.3; Brown 2000) and different from those in other sites (see also Camargo 1999; Camargo and Becker 1999). When typical species in a local community disappear, it may be suspected that the use of the landscape is no longer sustainable. If they continue to be present, their habitats will probably remain adequately conserved, unless systemic poisons are introduced into the landscape (even excess fertilizer can greatly change the vegetation along watercourses, and pesticides affect all animals).

In this way, the "bewildering complexity" of soil and vegetation mosaics in the Planalto can continue to be a stimulating source of resources, both economic and intellectual, for humans during and beyond the present occupation of the region by agroindustry.

PRIORITIES FOR CONTINUING RESEARCH

The cerrado region has been extensively transformed by large-scale cattle ranching and industrial plantations of soybeans, maize, and other crops in the past 30 years (chapter 5). Most of the data on Lepidoptera used here were gathered before. Recent visits suggest that the flora and fauna continue to persist, at least in steeper areas and in reserves of various sorts. It is still necessary to evaluate the effects of anthropic disturbance in areas adjacent to agriculture, especially those with open native savanna vegetation; the diversity can be increased (due to edge effects) but is often decreased due to agrochemicals and leveling, greatly affecting small marshes and other open microhabitats.

The data presented here need to be updated both in preserved areas and in those adjacent to human occupation, whose effects need to be recognized and separated from those of fires, excess seasonality (as in 1998–2000, with greatly reduced rainfall), population fluctuations of host plants, parasites, and predators, and the long-term dynamics of natural savannas. New censuses of Lepidoptera and other insects, more complete and effective than past data, could help in the formulation of effective management protocols for the singular landscapes characteristic of the central Brazil plateau, a rich biological resource whose description and recognition have lagged behind those for forests and more homogeneous vegetation in other parts of the Neotropics. The baselines established for Lepidoptera previous to extensive human occupation should be useful in the design and monitoring of sustainable programs for human use of the region and its resources.

ACKNOWLEDGMENTS

We are grateful to J. Ratter and A. Oliveira-Filho for orientation on the analysis of cerrado vegetation, and to A. V. L. Freitas for substantial contributions in data analysis. DRG received support from the Royal Society and the Royal Geographical Society in the Xavantina-Cachimbo expedition (1967–1968), the Royal Society and the CNPq (1971), the Brazilian Academy of Sciences (1972, 1978–1979), the Universities of Edinburgh and Brasília, FINEP (field study, 1976–1979), and the IBDF (Xingu, 1978; Bananal, 1979). Drs. G. P. Askew and R. F. Montgomery also participated in these field projects. Support from the BIOTA/FAPESP program contributed to the data analysis in 2000; data collection by KSB in the 1960s and 1970s was supported by many of the above agencies and a fellowship from the CNPq ("Pesquisador-Conferencista"). O. H. H. Mielke, H. and K. Ebert, S. Nicolay, and N. Tangerini made significant contributions to the database, and R. Marquis and P. S. Oliveira gave many suggestions for the text and figures.

REFERENCES

Brown Jr., K. S. 1972. Maximizing daily butterfly counts. *J. Lep. Soc.* 26:183–196.

Brown Jr., K. S. 1979. *Ecologia Geográfica e Evolução nas Florestas Neotropicais.* Campinas: Universidade Estadual de Campinas, Brazil.

Brown Jr., K. S. 1982a. Historical and ecological factors in the biogeography of aposematic Neotropical Lepidoptera. *Amer. Zool.* 22:453–471.

Brown Jr., K. S. 1982b. Paleoecology and regional patterns of evolution in Neotropical butterflies. In G.T. Prance, ed., *Biological Diversification in the Tropics*, pp. 255–308. New York: Columbia University Press.

Brown Jr., K. S. 1987a. Zoogeografia da região do Pantanal Matogrossense. In *Anais do 1° Simpósio sobre Recursos Naturais e Sócio-Econômicos do Pantanal (Corumbá, Mato Grosso)*, pp. 137–178. Brasília: Empresa Brasileira de Pesquisa Agropecuária.

Brown Jr., K. S. 1987b. Biogeography and evolution of Neotropical butterflies. In T. C. Whitmore and G. T. Prance, eds., *Biogeography and Quaternary History in Tropical America*, pp. 66–104. Oxford, England: Clarendon Press.

Brown Jr., K. S. 1987c. Conclusions, synthesis, and alternative hypotheses. In T. C. Whitmore and G. T. Prance, eds., *Biogeography and Quaternary History in Tropical America*, pp. 175–196. Oxford, England: Clarendon Press.

Brown Jr., K. S. 1991. Conservation of Neotropical environments: Insects as indicators. In N. M. Collins and J. A. Thomas, eds., *The Conservation of Insects and their Habitats*, pp. 349–404. London: Academic Press.

Brown Jr., K. S. 1997. Diversity, disturbance, and sustainable use of Neotropical forests: Insects as indicators for conservation monitoring. *J. Insect Cons.* 1:25–42.

Brown Jr., K. S. 2000. Insetos indicadores da história, composição, diversidade, e integridade de matas ciliares tropicais. In R. R. Rodrigues and H. F. Leitão Filho, eds., *Matas Ciliares: Conservação e Recuperação*, pp. 223–232. São Paulo: Editora da Universidade de São Paulo and Fundação de Amparo à Pesquisa do Estado de São Paulo.

Brown Jr., K. S. and A. V. L. Freitas. 2000. Atlantic Forest Butterflies: Indicators for landscape conservation. *Biotropica* 32:934–956.

Brown Jr., K. S. and O. H. H. Mielke. 1967a. Lepidoptera of the Central Brazil Plateau:. I. Preliminary list of Rhopalocera: Introduction, Nymphalidae, Libytheidae. *J. Lep. Soc.* 21:77–106.

Brown Jr., K. S. and O. H. H. Mielke. 1967b. Lepidoptera of the Central Brazil Plateau: I. Preliminary list of Rhopalocera (continued): Lycaenidae, Pieridae, Papilionidae, Hesperiidae. *J. Lep. Soc.* 21:145–168.

Brown Jr., K. S. and O. H. H. Mielke. 1968. Lepidoptera of the Central Brazil Plateau: III. Partial list for the Belo Horizonte area, showing the character of the southeastern "blend zone." *J. Lep. Soc.* 22:147–157.

Brown Jr., K. S. and O. H. H. Mielke. 1972. Lepidoptera of the Central Brazil Plateau: VI. Preliminary list of Saturniidae and Sphingidae; VII. Supplementary list of Rhopalocera. Unpublished manuscripts.

Brown Jr., K. S. and J. Vasconcellos-Neto. 1976. Predation on aposematic ithomiine butterflies by tanagers (*Pipraeidea melanonota*). *Biotropica* 8:136–141.

Callaghan, C. J., and K. S. Brown Jr. In prep. An annotated list of the Riodinine butterflies of the Brazilian Planalto.

Camargo, A. J. A. 1999. Estudo comparativo sobre a composição e a diversidade de lepidópteros noturnos em cinco áreas da Região dos Cerrados. *Rev. Bras. Zool.* 16:368–380.

Camargo, A. J. A. and V. O. Becker. 1999. Saturniidae (Lepidoptera) from the Brazilian Cerrado: Composition and biogeographic relationships. *Biotropica* 31:696–705.

Castro, A. A. J. F., F. R. Martins, J. Y. Tamashiro, and G. J. Shepherd. 1999. How rich is the flora of the Brazilian Cerrados? *Ann. Miss. Bot. Gard.* 86:192–224.

DNMet. 1992. *Normais Climatográficos, 1961–1990.* Brasília: Ministério da Agricultura, Serviço Nacional de Irrigação.

Gifford, D. R. 1979. Edaphic factors in dispersion of Heliconian and Ithomiine butterflies in the Araguaia region, Central Brazil. Unpublished manuscript.

Ledru, M.-P. 1993. Late Quaternary environmental and climatic changes in central Brazil. *Quat. Res.* 39:90–98.

Mielke, O. H. H., and M. M. Casagrande. 1998. Papilionoidea e Hesperioidea (Lepidoptera) do Parque Estadual do Morro do Diabo, Teodoro Sampaio, São Paulo, Brasil. *Rev. Bras. Zool.* 14:967–1001.

Motta, P. C. 2002. Butterfly diversity in Uberlândia, central Brazil: Species list and biological comments. *Rev. Bras. Biol.* 62:151–163.

Oliveira-Filho, A. T. and M. A. L. Fontes. 2000. Patterns of floristic differentiation among Atlantic Forests in south-eastern Brazil, and the influence of climate. *Biotropica* 32:793–810.

Oliveira-Filho, A. T. and J. A. Ratter. 1995. A study of the origin of central Brazilian forests by the analysis of plant species distribution patterns. *Edinb. J. Bot.* 52:141–194.

Oliveira-Filho, A. T. and J. A. Ratter. 2000. Padrões florísticos das matas ciliares na região do Cerrado e a evolução das paisagens do Brasil Central durante o Quaternário tardio. In R. R. Rodrigues and H. F. Leitão Filho, eds., *Matas Ciliares: Conservação e Recuperação,* pp. 73–89. São Paulo: Editora da Universidade de São Paulo and Fundação de Amparo à Pesquisa do Estado de São Paulo.

Pinheiro, C. E. G. and J. V. C. Ortiz. 1992. Communities of fruit-feeding butterflies along a vegetation gradient in central Brazil. *J. Biogeogr.* 19:505–511.

RADAMBRASIL, Projeto (Ministério de Minas e Energia). 1978–1983. *Levantamento de Recursos Naturais* (folhas Porto Velho, Goiás, Cuiabá, Corumbá, Campo Grande, Brasília, Goiânia, and Rio de Janeiro). Volumes 16, 25–29, 31–32. Salvador and Rio de Janeiro: Departamento

Nacional de Produção Mineral and Instituto Brasileiro de Geografia e Estatística.

Ratter, J. A., G. P. Askew, R. F. Montgomery, and D. R. Gifford. 1978. Observations on forests of some mesotrophic soils in central Brasil. *Rev. Bras. Bot.* 1:47–58.

Ratter, J. A., J. F. Ribeiro, and S. Bridgewater. 1997. The Brazilian cerrado vegetation and threats to its biodiversity. *Ann. Bot.* 80:223–230.

Ratter, J. A., P. W. Richards, G. Argent, and D. R. Gifford. 1973. Observations on the vegetation of northeastern Mato Grosso: I. The woody vegetation types of the Xavantina-Cachimbo Expedition area. *Phil. Trans. Royal. Soc. Lond.* 266B:449–492.

Rodrigues, R. R., and H. F. Leitão-Filho, eds. 2000. *Matas Ciliares: Conservação e Recuperação*. São Paulo: Editora da Universidade de São Paulo and Fundação de Amparo à Pesquisa do Estado de São Paulo.

Shepherd, G. J. 1995. *FITOPAC: Manual de Usuários*. Campinas: Departamento de Botânica, Universidade Estadual de Campinas.

StatSoft, Inc. 1995. *Statistica for Windows*. Tulsa, Oklahoma.

Ter Braak, C. J. F. 1987–1992. *CANOCO: A FORTRAN program for Canonical Community Ordination*. Ithaca: Microcomputer Power.

Tyler, H. A., K. S. Brown Jr., and K. H. Wilson. 1994. *Swallowtail Butterflies of the Americas: A Study in Biological Dynamics, Ecological Diversity, Biosystematics, and Conservation*. Gainesville: Scientific Publishers.

Appendix Environmental and Biological Characteristics of 45 Sampling Sites in the Brazilian Cerrado. Conventions for values of variables below

No. Site and State	Vegetation Type	Lat	Lon	Alt	Sur	Top	Wat	Veg	Vmo	Bam	Soil	Smo	Sba	Plu	Dry	Tem	Tva	Heli	Itb	Bait	Satur
01 Ilha do Bananal, TO	SDF	12.0	50.0	190	5	1	4	3	4	3	2	2	2	17	110	25.6	3.0	10	14	x	x
02 Barreiras, BA	CaCe	12.1	45.0	440	4	3	4	5	4	3	4	4	4	10	150	24.3	3.5	x	x	x	25
03 Vilhena, RO	FCe	12.6	60.1	615	4	2	2	5	3	3	2	2	3	20	120	23.0	3.0	x	x	x	29
04 Chapada dos Veadeiros, GO	Ce	14.0	47.7	1100	3	4	4	3	3	2	3	5	3	15	130	21.1	4.0	x	9	31	17
05 *Chapada dos Guimarães, MT	FCe	15.5	55.7	700	3	3	3	4	5	4	3	5	4	17	120	23.0	4.0	15	24	82	34
06 Alto Rio Maranhão, DF	FCe	15.5	47.7	650	4	4	4	5	3	3	3	3	5	16	120	22.0	5.0	9	14	33	37
07 Itiquira + Formosa, DF/GO	Ce	15.5	47.4	1000	4	5	4	3	4	2	3	3	2	16	130	21.5	5.0	5	10	x	64
08 FERCAL/Chap. Contagem, DF	Ce	15.6	47.9	1200	2	4	2	3	2	1	3	3	2	16	120	21.2	5.0	7	8	29	31
09 *Planaltina (EMBRAPA), DF	FCe	15.6	47.7	960	3	2	4	2	2	1	5	4	3	16	120	21.2	5.0	x	x	x	98
10 Sobradinho, DF	Ce	15.7	47.8	1050	2	2	2	3	3	1	4	2	2	16	120	21.2	5.0	10	12	38	39
11 *Brasília Microwave Tower, DF	Ce	15.8	47.9	1100	3	2	5	3	3	1	4	2	3	16	130	21.2	5.0	x	x	x	67
12 Goiás Velho, GO	SDF	15.9	50.1	490	5	3	3	5	2	3	4	4	5	18	140	24.7	4.0	8	20	x	x
13 *Jardim Zoobotânico, DF	Ce	15.9	48.0	1000	3	1	4	6	3	1	4	2	3	16	120	21.2	5.0	9	17	24	25

Appendix *(continued)*

No. Site and State	Vegetation Type	Lat	Lon	Alt	Sur	Top	Wat	Veg	Vmo	Bam	Soil	Smo	Sba	Plu	Dry	Tem	Tva	Heli	Itb	Bait	Satur
14 Est. Exp. Cabeça do Veado, DF	FCe	15.9	47.8	1050	3	2	4	4	2	2	4	2	2	16	120	21.2	5.0	11	21	44	43
15 Brasília Country Club, DF	FCe	16.0	48.0	1200	2	2	3	4	2	3	4	2	2	16	120	21.2	5.0	10	11	37	51
16 Parque do Gama, DF	FCe	16.0	48.1	900	4	4	2	5	4	4	4	3	3	16	120	21.5	5.0	11	17	64	72
17 Pirenópolis, GO	SDF	16.0	49.1	900	4	4	3	5	3	2	4	4	5	18	120	22.5	3.6	5	17	26	x
18 Santa Maria MW Tower, DF	Ce	16.1	48.0	1240	1	1	4	3	3	1	4	3	3	16	20	21.2	5.0	x	x	x	27
19 Anápolis, incl. MW Tower, GO	Ce	16.3	49.0	1020	3	2	3	5	2	3	4	2	4	16	120	21.4	4.0	x	11	32	38
20 Unaí de Minas, MG	CaCe	16.4	46.9	580	4	2	3	5	3	2	4	3	3	13	140	22.0	4.0	x	x	x	28
21 Cilú Microwave Tower, GO	Ce	16.4	48.2	900	4	3	2	4	2	2	5	2	4	15	130	21.2	5.0	x	x	x	48
22 Iporá to Piranhas, GO	SDF	16.4	51.3	550	5	3	2	5	3	3	5	4	4	15	140	23.0	4.0	7	11	x	x
23 Leopoldo Bulhões, GO	Ce	16.6	48.8	1030	4	2	2	5	3	3	5	5	4	16	130	21.2	4.0	x	11	26	x
24 *Goiânia + Campinas, GO	SDF	16.7	49.3	750	5	3	4	5	2	3	4	4	4	16	150	23.2	4.0	5	17	47	26
25 Vianópolis+ Ponte Funda MWT	Ce	16.7	48.5	1000	4	2	2	5	2	3	4	4	4	16	130	21.2	5.0	x	12	x	88
26 Paracatu-K231 Bras.-Belo,MG	FCe	17.2	46.9	920	4	4	2	4	3	5	3	3	3	15	130	22.6	5.0	6	x	36	44
27 Mineiros-K163 Jataí-A.Arag.	FCe	17.4	52.5	830	4	2	3	5	3	2	4	4	4	15	120	22.5	4.0	x	x	26	x

Appendix *(continued)*

No. Site and State	Vegetation Type	Lat	Lon	Alt	Sur	Top	Wat	Veg	Vmo	Bam	Soil	Smo	Sba	Plu	Dry	Tem	Tva	Heli	Itb	Bait	Satur
28 Mineiros-K123 Jataí-A.Arag.	FCe	17.4	52.5	830	4	2	3	4	2	2	4	3	4	14	120	22.5	4.0	4	10	x	x
29 Piracanjuba, GO	SDF	17.4	49.2	750	4	4	2	5	2	5	5	3	6	15	150	22.0	4.0	x	x	28	x
30 Morrinhos, GO	FCe	17.7	49.1	770	4	3	2	5	3	4	5	3	4	15	150	22.0	4.0	x	14	27	24
31 Três Marias, MG	Ce	18.2	45.2	540	5	2	5	4	2	3	3	2	2	12	120	22.0	3.0	x	x	x	19
32 Felixlândia-K222 Belo-Bras	Fce	18.7	44.9	630	4	2	4	5	3	2	4	3	3	12	150	21.0	4.0	x	8	x	x
33 Córrego Boa Vista,M.Alegre	FCe	18.9	49.0	780	3	2	3	4	2	3	4	3	3	16	130	21.4	5.1	x	15	x	25
34 Cabeceira Córrego Leitão, MG	FCe	19.1	44.6	800	3	2	3	5	3	3	3	3	4	12	150	21.0	5.0	6	9	29	x
35 *Uberlândia, MG	FCe	19.2	48.4	800	3	3	4	5	3	3	4	3	4	15	130	21.4	5.1	8	18	47	x
36 *Paraopeba, MG	SDF	19.3	44.4	730	5	2	2	6	3	3	5	3	4	12	150	20.9	5.5	6	16	35	42
37 *Belo Horizonte, MG	SDF	19.9	43.9	1000	2	5	4	6	3	3	5	5	3	15	150	21.1	5.1	12	21	61	40
38 *Salobra, MS	ChCe	20.3	56.7	180	4	3	3	5	5	3	5	3	6	10	150	25.0	6.1	x	x	x	27
39 *Mirassol, SP	SDF	20.8	49.6	470	4	3	2	6	2	3	5	2	4	13	120	24.0	5.0	x	19	31	x
40 Rio Brilhante, MS	ChCe	21.8	54.5	400	4	3	3	5	5	3	6	2	4	13	90	21.9	8.2	x	x	x	26
41 Itirapina, SP	Ce	22.2	48.0	600	3	2	3	3	2	3	3	2	2	15	50	22.3	6.4	x	17	x	x
42 *Faz.Camp-ininha, Mogi-Guaáu	Ce	22.3	47.2	530	3	3	4	3	3	2	3	2	2	15	70	20.5	6.0	6	22	x	x

Appendix (*continued*)

No. Site and State	Vegetation Type	Lat	Lon	Alt	Sur	Top	Wat	Veg	Vmo	Bam	Soil	Smo	Sba	Plu	Dry	Tem	Tva	Heli	Ith	Bait	Satur
43 Parque Est. Morro do Diabo, SP	DF	22.6	52.3	350	4	2	4	5	3	4	4	3	3	13	150	24.0	8.0	10	20	64	x
44 *Horto Florestal de Sumaré, SP	SDF	22.8	47.3	600	5	2	2	6	2	1	5	2	4	14	60	20.7	5.7	x	22	x	x
45 *Santa Genebra, Campinas, SP	SDF	22.9	47.1	600	5	2	3	6	2	4	5	2	4	14	60	20.7	5.7	13	28	83	x

* = Sites with considerable anthropic disturbance at some time or sectors. Site numbers, in order of latitude and then longitude, are those used in figures 11.1 and 11.2.

Vegetation types: Ca = Caatinga, Ce = Cerrado, Ch = Chaco, DF = Deciduous or dry forest, FCe = Forest mixed with Cerrado, SDF = Semi-deciduous forest.
Latitude and **Longitude** (0.1°), **Altitude** (m), **Pluviosity** = Annual rainfall (dm), average length of the **Dry** season (in days), **Temperature** (annual mean), and **Temperature Variation** (between means for warmest and coldest months, usually September and June in the Cerrado region) (in °C) are quantitative. Many of the climatic data came from Oliveira-Filho and Fontes (2000), DNMet (1992), or RADAMBRASIL (1978—83); most soil and vegetation data are from the latter.
Explanation of codes for categorical variables in matrix (ordered to show increasing humidity, richness, or complexity; see also Brown and Freitas 2000): **Surface** (Geomorphic, see chapter 2) is coded as 1 = highest (oldest) Surface 1, 2 = transition 1—2, 3 = Surface 2, 4 = transition 2—3, 5 = Surface 3 (youngest).
Topography is given as relief type, 1 = level, 2 = depression or gently rolling, 3 = rolling, 4 = strongly rolling, and 5 = steep, mountainous or escarpment.
Permanent **Water** is coded as 1 = dry landscape, 2 = small rivulets, 3 = broader streams, 4 = river and larger swamps, 5 = large open water bodies in site.
Vegetation categories correspond to inceasing humid arboreal physiognomies in the region, and are averaged for a site; base numbers are 1 = *campo* (open grassland), 2 = *campo cerrado* (sparse small trees), 3 = cerrado or *caatinga* (many trees, many grasses), 4 = *cerradão* or *agreste* (dominant higher woody layer, sparse grasses), 5 = dry or deciduous forest, or a mixture of cerrado and gallery forest, 6 = semi-deciduous, headwater, or very broad gallery forest.
Vegetation **mosaic** is the number of different vegetation types (as categorized above) prominent and interdigitating in the site. See also chapter 6.
Bamboo abundance in the site is coded as 1 = essentially absent to 6 = present in many large patches, up to 30% of the vegetation stems at ground level.
Soils are averaged among the following categories: 1 = rock, hardpan, laterite; 2 = sandy or concretionary, 3 = cambisols, plinthic soils, 4 = moderate-texture latosols, 5 = moderate-textured podzolized soils, and 6 = humic or very argyllic soils.
Soil mosaic is the number of different soil types (as categorized above) in the site, with 1 = over 80% of a single class of soils, 2 = 50—80% of a single class,

Appendix *(continued)*

3 = three types, 4 = four types, 5 = five or more soil types present in a fine mosaic.
Soil bases (fertility) is as 1 = hardpan or coarse sand, 2 = alic, 3 = alic + dystrophic, 4 = dystrophic, 5 = dystrophic + eutrophic, and 6 = eutrophic soils.
Heli = Number of distinct Heliconiini taxa recorded in the site, **Ith** = Ithomiinae, **Bait** = Bait-attracted Nymphalidae except Satyrinae, **Satur** = Saturniidae.

Additional data for Lepidoptera lists of various sites were obtained from N. Tangerini (material from the Microwave stations), Brown and Vasconcellos-Neto (1975, Sumaré), Brown (1987a, central Mato Grosso), Mielke and Casagrande (1998, Morro do Diabo), P. C. Motta (2002, Uberlândia), E.G. Munroe, C. E. G. Pinheiro and H. C. Morais (unpublished lists, Cabeça do Veado and other Federal District sites), and Fernando Corrêa Campos (Belo Horizonte).

12

The Character and Dynamics
of the Cerrado Herpetofauna

Guarino R. Colli, Rogério P. Bastos,
and Alexandre F. B. Araujo

THE EARLIEST WORK ON THE HERPETOFAUNA OF THE CERRADO IS
a list of 54 reptiles and amphibians from Lagoa Santa, state of Minas
Gerais, prepared by Warming (1892). More than 50 years later, Vanzolini
(1948) presented an annotated list of 22 snake and 11 lizard species from
Pirassununga, state of São Paulo. Moreover, Vanzolini (1974, 1976,
1988) examined the distribution patterns of cerrado and *caatinga* lizards,
concluding that no characteristic lizard fauna is harbored by either biome,
both biomes belonging to a corridor of open vegetation ranging from
northwestern Argentina to northeastern Brazil (see chapter 6). According
to Vanzolini (1976), the lack of differentiation of the herpetofauna along
his "great diagonal belt of open formations" results from present-day
ample, interdigitating contacts between the *caatinga* and the cerrado and
past climatic cycles that promoted the expansion and retraction of both
biomes, leading to extensive faunal mixing. This idea received further
support from Duellman (1979), Webb (1978), and Silva and Sites (1995).
More recently, in a preliminary survey conducted near Alto Araguaia,
Mato Grosso, Vitt (1991) concluded that the cerrado lizard fauna is
depauperate when compared to its *caatinga* and Amazonia counterparts,
most noticeably with respect to the Gekkonidae, and that ecological factors such as low habitat structural diversity and the occurrence of fires
may account for the low lizard diversity.

In sum, the aforementioned works indicate that the cerrado herpetofauna (1) has low levels of species diversity and endemism; (2) lacks a
character; and (3) is more similar to the *caatinga* than to other South

American biomes. We advance, however, that these claims, largely the outcome of inadequate sampling and/or analyses, do not adequately describe the nature of the cerrado herpetofauna. For instance, Heyer (1988) analyzed the adequacy of the distributional database of 10 species-groups of frogs east of the Andes and concluded that frog distributional data were largely inadequate for biogeographical analysis, that the cerrado and *caatinga* were the poorest-sampled biomes, and that the question whether the cerrado and *caatinga* share a common frog fauna was unanswerable with the data available at the time.

In a similar vein, of the 101 cerrado lizard localities we obtained from three Brazilian institutions, 97% contain fewer than 12 species (see fig. 12.1). Records from the three most extensively sampled localities (Brasília, Distrito Federal; Minaçu, state of Goiás; and Chapada dos Guimarães, state of Mato Grosso) suggest that the local lizard diversity in the cerrado is around 25 species. Therefore, most collecting localities have been inadequately sampled in the cerrado, and our estimates indicate that, contrary to Vitt's (1991) impression, the local diversity of lizards in the cerrado (around 25 species) is greater than that of the *caatinga* (18

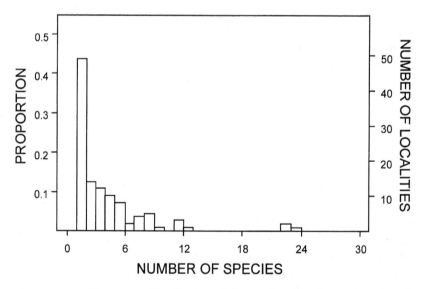

Figure 12.1 Frequency distribution of the cerrado localities where lizards have been collected according to the number of lizard species. Data from the Museu Paraense Emílio Goeldi (MPEG), Museu Nacional do Rio de Janeiro (MNRJ), and Coleção Herpetológica da Universidade de Brasília (CHUNB).

species; Vitt 1995). However, it is possible that the three best-studied localities are unusual and that, in fact, the other 98 studied localities, reported to have 12 or fewer species, do have low lizard diversity. This latter view implies that species diversity in the cerrados is highly variable on a regional scale. Further studies are necessary to clarify this issue and identify possible determinants of lizard diversity in the cerrado region.

Several species of reptiles and amphibians of the cerrado have been described recently, and several undescribed species still await adequate studies. At present, 10 species of turtles, 5 crocodilians, 15 amphisbaenians, 47 lizards, 107 snakes, and 113 amphibians are known to occur in the Cerrado Biome (appendix). Among these species, some are typical of the Atlantic forest or Amazonia biomes but often occur deep in the cerrados, along gallery forests. As an example, *Anilius scytale* has been recorded in gallery forests of the Tocantins river drainage, at Minaçu, state of Goiás, in the core region of the cerrados. Other species, such as *Leptodactylus petersii* (Heyer 1994), have a more marginal distribution in the biome. The species list we present roughly corresponds to 20% of the Brazilian species of amphibians and 50% of the reptiles. The estimates for lizards and snakes are almost twice those previously recorded by Vanzolini (1988) and Silva and Sites (1995). Moreover, contrary to early beliefs, the cerrados harbor a large number of endemics: 8 species of amphisbaenians (50% of the total amphisbaenian species), 12 species of lizards (26%), 11 species of snakes (10%), and 32 (28%) species of amphibians (appendix). This level of endemicity differs sharply from the 3.8% recorded for cerrado birds (Silva 1995) and, at least for amphisbaenians, is comparable to that registered for the cerrado flora, which is approximately 50% (Heringer et al. 1977; chapter 6). The cerrado herpetofauna also includes three endangered species of anurans, four turtles, five crocodilians, five lizards, and six snakes (see appendix). With the exception of *Caiman niger* (*sensu* Poe 1996), listed in appendix I of CITES, the remaining endangered species are listed in appendix II.

It is commonly assumed that savanna biomes, due to their lower habitat heterogeneity, harbor an impoverished herpetofauna relative to forest biomes (e.g., Duellman 1993; Lamotte 1983). Further, it seems to be taken for granted that species richness in South America is centered in Amazonia. Taking lizards as an example, there are about 100 species in Brazilian Amazonia (Ávila-Pires 1995), whereas our list for the cerrados contains about half that number. Nevertheless, Brazilian Amazonia covers an area of approximately 4 million km^2, roughly twice the area covered by the cerrados. Therefore, the difference in species richness between the two biomes may stem from an area effect, rather than from a habitat heterogeneity effect. In terms of local species richness, moreover, there is no difference in the mean number of lizard species between cerrado and neotropical forest sites (tables 12.1, 12.2). This is a remarkable and apparently counterintuitive result. If local species assemblages are but a random sample of the regional pool of species, it follows that neotropical forest sites should be richer than cerrado sites (Schluter and Ricklefs 1993). We advance that the similar richness in local lizard assemblages between cerrado and neotropical forest sites results from two factors. First, the pronounced horizontal variability in the cerrado landscape balances the predominantly vertical variability typical of forested biomes when it comes to allowing the coexistence of species. The horizontal variability of the landscape may be further enhanced by fire history, as documented in other open vegetation regions (e.g., Pianka 1989). Second, there is a greater regional differentiation of the herpetofauna in neotropical forests relative to the cerrados: the average species turnover (beta diversity) among the five forest sites listed in Duellman (1990) is significantly higher than that among the five cerrado sites in table 12.1 (forest: $\bar{x} = 0.78 \pm 0.19$; cerrado: $\bar{x} = 0.27 \pm 0.09$; Mann-Whitney $U = 0.00$, $p < .001$). Thus, even though neotropical forests as a whole harbor a richer herpetofauna, the local lizard diversity is comparable to that recorded in the cerrado.

The much lower species turnover in the cerrados, relative to Amazonia, suggests that the high regional heterogeneity of the cerrado has little effect upon the composition of the herpetofauna. However, some areas, such as high altitude habitats, harbor an unusually high number of endemics. For instance, the *campos rupestres* (rocky grasslands; see chapter 6) along the Espinhaço Range include large numbers of endemic amphibians and reptiles (Heyer 1999; Rodrigues 1988). There may be other areas of high endemism in the biome, such as the Jalapão sand dunes in the state of Tocantins, but this region as well as large tracts of the cerrados have never been adequately sampled.

NATURAL HISTORY

Most of the available information on the natural history of the cerrado herpetofauna is restricted to lizards and anurans. Salamanders are absent from the cerrados, and practically no information is available for caecilians, amphisbaenians, crocodiles, and testudines. Our coverage, therefore, is concentrated on lizards and anurans, reflecting this knowledge bias. A striking feature of the Cerrado Biome is its marked seasonality,

Table 12.1   Composition of Lizard Assemblages from Cerrado Sites

Taxon	CG	BG	MI	PI	BR	Diel	Habitat	Micro-habitat
Hoplocercidae								
Hoplocercus spinosus	X	X	X	—	X	D	CE	G
Iguanidae								
Iguana iguana	X	X	X	—	—	D	F	T
Polychrotidae								
Anolis meridionalis	X	X	X	X	X	D	CE	G,T
Anolis nitens	X	X	X	X	X	D	F,CE	G,T
Enyalius bilineatus	—	—	—	—	X	D	F	G,T
Polychrus acutirostris	X	X	X	X	X	D	CE	T
Tropiduridae								
Stenocercus caducus	—	X	—	—	—	D	CE,F	G,T
Tropidurus guarani	X	—	—	—	—	D	CE,F	T
Tropidurus itambere	—	—	—	X	X	D	CE	G
Tropidurus montanus	X	X	X	—	—	D	CE	G
Tropidurus oreadicus	X	X	X	X	X	D	CE	G
Tropidurus torquatus	X	X	X	—	X	D	F	G,T
Gekkonidae								
Coleodactylus brachystoma	—	—	X	—	—	D	F,CE	L
Coleodactylus meridionalis	X	—	X	—	—	D	F	L
Gymnodactylus geckoides	X	X	X	X	—	N(?)	CE	G
Hemidactylus mabouia	X	—	—	—	X	N	CE	G,T
Phyllopezus pollicaris	X	—	X	—	—	N(?)	CE	G
Scincidae								
Mabuya dorsivittata	—	—	—	—	X	D	CE,F	G,L
Mabuya frenata	X	X	X	X	X	D	CA,CE	G,L
Mabuya guaporicola	—	—	—	—	X	D	CE,F	G,L
Mabuya nigropunctata	X	X	X	X	X	D	CE	G,L

Table 12.1 (*continued*)

Taxon	CG	BG	MI	PI	BR	Diel	Habitat	Micro-habitat
			Site				*Ecology*	
Gymnophthalmidae								
Bachia bresslaui	X	—	—	—	X	D	CE	F
Cercosaura ocellata	—	X	X	—	X	D	E,CA,F	L
Colobosaura modesta	—	X	X	—	X	D	CE,F	L
Micrablepharus atticolus	—	X	X	X	X	D	CE	L
Micrablepharus maximiliani	X	—	X	X	X	D	CE	L
Pantodactylus schreibersii	X	X	X	X	X	D	CE	L
Teiidae								
Ameiva ameiva	X	X	X	X	X	D	CE,F	G
Cnemidophorus ocellifer	X	X	X	X	X	D	CE	G
Kentropyx paulensis	X	X	X	—	X	D	CE,CA	G
Tupinambis duseni	—	—	—	—	X	D	CE,CA	G
Tupinambis merianae	X	X	X	X	X	D	CE,F	G
Tupinambis quadrilineatus	X	X	X	—	—	D	F,CE	G
Anguidae								
Ophiodes striatus	X	X	X	—	X	D	CE,F,CA	F
Total number of species	**24**	**22**	**25**	**14**	**25**			

Sources: Araujo and records from the Coleção Herpetológica da Universidade de Brasília (CHUNB). Note: CG = Chapada dos Guimarães, Mato Grosso; BG = Barra do Garças, Mato Grosso; MI = Minaçu, Goiás; PI = Pirenópolis, Goiás; BR = Brasília, Distrito Federal. Diel: D = diurnal, N = nocturnal. Habitat: CA = *campo cerrado, campo limpo,* or *campo sujo,* CE = cerrado *sensu stricto,* F = forest. Microhabitat: F = fossorial, G = ground, L = litter, T = tree. In columns under sites, X = present, — = presumably absent. See chapter 6 for descriptions of cerrado physiognomies.

with well-defined dry and rainy seasons. Moreover, the cerrado is a highly heterogeneous landscape, with a physiognomic mosaic that ranges from treeless grass fields to relatively dense gallery forests, with a predominance of open vegetation physiognomies (chapter 6). These two aspects, marked seasonality and high spatial heterogeneity, have profound effects on the ecology of the herpetofauna in general, since these animals are ectothermic, and amphibians in particular, because of their highly permeable skin.

Table 12.2 Ecology and Families of Lizards from Cerrado and Neotropical Forest Sites

	Cerrado						Neotropical rainforest					
	CG	BG	MI	PI	BR	mean ± sd	LS	BCI	MN	MA	SC	mean ± sd
Diel activity												
Diurnal	87.5	95.5	92.0	92.9	96.0	92.8 ± 3.4	92.0	92.3	95.8	93.7	96.7	94.1 ± 2.1
Nocturnal	12.5	4.5	8.0	7.1	4.0	7.2 ± 3.4	8.0	7.7	4.2	6.3	3.3	5.9 ± 2.1
Habitat												
Fossorial	8.3	4.5	4.0	-	8.0	5.0 ± 3.4	-	3.8	8.3	6.3	3.3	4.3 ± 3.2
Ground[a]	66.7	63.6	56.0	64.3	64.0	62.9 ± 4.1	40.0	15.4	25.0	43.8	26.7	30.1 ± 11.7
Litter	20.8	27.3	36.0	35.7	36.0	31.2 ± 6.9	44.0	30.8	33.3	12.5	36.7	31.5 ± 11.7
Trees[a,b]	29.2	27.3	20.0	21.4	24.0	24.4 ± 3.9	52.0	57.7	45.8	56.3	43.3	51.0 ± 6.3
Family												
Hoplocercidae	4.2	4.5	4.0	-	4.0	3.3 ± 1.9	-	7.7	-	6.3	6.7	2.6 ± 3.6
Corytophanidae	-	-	-	-	-	-	12.0	7.7	-	-	-	3.9 ± 5.6
Iguanidae	4.2	4.5	4.0	-	-	2.5 ± 2.3	4.0	3.8	4.2	-	-	2.4 ± 2.2
Polychrotidae[c]	12.5	13.6	12.0	21.4	16.0	15.1 ± 3.8	36.0	46.2	16.7	18.8	23.3	28.2 ± 12.5
Tropiduridae	16.7	18.2	12.0	14.3	12.0	14.6 ± 2.8	-	-	16.7	18.8	6.7	8.4 ± 9.0
Gekkonidae	16.7	4.5	16.0	7.1	4.0	9.7 ± 6.2	16.0	15.4	12.5	18.8	13.3	15.2 ± 2.5
Scincidae[c]	8.3	9.1	8.0	14.3	16.0	11.1 ± 3.7	8.0	3.8	4.2	6.3	3.3	5.12 ± 2.0
Xantusiidae	-	-	-	-	-	-	4.0	3.8	-	-	-	1.6 ± 2.1
Gymnophthalmidae	12.5	18.2	20.0	21.4	24.0	19.2 ± 4.3	-	7.7	33.3	6.3	33.3	16.1 ± 16.0
Teiidae[c]	20.8	22.7	20.0	21.4	20.0	21.0 ± 1.1	8.0	11.5	12.5	18.8	13.3	12.8 ± 3.9
Anguidae	4.2	4.5	4.0	0.00	4.0	3.3 ± 1.9	12.0	-	-	6.3	-	3.7 ± 5.4
Number of species	24	22	25	14	25	22.0 ± 4.6	25	26	24	16	30	24.2 ± 5.1

Sources: Cerrado sites are from Araujo (1992) and records from the Coleção Herpetológica da Universidade de Brasília (CHUNB); neotropical rainforest sites are from Duellman (1990).

Note: CG = Chapada dos Guimarães, Mato Grosso; BG = Barra do Garças, Mato Grosso; MI = Minaçu, Goiás; PI = Pirenópolis, Goiás; BR = Brasília, Distrito Federal. Numbers indicate percentage of lizard species occurring in each category. Some species are present in more than one category.

[a] Means of arcsin transformed values are statistically different between cerrado and neotropical rainforest (Ground: $F = 31.4$, $p < .001$; Trees: $F = 64.7$, $p < .001$).

[b] Species in "bushes" category of Duellman (1990) were placed in the "trees" category.

[c] Means of arcsin transformed values are statistically different between cerrado and neotropical rainforest (Polychrotidae: $F = 5.5$, $p < .05$; Scincidae: $F = 11.8$, $p < .01$; Teiidae: $F = 17.7$, $p < .01$).

Below, we review different aspects of the ecology of the cerrado herpeto-fauna, such as habitat and microhabitat preferences, diel activities, and reproductive cycles.

Habitat, Microhabitat, and Diel Activity

Considering the pronounced seasonality and heterogeneity of the cerrado landscape (chapter 11), the restriction of moist habitats to the periphery of river drainages, and their great physiological requirements, it is expected that cerrado anurans should adopt one or more of the following strategies: (a) to be nocturnal; (b) to live close to or in moist places, like swamps; (c) to show activity only during the rainy season; or d) to live permanently buried in the soil (Lamotte 1983). Conversely, because of their highly impermeable skin, strictly pulmonary respiration, and amniotic eggs, reptiles can explore a broader range of environments.

The anuran assemblage at the Estação Ecológica de Águas Emendadas, Brasília (Distrito Federal), contains 26 species distributed in four habitats: *campo-úmido/vereda* (22 species), *vereda* (16), cerrado/*cerradão* (12), and gallery forest (10) (Brandão and Araujo 1998; chapter 6). Most species use two or more habitats and the *campo-úmido/vereda*, a seasonally flooded field with buriti-palms (*Mauritia flexuosa*) harboring the greatest number of species, including adults and tadpoles. Only two species are restricted to gallery forest: *Hyla biobeba* and *Aplastodiscus perviridis*. Motta (1999) compared the leaf-litter anuran assemblages of cerrados and gallery forest, by using pitfall traps and drift fences, at the Estação Florestal de Experimentação, Silvânia, state of Goiás. *Barycholos savagei* and *Physalaemus cuvieri* were the most abundant species in both habitats. Despite the greater abundance and diversity of anurans in the gallery forest, there was a high similarity between the two assemblages.

A majority of the lizard species distributed within the cerrado region occurs primarily in "cerrado-type" physiognomies (see chapter 6), whereas *Iguana iguana, Enyalius bilineatus, Anolis chrysolepis brasiliensis, Tropidurus torquatus, Coleodactylus brachystoma,* and *Tupinambis quadrilineatus* are mainly restricted to forested habitats (table 12.1). However, apparently there is no difference in the mean richness of lizards between cerrado and forested habitats (Araujo 1992). Some species are strongly associated with specific habitat characteristics. For example, *Phyllopezus pollicaris* and *Gymnodactylus geckoides* are found primarily in rock outcrops. On the other hand, *Bachia bresslaui* and *Cnemidophorus ocellifer* are strongly associated with sandy soils (Magnusson et al. 1986; Colli et al. 1998). Still other species, such as *Ameiva ameiva* (Colli 1991; Vitt and

Colli 1994; Sartorius et al. 1999) and *Tropidurus torquatus* (Wieder-hecker et al. 2002), seem to benefit from human disturbance, as they are very abundant in anthropic areas.

Among reptiles, only crocodilians, most turtles, some snakes such as *Eunectes murinus* and *Helicops modestus*, and the lizard *Iguana iguana* are strongly associated with water in the cerrados. Several species of cerrado reptiles are fossorial, including all amphisbaenids, all species of the gymnophthalmid genus *Bachia*, the anguid *Ophiodes striatus*, and several snake genera including *Liotyphlops*, *Leptotyphlops*, *Typhlops*, *Anilius*, *Phalotris*, *Apostolepis*, *Atractus*, *Tantilla*, and *Micrurus*. A number of species also uses cavities in trees, termite and ant nests, and armadillo burrows as shelter. Cavities in trees are often used by the arboreal *Tropidurus guarani* (Colli et al. 1992). Several snakes such as *Bothrops neuwiedi*, *Tantilla melanocephala*, and *Oxyrhopus trigeminus*, and the lizards *Anolis meridionalis* and *Gymnodactylus geckoides* are commonly found under termite nests. Apparently, the amphisbaenid *Amphisbaena alba* (Colli and Zamboni 1999) and the gymnophthalmid *Micrablepharus atticolus* (Vitt and Caldwell 1993) are associated with nests of leaf-cutter ants. The ability to use these microhabitats seems especially critical during the periodic fires.

Another behavioral adaptation of the herpetofauna to resist desiccation is nocturnal activity. Diurnal species of anurans, relatively common in Amazonic and Atlantic Forests due to high humidity, are relatively rare in the cerrados (Heyer et al. 1990; Hödl 1990). Some species, however, can call during the day after intense rains (Brandão and Araujo 1998). Vanzolini (1948) noticed a preponderance of nocturnal and fossorial species in a cerrado assemblage of squamates, suggesting that these traits would be an adaptation to high temperatures and visibility to predators at the soil surface, coupled with thermal stability, good aeration, and an abundance of prey beneath the surface. We evaluated Vanzolini's statement by classifying five lizard assemblages from the cerrado according to diel and microhabitat categories, following Duellman (1990). The number of nocturnal species in each site varied from one to three (table 12.1), and the percentage of nocturnal species averaged 7.2% (table 12.2). Curiously, we observed no significant difference in the average of nocturnal species between the cerrado and neotropical forest (table 12.2). The number of fossorial species ranged from zero to two species (table 12.1), averaging 5% of the species in the cerrado (table 12.2). Again, we observed no difference between the mean percentage of fossorial species between the two biomes (table 12.2). This pattern seemingly derives from a comparable representation of nocturnal (Gekkonidae) and fossorial (Anguidae, Gymnophthalmidae) lineages in both biomes. Thus, we find no evidence for Vanzolini's (1948) suggestion that there is a trend toward nocturnal activity and fossoriality in the cerrado herpetofauna.

Contrasting with the savannas of northern South America (Staton and Dixon 1977; Rivero-Blanco and Dixon 1979; Vitt and Carvalho 1995), the higher density of trees and shrubs in the cerrados supports a relatively large number of arboreal squamates. Among lizards, the most arboreal species is *Polychrus acutirostris*, which even possesses a prehensile tail (Vitt and Lacher 1981). The presence of a developed arboreal stratum allows the coexistence of divergent, congeneric species pairs, such as *Tropidurus guarani* (larger, arboreal) and *T. oreadicus* (smaller, ground) in the same area (Colli et al. 1992). Perhaps the same could be said about *Anolis chrysolepis brasiliensis* and *A. meridionalis*, but there are no detailed studies on their coexistence. The only difference in the distribution of lizard species across habitat categories between cerrado and neotropical rainforest assemblages concerns the proportion of arboreal versus ground species. In neotropical rainforest sites, the mean percentage of arboreal species is approximately twice that for cerrado sites, whereas the reverse is true concerning species that use primarily the ground (table 12.2). That difference stems from the higher number of polychrotids in neotropical rainforest sites and of skinks and teiids in cerrado sites (table 12.2). Polychrotids are primarily arboreal, whereas teiids and skinks are primarily ground-dwellers.

Life History

Apparently, most anurans of the cerrado region breed in areas of open vegetation (Haddad et al. 1988, Feio and Caramaschi 1995, Brandão and Araujo 1998). Conversely, in anuran assemblages from forested biomes, such as Amazonia and Atlantic forests, a much lower proportion of species reproduce in open vegetation (Crump 1974; Heyer et al. 1990; Hödl 1990; Bertoluci 1998). However, Manaus (state of Amazonas) and Belém (state of Pará) present percentages that are similar to the cerrados, perhaps because of the noticeable dry season they exhibit (Hödl 1990). The prevalence of species with foam-nest reproduction in Manaus and Belém may be an adaptation to the seasonal environment (Hödl 1990). Moreover, in open habitats, as opposed to forest formations, the number of species is usually much larger than the number of available microhabitats such as calling perches (Cardoso et al. 1989). Consequently, spatial overlapping is seemingly more extensive among cerrado anurans (Blamires et al. 1997).

In tropical regions, rainfall and temperature are the main variables affecting anuran reproduction, and most species breed in a restricted period, determining a reproductive cycle (Aichinger 1987; Rossa-Feres and Jim 1994; Pombal 1997; Bertoluci 1998). One of us (RPB) and some students monitored the breeding activity of one cerrado anuran assemblage, between June 1995 and May 1998, at the Estação Florestal de Experimentação (EFLEX), in Silvânia, state of Goiás (Blamires et al. 1997). Even though most species reproduced in the rainy season, between October and April, some were active even during the dry season, such as *Hyla albopunctata*, *H. biobeba*, *H. goiana*, and *Odontophrynus cultripes*. Like several other neotropical anurans (Pombal et al. 1994; Bastos and Haddad 1996, 1999), these species can be regarded as prolonged breeders. At EFLEX, males of most species (e.g., *Hyla albopunctata*, *H. biobeba*, *H. cruzi*, and *H. goiana*) aggregate during several nights, forming choruses and leks (exhibition areas) and defending territories with resources to oviposition.

A common cerrado leptodactylid, *Physalaemus cuvieri*, breeds exclusively in the rainy season throughout its range (Barreto and Andrade 1995; Moreira and Barreto 1997). A detailed study on the reproductive ecology of *Scinax centralis* was also carried out at EFLEX (Bugano 1999). This species belongs to the *catharine* group, typical of the Atlantic forest, and restricted to gallery forest in the cerrado region (Pombal and Bastos 1996). Leks occur from November to July; males always outnumber females, showing a uniform spatial pattern maintained by acoustic interactions (three call types were observed) or fights; and oviposition occurs away from calling perches.

With the exception of species of *Mabuya* and *Ophiodes striatus*, all lizard species of the cerrado are oviparous. All South American *Mabuya* have a long gestation period, ovulate small ova, form a chorioallantoic placenta, and have placental provision of practically all the energetic requirements for development (Blackburn and Vitt 1992). Clutch size is practically fixed in some lineages, with *Anolis* producing predominantly one egg, whereas gekkonids and gymnophthalmids produce clutches of usually two eggs (Vitt 1991). In the remaining species, clutch size shows a positive correlation with body size. The species so far studied are cyclic breeders, including *Ameiva ameiva* (Colli 1991), *Amphisbaena alba* (Colli and Zamboni 1999), *Mabuya frenata* (Vrcibradic and Rocha 1998; Pinto 1999), *M. nigropunctata* (Pinto 1999), *Tropidurus itambere* (Van Sluys 1993), and *T. torquatus* (Wiederhecker et al. 2002). With the exception of *A. alba*, all other species studied to date reproduce primarily during the rainy season. Furthermore, with the exception of the species of *Mabuya*, the remaining species produce multiple clutches during the breeding season. It has been repeatedly suggested that reproductive seasonality in the neotropical region is largely determined by fluctuations in arthropod abundance which, in their turn, are associated with fluctuations in rainfall. However, by using fat body mass as an indicator of food availability, Colli et al. (1997) concluded that lizard reproductive seasonality is not constrained by food limitation in the cerrado. Long-term studies and comparisons among populations of the same species in distinct regions of the Cerrado Biome, however, are still lacking.

SUGGESTIONS FOR FUTURE RESEARCH

The knowledge about the cerrado herpetofauna is in its early infancy. New species are still being described, and large tracts of the biome have never been adequately sampled. Meaningful biogeographic analyses of the herpetofauna will depend heavily on museum data. Most large collections throughout the world have few specimens from the cerrado region, while several small collections in Brazilian institutions, housing potentially useful material, usually go unnoticed due to lack of integration and institutional support. The organization of such collections and their integration with larger institutions would be very helpful in future systematic and ecological work.

Basic natural history information is currently lacking for most species of the cerrado herpetofauna. This kind of descriptive work, often undervalued by funding agencies and reviewers of scientific journals, is essential for the formulation of generalizations and testing of hypotheses. Species which are locally abundant and/or have broad distributions in the cerrados, such as *Bufo paracnemis*, *Hyla albopunctata*, *H. minuta*, *Leptodactylus ocellatus*, *Physalaemus cuvieri*, *Cnemidophorus ocellifer*, and *Tropidurus oreadicus*, provide model organisms for basic studies of natural history. Finally, detailed demographic and community-level studies on the cerrado herpetofauna are badly needed. Most importantly, they should be analyzed in concert with phylogenetic hypotheses to answer questions adequately such as, "To what extent do current environmental conditions and lineage effects determine the natural history characteristics of the herpetofauna?" A promising group to be investigated is the genus *Tropidurus*, with several species in open and forested biomes, for which well-resolved phylogenetic hypotheses are available. Another interesting venue of research is the investigation of the relative roles of structural features of the environment and fire history versus past geological and climatic events on the distribution and composition of amphibian and reptile assemblages. For example, are species often associated with rock outcrops, such as *Leptodactylus syphax* and *Phyllopezus pollicaris*, dependent on some kind of resource only available at those sites, or do they represent a relictual distribution on ancient and dissected plateaus? The predictive power that could be gained with this kind of research would be invaluable in assisting conservation efforts for the biome, considering the accelerated pace of its destruction.

ACKNOWLEDGMENTS

We thank R. A. Brandão, W. Ronald Heyer, R. F. Juliano, D. Melo e Silva, C. C. Nogueira, P. S. Oliveira, P. H. Valdujo, and L. J. Vitt for insightful criticisms on earlier versions of the manuscript. We also thank IBAMA for granting access to the EFLEX study site. This study was partially supported by grants from the Brazilian Research Council (CNPq) to GRC (no. 302343/88-1), and RPB (no. 400381-97.4).

REFERENCES

Aichinger, M. 1987. Annual activity patterns of anurans in a seasonal neotropical environment. *Oecologia* 71:583–592.

Araujo, A. F. B. 1992. "Estrutura Morfométrica de Comunidades de Lagartos de Áreas Abertas do Litoral Sudeste e Brasil Central." Ph.D. thesis, Universidade Estadual de Campinas, Campinas, Brazil.

Ávila-Pires, T. C. S. 1995. Lizards of Brazilian Amazonia (Reptilia: Squamata). *Zool. Verh.* 1995:3–706.

Barreto, L. and G. V. Andrade. 1995. Aspects of the reproductive biology of *Physalaemus cuvieri* (Anura: Leptodactylidae) in northeastern Brazil. *Amphibia-Reptilia* 16:67–76.

Bastos, R. P. and C. F. B. Haddad. 1996. Breeding activity of the neotropical treefrog *Hyla elegans* (Anura, Hylidae). *J. Herpetol.* 30:355–360.

Bastos, R. P. and C. F. B. Haddad. 1999. Atividade reprodutiva de *Scinax ribilis* (Bokermann) (Anura, Hylidae) na Floresta Atlântica, sudeste do Brasil. *Rev. Bras. Zool.* 16:409–421.

Bertoluci, J. 1998. Annual patterns of breeding activity in Atlantic rainforest anurans. *J. Herpetol.* 32:607–611.

Blackburn, D. G. and L. J. Vitt. 1992. Reproduction in viviparous South American lizards of the genus *Mabuya*. In W. C. Hamlett, ed., *Reproductive Biology of South American Vertebrates*, pp. 150–164. Berlin: Springer-Verlag.

Blamires, D., J. A. O. Motta, K. G. Sousa, and R. P. Bastos. 1997. Padrões de distribuição e análise de canto em uma comunidade de anuros no Brasil central. In L. L. Leite and C. H. Saito, eds., *Contribuição ao Conhecimento Ecológico do Cerrado*, pp. 185–190. Brasília: ECL/Universidade de Brasília.

Brandão, R. A. and A. F. B. Araujo. 1998. A herpetofauna da Estação Ecológica de Águas Emendadas. In J. S. Marinho-Filho, ed., *Vertebrados da Estação Ecológica de Águas Emendadas: História Natural e Ecologia de um Fragmento de Cerrado do Brasil Central*, pp. 9–21. Brasília: Instituto de Ecologia e Meio Ambiente do Distrito Federal.

Bugano, M. A. 1999. "Ecologia Reprodutiva de Scinax centralis (Anura: Hylidae) em uma Região do Brasil Central." Master's thesis, Universidade Federal de Goiás, Goiânia, Brazil.

Cardoso, A. J., G. V. Andrade, and C. F. B. Haddad. 1989. Distribuição espacial em comunidades de anfíbios (Anura) no sudeste do Brasil. *Rev. Bras. Biol.* 49:241–249.

Colli, G. R. 1991. Reproductive ecology of *Ameiva ameiva* (Sauria: Teiidae) in the Cerrado of central Brazil. *Copeia* 1991:1002–1012.

Colli, G. R., A. F. B. de Araújo, R. da Silveira, and F. Roma. 1992. Niche partitioning and morphology of two syntopic *Tropidurus* (Sauria: Tropiduridae) in Mato Grosso, Brazil. *J. Herpetol.* 26:66–69.

Colli, G. R., A. K. Péres, Jr., and M. G. Zatz. 1997. Foraging mode and reproductive seasonality in tropical lizards. *J. Herpetol.* 31:490–499.

Colli, G. R., and D. S. Zamboni. 1999. Ecology of the worm-lizard *Amphisbaena alba* in the Cerrado of central Brazil. *Copeia* 1999:733–742.

Colli, G. R., M. G. Zatz, and H. J. da Cunha. 1998. Notes on the ecology and geographical distribution of the rare gymnophthalmid lizard, *Bachia bresslaui*. *Herpetologica* 54:169–174.

Crump, M. L. 1974. Reproductive strategies in a tropical anuran community. *Misc. Publ. Mus. Nat. Hist. Univ. Kansas* 61:1–68.

Duellman, W. E. 1979. The South American herpetofauna: A panoramic view. In W. E. Duellman, ed., *The South American Herpetofauna: Its Origin, Evolution, and Dispersal*, pp. 1–28. Lawrence: The Museum of Natural History, The University of Kansas.

Duellman, W. E. 1990. Herpetofaunas in neotropical rainforests: Comparative composition, history, and resource use. In A. H. Gentry, ed., *Four Neotropical Rainforests*, pp. 455–505. New Haven: Yale University Press.

Duellman, W. E. 1993. Amphibians in Africa and South America: Evolutionary history and ecological comparisons. In P. Goldblatt, ed., *Biological Relationships Between Africa and South America*, pp. 200–243. New Haven: Yale University Press.

Feio, R. N. and U. Caramaschi. 1995. Aspectos zoogeográficos dos anfíbios do médio Rio Jequitinhonha, nordeste de Minas Gerais. *Revista Ceres* 42:53–61.

Haddad, C. F. B., G. V. Andrade, and A. J. Cardoso. 1988. Anfíbios anuros no Parque Nacional da Serra da Canastra, estado de Minas Gerais. *Brasil Florestal* 64:9–20.

Heringer, E. P., G. M. Barroso, J. A. Rizzo, and C. T. Rizzini. 1977. A flora do Cerrado. In M. G. Ferri, ed., *IV Simpósio sobre o Cerrado*, pp. 211–232. Belo Horizonte and São Paulo: Livraria Itatiaia Editora and Editora da Universidade de São Paulo.

Heyer, W. R. 1988. On frog distribution patterns east of the Andes. In P. E. Vanzolini and W. R. Heyer, eds., *Proceedings of a Workshop on Neotropical Distribution Patterns*, pp. 245–273. Rio de Janeiro: Academia Brasileira de Ciências.

Heyer, W. R. 1994. Variation within the *Leptodactylus podicipinus-wagneri* complex of frogs (Amphibia: Leptodactylidae). *Smithsonian Contrib. Zool.* 546:1–124.

Heyer, W. R. 1999. A new genus and species of frog from Bahia, Brazil (Amphibia: Anura: Leptodactylidae) with comments on the zoogeography of the Brazilian campos rupestres. *Proc. Biol. Soc. Washington* 112:19–39.

Heyer, W. R., A. S. Rand, C. A. G. d. Cruz, O. L. Peixoto, and C. E. Nelson. 1990. Frogs of Boracéia. *Arq. Zool. São Paulo* 31:231–410.

Hödl, W. 1990. Reproductive diversity in Amazonian lowland frogs. *Fortschr. D. Zool.* 38:41–60.

Lamotte, M. 1983. Amphibians in savanna ecosystems. In F. Bourlière, ed., *Ecosystems of the World: 13. Tropical Savannas*, pp. 313–323. Amsterdam: Elsevier Scientific Publishing Company.

Magnusson, W. E., C. R. Franke, and L. A. Kasper. 1986. Factors affecting densities of *Cnemidophorus lemniscatus*. *Copeia* 1986:804–807.

Moreira, G. and L. Barreto. 1997. Seasonal variation in nocturnal calling activity of a savanna anuran community in central Brazil. *Amphibia-Reptilia* 18:49–57.

Motta, J. A. O. 1999. "A Herpetofauna no Cerrado: Composição de Espécies, Sazonalidade e Similaridade." Master's thesis, Universidade Federal de Goiás, Goiânia, Brazil.

Pianka, E. R. 1989. Desert lizard diversity: Additional comments and some data. *Amer. Nat.* 134:344–364.

Pinto, M. G. M. 1999. "Ecologia das Espécies de Lagartos Simpátricos *Mabuya nigropunctata* e *Mabuya frenata* (Scincidae), no Cerrado de Brasília e Serra da Mesa (GO)." Master's thesis, Universidade de Brasília, Brasília, Brazil.

Poe, S. 1996. Data set incongruence and the phylogeny of crocodilians. *Syst. Biol.* 45:393–414.

Pombal, J. P., Jr. 1997. Distribuição espacial e temporal de anuros (Amphibia) em uma poça permanente na Serra de Paranapiacaba, sudeste do Brasil. *Rev. Bras. Biol.* 57:583–594.

Pombal, J. P., Jr. and R. P. Bastos. 1996. Nova espécie de *Scinax* Wagler, 1830 do Brasil central (Amphibia, Anura, Hylidae). *Bol. Mus. Nac., nova sér., Zool., Rio de Janeiro* 371:1–11.

Pombal, J. P., Jr., I. Sazima, and C. F. B. Haddad. 1994. Breeding behavior of the pumpkin toadlet, *Brachycephalus ephippium* (Brachycephalidae). *J. Herpetol.* 28:516–519.

Rivero-Blanco, C. and J. R. Dixon. 1979. Origin and distribution of the herpetofauna of the dry lowland regions of northern South America. In W. E. Duellman, ed., *The South American Herpetofauna: Its Origin, Evolution, and Dispersal*, pp. 281–298. Lawrence: The Museum of Natural History, The University of Kansas.

Rodrigues, M. T. 1988. Distribution of lizards of the genus *Tropidurus* in Brazil (Sauria, Igaunidae). In W. R. Heyer and P. E. Vanzolini, eds., *Proceedings of a Workshop on Neotropical Distribution Patterns*, pp. 305–315. Rio de Janeiro: Academia Brasileira de Ciências.

Rossa-Feres, D. C. and J. Jim. 1994. Distribuição sazonal em comunidades de anfíbios anuros na região de Botucatu, São Paulo. *Rev. Bras. Biol.* 54:323–334.

Sartorius, S. S., L. J. Vitt, and G. R. Colli. 1999. Use of naturally and anthropogenically disturbed habitats in Amazonian rainforests by the teiid lizard *Ameiva ameiva*. *Biol. Conserv.* 90:91–101.

Schluter, D., and R. E. Ricklefs. 1993. Convergence and the regional component of species diversity. In R. E. Ricklefs and D. Schluter, eds., *Species Diversity in Ecological Communities: Historical and Geographical Perspectives*, pp. 230–240. Chicago: The University of Chicago Press.

Silva, J. M. C. 1995. Birds of the Cerrado Region, South America. *Steenstrupia* 21:69–92.

Silva, N. J., Jr. and J. W. Sites, Jr. 1995. Patterns of diversity of neotropical squamate reptile species with emphasis on the Brazilian Amazon and the conservation potential of indigenous reserves. *Conserv. Biol.* 9:873–901.

Staton, M. A. and J. R. Dixon. 1977. The herpetofauna of the central llanos of Venezuela: Noteworthy records, a tentative checklist and ecological notes. *J. Herpetol.* 11:17–24.

Van Sluys, M. 1993. The reproductive cycle of *Tropidurus itambere* (Sauria: Tropiduridae) in southeastern Brazil. *J. Herpetol.* 27:28–32.

Vanzolini, P. E. 1948. Notas sôbre os ofídios e lagartos da Cachoeira de Emas, no município de Pirassununga, estado de São Paulo. *Rev. Bras. Biol.* 8:377–400.

Vanzolini, P. E. 1974. Ecological and geographical distribution of lizards in Pernambuco, northeastern Brazil (Sauria). *Pap. Avul. Zool., S. Paulo* 28:61–90.

Vanzolini, P. E. 1976. On the lizards of a Cerrado-caatinga contact, evolutionary and zoogeographical implications (Sauria). *Pap. Avul. Zool., S. Paulo* 29:111–119.

Vanzolini, P. E. 1988. Distribution patterns of South American lizards. In P. E. Vanzolini and W. R. Heyer, eds., *Proceedings of a Workshop on*

Neotropical Distribution Patterns, pp. 317–343. Rio de Janeiro: Academia Brasileira de Ciências.

Vitt, L. J. 1991. An introduction to the ecology of Cerrado lizards. *J. Herpetol.* 25:79–90.

Vitt, L. J. 1995. The ecology of tropical lizards in the Caatinga of northeast Brazil. *Occ. Pap. Oklahoma Mus. Nat. Hist.* 1:1–29.

Vitt, L. J. and J. P. Caldwell. 1993. Ecological observations on Cerrado lizards in Rondônia, Brazil. *J. Herpetol.* 27:46–52.

Vitt, L. J. and C. M. Carvalho. 1995. Niche partitioning in a tropical wet season: Lizards in the Lavrado area of northern Brazil. *Copeia* 1995:305–329.

Vitt, L. J. and G. R. Colli. 1994. Geographical ecology of a neotropical lizard: *Ameiva ameiva* (Teiidae) in Brazil. *Can. J. Zool.* 72:1986–2008.

Vitt, L. J. and T. E. Lacher, Jr. 1981. Behaviour, habitat, diet and reproduction of the iguanid lizard *Polychrus acutirostris* in the Caatinga of northeastern Brazil. *Herpetologica* 37:53–63.

Vrcibradic, D., and C. F. D. Rocha. 1998. Reproductive cycle and life-history traits of the viviparous skink *Mabuya frenata* in southeastern Brazil. *Copeia* 1998:612–619.

Warming, E. 1892. *Lagoa Santa. Et bidrag til den biologiske plantegeographi.* Copenhagen: K. danske vidensk Selsk., 6.

Webb, S. D. 1978. A history of savanna vertebrates in the New World: Part II, South America and the Great Interchange. *Annu. Rev. Ecol. Syst.* 9:393–426.

Wiederhecker, H. C., A. C. S. Pinto, and G. R. Colli. 2002. Reproductive ecology of *Tropidurus torquatus* (Squamata: Tropiduridae) in the highly seasonal Cerrado biome of central Brazil. *J. Herpetol.* 36:82–91.

Appendix   Checklist of the Cerrado Amphibians and Reptiles

LISSAMPHIBIA: APODA: Caeciliidae: *Siphonops annulatus, S. paulensis*. Total species of caecilians: 2; total endemics: 0. ANURA: Bufonidae: *Bufo crucifer, B. granulosus, B. guttatus, B. ocellatus,*[E] *B. paracnemis, B. rufus, B. typhonius*; Centrolenidae: *Hyalinobatrachium eurygnathum*; Dendrobatidae: *Colostethus goianus,*[E] *Epipedobates braccatus,*[E,C2] *E. flavopictus,*[C2] *E. pictus*[C2]; Hylidae: *Aplastodiscus perviridis, Corythomantis greeningi, Hyla albopunctata, H. alvarengai,*[E] *H. anataliasiasi,*[E] *H. biobeba,*[E] *H. branneri, H. cipoensis,*[E] *H. circumdata, H. crepitans, H. faber, H. ibitiguara, H. melanargyrea, H. microcephala, H. minuta, H. multifasciata, H. nana, H. nanuzae,*[E] *H. pardalis, H. pinima, H. polytaenia, H. pseudopseudis,*[E] *H. pulchella, H. punctata, H. raniceps, H. rubicundula,*[E] *H. saxicola,*[E] *H. sazimai,*[E] *H. tritaeniata,*[E] *Phasmahyla jandaia,*[E] *Phrynohyas venulosa, Phyllomedusa burmeisteri, P. centralis,*[E] *P. hypochondrialis, P. megacephala, Scinax acuminatus, S. canastrensis,*[E] *S. centralis,*[E] *S. duartei, S. fuscomarginatus, S. fuscovarius, S. luizotavioi, S. machadoi,*[E] *S. maracaya,*[E] *S. nebulosus, S. squalirostris, Trachycephalus nigromaculatus*; Leptodactylidae: *Adenomera bokermanni, A. martinezi, Barycholos savagei,*[E] *Crossodactylus bokermanni, C. trachystomus, Eleutherodactylus dundeei, E. fenestratus, E. juipoca, Hylodes otavioi, Leptodactylus camaquara,*[E] *L. chaquensis, L. cunicularius,*[E] *L. furnarius, L. fuscus, L. jolyi,*[E] *L. labyrinthicus, L. mystaceus, L. mystacinus, L. ocellatus, L. petersii, L. podicipinus, L. pustulatus, L. syphax, L. troglodytes, L. tapiti,*[E] *Odontophrynus americanus, O. cultripes, O. moratoi,*[E] *O. salvatori,*[E] *Physalaemus albonotatus, P. centralis, P. cuvieri, P. deimaticus,*[E] *P. evangelistai,*[E] *P. fuscomaculatus, P. nattereri, Proceratophrys cururu,*[E] *P. goyana,*[E] *Pseudopaludicola boliviana, P. falcipes, P. mineira,*[E] *P. mystacalis, P. saltica, P. ternetzi, Thoropa megatympanum,* Microhylidae: *Chiasmocleis albopunctata, C. centralis,*[E] *C. mehelyi, Dermatonotus muelleri, Elachistocleis bicolor, E. ovalis*; Pseudidae: *Lysapsus caraya, L. limellus, Pseudis paradoxa*; Ranidae: *Rana palmipes*. Total species of anurans: 113; total endemics: 32.

REPTILIA: TESTUDINES: Pelomedusidae: *Podocnemis expansa,*[C2] *Podocnemis unifilis*[C2]; Chelidae: *Chelus fimbriatus, Phrynops geoffroanus, P. gibbus, P. vanderhaegei, Platemys platycephala*; Kinosternidae: *Kinosternon scorpioides*; Testudinidae: *Geochelone carbonaria,*[C2] *G. denticulata.*[C2] Total species of turtles: 10; total endemics: 0. CROCODYLIA: Alligatoridae: *Caiman crocodilus crocodilus,*[C2] *C. crocodilus yacare,*[C2] *C. latirostris,*[C1] *C. niger,*[C1] *Paleosuchus palpebrosus,*[C2] *P. trigonatus.*[C2] Total species of crocodilians: 5; total endemics: 0. SQUAMATA: Amphisbaenidae: *Amphisbaena alba, A. anaemariae,*[E] *A. crisae,*[E] *A. fuliginosa, A. leeseri, A. miringoera,*[E] *A. neglecta,*[E] *A. sanctaeritae,*[E] *A. silvestrii,*[E] *A. talisiae,*[E] *A. vermicularis, Bronia kraoh,*[E] *Cercolophia roberti, C. steindachneri, Leposternon infraorbitale, L. microcephalum.* Total amphisbaenian species: 16; total endemics: 8. Hoplocercidae: *Hoplocercus spinosus*[E]; Iguanidae: *Iguana iguana*[C2]; Polychrotidae: *Anolis chrysolepis brasiliensis, A. meridionalis,*[E] *Enyalius bilineatus, E. brasiliensis, E. catenatus, Polychrus acutirostris*; Tropiduridae: *Stenocercus caducus, Tropidurus etheridgei, T. hispidus, T. itambere,*[E] *T. montanus,*[E] *T. oreadicus, T. gnarani, T. torquatus*; Gekkonidae: *Coleodactylus brachystoma,*[E] *C. meridionalis, Gonatodes humeralis, Gymnodactylus geckoides, Hemidactylus mabouia, Phyllopezus pollicaris*; Teiidae: *Ameiva ameiva, Cnemidophorus ocellifer, Kentropyx calcarata, K. paulensis,*[E] *K. striata, K. vanzoi,*[E] *K. viridistriga, Teius teyou, Tupinambis duseni,*[C2] *T. merianae,*[C2] *T. quadrilineatus,*[E,C2] *T. teguixin*[C2]; Gymnophthalmidae: *Bachia bresslaui,*[E] *B. scolecoides,*[E] *B. cacerensis,*[E] *Cercosaura ocellata, Colobosaura modesta, Micrablepharus atticolus,*[E] *M. maximiliani, Pantodactylus schreibersii,*

Vanzosaura rubricauda; Scincidae: *Mabuya dorsivittata, M. frenata, M. guaporicola, M. nigropunctata (= bistriata);* Anguidae: *Ophiodes striatus.* Total species of lizards: 47; total endemics: 12. Anomalepididae: *Liotyphlops beui, L. ternetzii;* Leptotyphlopidae: *Leptotyphlops albifrons, L. koppesi, L. munoai;* Typhlopidae: *Typhlops brongersmianus;* Aniliidae: *Anilius scytale;* Boiidae: *Boa constrictor,*[C2] *Corallus caninus,*[C2] *C. hortulanus,*[C2] *Epicrates cenchria,*[C2] *Eunectes murinus*[C2]; Colubridae: *Apostolepis assimilis, A. dimidiata, A. flavotorquata,*[E] *A. goiasensis,*[E] *A. lineata,*[E] *A. vittata*[E]*, Atractus pantostictus, A. reticulatus, A. taeniatus, Boiruna maculata, Chironius bicarinatus, C. exoletus, C. flavolineatus, C. laurenti, C. quadricarinatus, Clelia bicolor, C. plumbea, C. quimi, C. rustica, Dipsas indica, Drymarchon corais, Drymoluber brazili, Echinanthera occipitalis, Erythrolamprus aesculapii, Gomesophis brasiliensis, Helicops angulatus, H. carinicaudus, H. gomesi, H. leopardinus, H. modestus, Hydrodynastes bicinctus, H. gigas,*[C2] *Imantodes cenchoa, Leptodeira annulata, Leptophis ahaetulla, Liophis almadensis, L. cobellus, L. dilepis, L. longiventris, L. maryellenae, L. meridionalis, L. miliaris, L. paucidens, L. poecilogyrus, L. reginae, L. typhlus, Lystrophis dorbignyi, L. histricus, L. mattogrossensis, L. nattereri, Mastigodryas bifossatus, M. boddaerti, Oxybelis aeneus, Oxyrhopus guibei, O. petola, O. rhombifer, O. trigeminus, Phalotris concolor,*[E] *P. lativittatus,*[E] *P. mertensi, P. multipunctatus,*[E] *P. nasutus,*[E] *P. tricolor, Philodryas aestivus, P. livida,*[E] *P. mattogrossensis, P. nattereri, P. olfersii, P. psammophideus, P. patagoniensis, Phimophis guerini, Pseudablabes agassizii, Pseudoboa neuwiedii, P. nigra, Pseudoeryx plicatilis, Psomophis genimaculatus, P. joberti, Rhachidelus brazili, Sibynomorphus mikanii, S. turgidus, Simophis rhinostoma, Spilotes pullatus, Tantilla melanocephala, Thamnodynastes rutilus, T. strigilis, Waglerophis merremi, Xenopholis undulatus;* Elapidae: *Micrurus brasiliensis*[E]*, M. frontalis, M. lemniscatus;* Viperidae: *Bothrops alternatus, B. itapetiningae,*[E] *B. moojeni, B. neuwiedi, Crotalus durissus.* Total species of snakes: 107; total endemics: 11.

Note: We included in the list all species known to occur within the borders of the Cerrado Biome as defined by the maps prepared during the workshop "Ações Prioritárias para a Conservação da Biodiversidade do Cerrado e Pantanal," held in Brasília, Distrito Federal in 1999 (see chapter 18). [C1] = listed in Appendix I of CITES, [C2] = listed in Appendix II of CITES, [E] = endemic.

13

The Avifauna: Ecology, Biogeography, and Behavior

Regina H. F. Macedo

ONE OF THE EARLIEST STUDIES OF THE BIRDS IN THE CERRADO region was that of Sick (1955), whose records date to an expedition to Rio das Mortes in central Brazil in 1944. Subsequently, Sick's (1965, 1966) field work in Goiás, Mato Grosso and Pará, at the time almost untouched by civilization, led him to conclude that it was difficult to describe a "typical" avifauna for the cerrado region, and that it was relatively poor (in this chapter "cerrado region" or "Brazilian cerrados" refer to the biome itself, while "cerrado" refers to the *sensu stricto* vegetation type. See chapter 6 for detailed descriptions of cerrado physiognomies). He estimated a diversity of only 200 species for the biome, of which approximately 11% was endemic. Since then, other studies gradually increased the number of species to 837 (see table 13.1), and also identified typical species for the different cerrado region formations. Most of these species (759, 90.7%) are known or assumed to breed in the region (see appendix). Considering that Brazil has 1,590 resident bird species, the conservation of the Brazilian cerrados should be considered a top priority (chapter 18), since the region harbors approximately 48% of the country's avifauna. Of the region's species, approximately 30 are considered endemic. The habitat transformations currently underway in the Brazilian cerrados (chapter 5) will undoubtedly limit the breeding ranges of most of these species, while reducing or completely eliminating endemic bird populations.

Avian community structure results from several complex factors, all interacting at various levels. I have grouped information on cerrado region birds into sections that reflect some of these elements: community

Table 13.1   Records of Bird Species Richness
in the Cerrado Region

Reference	Number of species	Increase over previous total (%)
Sick 1955	245	
Fry 1970	263	18 (7)
Müller 1973	357	94 (36)
Negret et al. 1984	429	72 (20)
Vuilleumier 1988[a]	454	25 (5.8)
Silva 1995	837	383 (84.4)

[a]Includes birds from cerrado, *chaco*, and *caatinga* regions (see chapter 6).

composition and biogeography, vegetation structure, seasonality effects, foraging and food resources, and impacts of fire. Additionally, I have focused upon the breeding patterns of a few species, including some socially breeding birds. In conclusion, I briefly discuss promising avenues for future work.

COMMUNITY COMPOSITION AND BIOGEOGRAPHY

The most species-rich families in the breeding avifauna are the Tyrannidae (111 species), Emberizidae (87 species), Formicariidae (58 species: in recent treatments subdivided into two families, with most of the cerrado region species included in the Thamnophilidae), Furnariidae (41 species), Trochilidae (36 species), and Psittacidae (33 species) (see appendix). Levels of endemicity range from a low estimate of 3.8% (Silva 1995) to higher estimates of 12% (Müller 1973) and 11% (Sick 1965; see table 13.2).

Naturally, some species exhibit an abundance of individuals, while most species are represented by fewer individuals. In a study by Negret (1983) the 10 most abundant species in the Brazilian cerrados were: *Aratinga aurea, Streptoprocne zonaris, Colibri serrirostris, Colaptes campestris, Elaenia flavogaster, Suiriri suiriri, Camptostoma obsoletum, Cyanocorax cristatellus, Neothraupis fasciata,* and *Ammodramus humeralis.* Some species are generalists and can be found in most habitats within the cerrado region, for example: *Colibri serrirostris, Milvago chimachima, Sporophila nigricollis, Piaya cayana, Nyctidromus albicollis, Galbula ruficauda, Chelidoptera tenebrosa, Turdus amaurochalinus, Basileuterus flaveolus,* and *Ramphocelus carbo* (Fry 1970; Negret 1983).

Table 13.2   Endemic Birds of the Cerrado Region
by Habitat and Range

Family	Species	Habitat[a]	Range
Tinamidae	*Nothura minor*	Cerrado	W
	Taoniscus nanus	Cerrado	W
Cracidae	*Penelope ochrogaster*	Forest	W
Columbidae	*Columbina cyanopis*	Cerrado	W
Psittacidae	*Pyrrhura pfrimeri*	Forest	R
	Amazona xanthops	Cerrado	W
Caprimulgidae	*Caprimulgus candicans*	Cerrado	W
Trochilidae	*Augastes scutatus*	Cerrado	R
Furnariidae	*Geobates poecilopterus*	Cerrado	W
	Philydor dimidiatus	Forest	W
	Synallaxis simoni	Forest	R
	Asthenes luizae	Cerrado	R
	Automolus rectirostris	Forest	W
Formicariidae	*Herpsilochmus longirostris*	Forest	W
	Cercomacra ferdinandi	Forest	R
Rhinocryptidae	*Melanopareia torquata*	Cerrado	W
	Scytalopus novacapitalis	Forest	R
Tyrannidae	*Phyllomyias reiseri*	Forest	W
	Polystictus superciliaris	Cerrado	R
	Knipolegus franciscanus	Forest	R
Pipridae	*Antilophia galeata*	Forest	W
Emberizidae	*Poospiza cinerea*	Cerrado	W
	Embernagra longicauda	Cerrado	R
	Sporophila melanops	Cerrado	R
	Charitospiza eucosma	Cerrado	W
	Paroaria baeri	Forest/Cerrado	R
	Saltator atricollis	Cerrado	W
	Porphyrospiza caerulescens	Cerrado	W
	Conothraupis mesoleuca	Forest	R
	Tachyphonus nattereri	Forest	R
Parulidae	*Basileuterus leucophrys*	Forest	W
Corvidae	*Cyanocorax cristatellus*	Cerrado	W

Sources: Data from Sick (1997), Silva (1995, 1997), and Cavalcanti (1999).
Note: W = widespread; R = restricted range.
[a]"Cerrado" refers to the following open physiognomies: cerrado *sensu stricto*, *campo cerrado*, *campo sujo*, *campo limpo*, *vereda*, and *campo rupestre*. "Forest" includes gallery forest, the forest-like *cerradão*, and the transitional zone between these formations. See chapter 6 for description of vegetation physiognomies.

However, many species (86 in Negret's 1983 study) are specialists and occur in only one habitat. The greatest number of these exclusive species are restricted to gallery forests, followed by the cerrado. Negret (1983) considered that of the 118 species he found in the cerrado, 102 were tree-dwelling birds, and 13 were essentially terrestrial. The latter include members of the family Tinamidae and also the very conspicuous *Cariama cristata*. In the *veredas* (chapter 6), the *Mauritia* palm groves are dominated by species such as *Thraupis palmarum*, *Reinarda squamatta* and *Phacellodomus ruber*. The Psittacidae are well represented: in the *veredas* of the region lives the world's largest macaw, *Anodorhynchus hyacinthinus*, as well as *Ara ararauna*, with a wider distribution (Sick 1965).

Most studies concerning the biogeography of the Brazilian cerrados avifauna were published in the last five years by Silva (1995a, b 1996, 1997) and thus constitute the basis for any discussion on the subject (see Nores 1992, 1994, for alternative discussions regarding biogeography of South American birds). Species diversity results from two processes: speciation within a region and biotic interchange between regions (Ricklefs 1990; Ricklefs and Schluter 1993). Species production in the Brazilian cerrados seems to have been low during the last two million years, and most of the endemic species are very old, dating from the middle and late Tertiary (Sick 1966; Silva 1995b). As for biotic interchanges, biogeographic analyses suggest that many species were received from neighboring biomes during the Pleistocene-Holocene cyclic climatic fluctuations (Silva 1995b) and were able to maintain viable populations because their habitats did not disappear entirely through successive changes in the vegetation (Silva 1997; see also chapters 3, 6).

Two of the three ancient savanna corridors that purportedly linked the central cerrado region with savannas of northern South America are congruent with present-day ranges of savanna birds. This led Silva (1995) to determine that the distribution patterns of these savanna-adapted birds do not conform to one of the most important assumptions of the refuge theory positing the existence of a savanna corridor right through central Amazonia during the dry and cold periods of the Quaternary (see Haffer 1969, 1974). The patterns suggest that connections between the avifaunas of the northern Amazonia savannas and those of central Brazil probably occurred along the eastern borders, and not through the center of the Amazon rainforest.

The headwaters of some major South American rivers (e.g., São Francisco, Tocantins, Araguaia, Paraguay) are located in central Brazil. Thus, the cerrado region is criss-crossed by gallery forests that run along rivers connecting the central Brazilian savanna with the Amazon or the Atlantic forests. Several researchers have suggested that these gallery forests may constitute mesic corridors through which forest-dependent organisms could have colonized the central savanna (chapters 6, 11, 14). Silva (1996) evaluated the influence of altitude and distance from the source areas on the distribution of birds within the Brazilian cerrados. Included in the analysis were 278 species, of which 200 were considered as Amazonian and 78 as Atlantic forest elements. Most Amazonian elements (86%) do not extend more than 250 km into the Brazilian cerrados region, and no species extends more than 750 km. In contrast, only 50% of Atlantic forest birds are restricted to less than 250 km, and 14% extend more than 1,000 km into the cerrado region. Silva (1996) also found that the altitudes of Amazonian elements are significantly lower than those of Atlantic forest elements. Thus, both distance from their centers of distribution as well as altitude appear to determine the distribution of birds in the gallery forest system of central Brazil.

VEGETATION STRUCTURE

The association between vegetation structure and number of species has been studied intensively (Orians 1969), especially since the significant contribution by MacArthur and MacArthur (1961), who postulated that bird species diversity could be predicted by the height profile of foliage density. Sick (1966) suggests that a large proportion of the typical species inhabiting the cerrado region are forest birds, living in trees, and cannot be considered savanna birds. In fact, approximately 70% of the breeding avifauna of the Brazilian cerrados is composed of species that are partially or totally dependent upon forests (gallery forests or tropical dry forests), a habitat that covers less than 15% of the region (Silva 1995a).

In Negret's (1983) study, 215 species were censused in all major vegetation types of the cerrados. Of these, 168 (64%) were observed within gallery forests; the poorest habitat was that of *campo limpo*, with only 31 species and, predictably, the simplest structural complexity (chapter 6). The greatest abundance of individuals, however, was registered for the cerrado *sensu stricto*. In a previous, smaller-scale study in the northeast of the state of Mato Grosso, only 78 species were recorded for gallery forests, of which 33 occurred exclusively in that vegetation (Fry 1970). Parrots were poorly represented, while kingfishers, toucans, woodcreepers, and manakins were all abundant in gallery forests, which boasted, however, many fewer flycatchers than the cerrado. Fry (1970) also recorded a higher species total for the cerrado than for gallery forest, an incongruous finding relative to other studies, probably due to the short period of the study and the difficulty of sampling middle storey and canopy birds in gallery forests.

A comparison of the more altered open physiognomies of the cerrado region (chapter 6) showed that, as expected, there is greater richness, diversity, and abundance of birds in the areas containing more shrubs and trees (Tubelius 1997). Canopy-dwelling species appear more vulnerable to changes in their habitat. The simplification of the habitat from cerrado *sensu stricto* to *campo sujo* and eventually *campo limpo* apparently favors granivorous species such as *Volatinia jacarina*, *Sicalis citrina*, *Nothura maculosa*, and *Myospiza humeralis*.

Some birds occupy different strata of gallery forests, thereby increasing species diversity. For example, three sympatric and partially syntopic *Basileuterus* warblers inhabit gallery forests of central Brazil (Marini and Cavalcanti 1993): *B. leucophrys*, *B. hypoleucus*, and *B. flaveolus*. The two species (*B. flaveolus* and *B. leucophrys*) that forage below 3 m height are not syntopic. *Basileuterus hypoleucus*, however, which occupies strata above 3 m, occurs syntopically with each of the other two species.

A structural component of landscapes that is often overlooked, and which may be important in the organization of grassland bird communities, is the existence of forest edges. In the cerrado region, gallery forest edges may provide crucial resources that decline in the more open habitats during the dry season. Some tyrannid flycatchers, which breed during the rainy season, migrate elsewhere during the dry season; however, others (e.g., *Tyrannus savana*) disappear during the dry season from the cerrado but may be found feeding on fruits along the edges of gallery forests (Cavalcanti 1992). This pattern of gallery forest edge use is evident for other families, particularly for hummingbirds. The reverse situation occurs during the rainy season, when forest birds (e.g., *Turdus* spp. and *Saltator similis*) forage in adjacent cerrado areas, particularly rich in alates of termites and ants (Cavalcanti 1992). Thus, in addition to harboring its own avian community, gallery forests provide cover and resources for cerrado birds in periods of stress (e.g., grassland fires, dry season, cover from predators).

SEASONALITY

The climate of the cerrado region is highly seasonal, with marked dry (May–August) and rainy (September–March) seasons. This seasonality is evident in the differences exhibited by the vegetation as well as by insect, fruit, and flower abundance, all of which profoundly affect bird communities. The rigor and periodicity of moisture gradients influence all aspects of community composition and organization in the Brazilian cerrados, and condition such phenomena as the timing of breeding, flock occurrence and composition, migration, shifts in foraging behavior, and competition for resources.

Studies in other tropical areas have shown that spatially and tempo-
rally variable resources are important factors that induce seasonal bird
movements as well as changes in bird abundances and in the composition
of local avian communities. For Costa Rican birds, diet and habitat are
strongly related to seasonal movements (Levey 1988; Levey and Stiles
1992; Stiles 1983), and the patchy and ephemeral nature of resources such
as fruits affects the structure of bird communities (Blake and Loiselle
1991). In the humid tropical forest of Barro Colorado Island, small-
seeded fruits ripen rather evenly throughout the year, whereas large-
seeded fruits tend to be seasonal, affecting seed dispersal patterns (Smythe
1970). In the cerrado region, it appears that the seasonal abundance of
insects contributes significantly to the variation of richness and abundance of birds throughout the year. The maximum peak of insect abundance, in the rainy season, coincides with the arrival of migrants and also
seems to affect patterns of movement within the region. The most abundant and well-known migratory species is *Tyrannus savanna*. This species'
range encompasses Mexico, Central America, and all of South America
except Chile (Sick 1997). It arrives in the cerrado region in August and
September, reproduces in the region in October and November, and leaves
in January and February, probably returning to the Amazon region
(Negret and Negret 1981).

Likewise, flowering and fruiting patterns affect the movements of
birds in the region. Ornithophilous plants flower almost continuously
within gallery forests, providing a source of nectar during the dry season,
while in *campo sujo* flowering happens only during the rainy season
(Oliveira 1998). Thus, a local movement of hummingbirds into gallery
forests is probably a general phenomenon during the dry season. In some
cases, specific plants flower only during very circumscribed periods and
provide much-needed resources. Such plants may be regarded as pivotal
species in their habitats but have been discussed only occasionally in the
literature. The leguminous tree *Bowdichia virgilioides* occurs in cerrado
habitat in various areas of Brazil, and flowers only at the end of the dry
season (Rojas and Ribon 1997). A small but diverse group of birds dependent upon this tree includes: six species of hummingbirds (*Colibri serrirostris, Chlorostilbon aureoventris, Eupetomena macroura, Calliphlox
amethystina, Amazilia lactea,* and perhaps *A. fimbriata*); two Coerebidae
(*Dacnis cayana* and *Coereba flaveola*); as well as some species that eat the
flowers (*Aratinga aurea,* Psittacidae; *Tangara cayana, Thraupis sayaca,*
Thraupidae; and *D. cayana*). A similar pivotal species phenomenon
was described for the rain forest tree *Casearia corymbosa* in the La Selva
Biological Station in Costa Rica: fruiting was restricted to an annual

period of fruit scarcity, thus supporting three obligate frugivores in the area (Howe 1977).

Another seasonal effect is that of mixed-species flock formation. Although these can be found year-round, their highest occurrence is early in the dry season, the non-reproductive period, with a sharp decline during the breeding season (Davis 1946; Silva 1980; Alves and Cavalcanti 1996). Not only the frequency of flocks, but also their composition, is affected by the seasonality in the region. Because of migrants that arrive during the rains, species such as *Elaenia chiriquensis*, *Myiarchus swainsoni*, and *Tyrannus savanna* participate in flocks only during the period from August to March (Alves and Cavalcanti 1996).

Breeding periods worldwide are associated with food availability for the young, which means that in temperate regions breeding occurs during the spring and summer. In the cerrados region, as in other savannas and areas with sharply defined rainy seasons, breeding coincides with peak vegetation growth and insect abundance. Thus, insectivores should breed somewhat earlier than granivorous species or those that depend upon tall grasses for nesting sites (Lack 1968; Young 1994). For the cerrado region, there have been only a few in-depth studies recording specific reproductive periods, courtship displays, nestling growth patterns, and reproductive success rates. The manakin *Antilophia galeata*, an inhabitant of gallery forests, has its peak sexual activity from August to November, coinciding with the beginning to middle of the rainy season, although males maintain their territory all year (Marini 1992b). For *Neothraupis fasciata*, nesting occurs from September to January or later (Alves and Cavalcanti 1990). The macaw *Ara ararauna*, a cavity nester which uses *Mauritia* palm trees and initiates courtship behavior in May, nests between August and December (Bianchi 1998). A few species, such as *Rhynchotus rufescens*, *Nothura maculosa*, and *Columbina talpacoti*, seem to breed year-round, although peaking during the rains (Soares 1983; Couto 1985; Setubal 1991). Most species appear to have their breeding periods restricted to the rainy season, and descriptive data (but few in-depth studies) are available for most (Sick 1997).

FORAGING AND FOOD RESOURCES

Foraging studies for tropical species have been conducted mostly for forest birds, where foraging specialization may promote coexistence of many species, leading to complex communities (Orians 1969). Less is known about the diet and foraging patterns of savanna birds. Food may be superabundant during breeding periods in some habitats and systems but limited during other periods, thus reducing the reproductive success of some individuals and affecting survival of both adults and nestlings (Martin 1987). In nearly all diet categories (insectivory, granivory, frugivory, and nectarivory), the cerrado region avifauna offers interesting possibilities for studies. Detailed descriptions of the diet of most species are not available, but some studies have described foraging specializations (see Alves 1991; Marini 1992a).

In the cerrado region communities there is an elevated number of insectivores (113 species of flycatchers alone), and some studies have examined how food availability allows the coexistence of so many species with broadly overlapping diets. For example, Negret (1978) considered eight flycatchers at Distrito Federal (central Brazil) to determine their ecological niche and trophic relations: *Xolmis cinerea, Xolmis velata, Knipolegus lophotes, Tyrannus savanna, Megarhynchus pitangua,Pitangus sulphuratus, Suiriri suiriri,* and *Leptopogon amaurocephalus.* These species vary in their habitats, which range from urban areas to gallery forest. The study clarified the relation between bill morphology and prey items. *Megarhynchus pitangua,* which has the widest bill, for instance, foraged upon larger insects. This characteristic also corresponded to the "flycatching" and "leaf-snatching" foraging modes. The shortest bills corresponded to the "leaf-gleaning" and "ground-feeding" behaviors. No tendency was found for predation of a particular group of insects. Instead, the study suggests an association between the feeding behavior of each species and the habitat occupied by its prey.

Depending on the habitat, from 50% to 90% of tropical shrubs and trees have their seeds dispersed by vertebrates. However, knowledge of the diet of tropical birds is still limited, and it is often difficult to determine which species are frugivorous (Herrera 1981; Fleming et al. 1987). The morphology of birds and plants affects which species are essentially frugivorous (Moermond and Denslow 1985), with additional important characteristics including spatio-temporal distribution and nutritional characteristics of the fruits. In the Brazilian cerrados, fruit-consuming birds are not primarily frugivorous, using fruits instead to complement their diets.

Roughly about 50% to 60% of plants in the cerrado region are dispersed by animals (Gottsberger and Silberbauer-Gottsberger 1983; Pinheiro 1999). In the Brazilian cerrados, birds constitute the greatest proportion of frugivorous animals, with 75 species that can be classified as frugivorous or partially frugivorous (Bagno 1998). A surprisingly large number of cerrado plants are potentially bird-dispersed, large even when compared to wet forests. For example, in the La Selva Biological Station,

Costa Rica, at least 137 plant species were found to be bird-dispersed, although the study only included plants that had fruits below 10 m (i.e., plants in the mid-canopy and canopy were not included; Loiselle and Blake 1991). Thus, this most likely is a low estimate of the number of plants dispersed by birds in that area. In the cerrado region, over 179 plants may be bird-dispersed. Important birds for seed dispersal include *Rhea americana*, *Neothraupis fasciata*, *Rhynchotus rufescens*, *Tinamus solitarius*, some species of *Nothura*, pigeons and doves (e.g., *Columbina cyanopis* and *Uropelia campestris*), parakeets, parrots, macaws, toucans, blackbirds of the family Icteridae, and several Tyrannidae and Furnariidae (Gottsberger and Silberbauer-Gottsberger 1983; Alves 1991; Paes 1993).

The frugivore guild associated with the *Mauritia flexuosa* palm is composed not only of birds from various families, but also of birds exhibiting a wide range of bill gape widths. Foraging behaviors also vary greatly: there are terrestrial birds that pick up fruits dropped at the base of the tree by canopy foragers; birds that consume the fruit on the tree; those that consume the fruit only partially; and birds that take it away in flight (Villalobos 1994). Over 90% of the fruit consumption is by three psittacids, *Ara manilata*, *A. ararauna*, and *Amazona aestiva*. They usually eat only part of the fruit pulp and then drop it at the base of the tree. Once the macaws drop the fruit, a number of mammals feed upon it, including some rodents, tapirs, maned wolves, and opossums. Other birds that also feed on the fruits from this palm include *Schistochlamys melanopis* and *Thraupis palmarum* (Thraupidae), *Gnorimopsar chopi* (Icteridae), *Polyborus plancus* (Falconidae), *Cyanocorax cristatellus* (Corvidae), and *Porzana albicollis* (Rallidae).

Nectarivory in the tropics has also attracted much interest. There are more than 300 species of hummingbirds (Trochilidae) distributed in the New World, with peak diversity occurring in the tropics. The combination of ecological constraints, usually in the form of interspecific competition and mutualistic coevolution with flowers, may be responsible for the diversity of morphologies and species of hummingbirds (Brown and Bowers 1985).

Stiles (1985) suggests the existence of a relation between the number of plant species with adequate resources and the total number of hummingbird species that can be supported in an area. He cites as an example the existence of about 20 hummingbird species and a community of 50 plants they pollinate in the La Selva Biological Station (Costa Rica). In contrast, in a southeastern site in Puerto Rico there are only three hummingbird species and 13 plants they pollinate (Kodric-Brown et al. 1984). These observations would lead one to expect but a small hummingbird community in the cerrado region, since the number of plants that present the ornithophily syndrome (tubular corollas, conspicuous colors, etc.) is small (Silberbauer-Gottsberger and Gottsberger 1988; see chapter 17). However, that is not the case, as the cerrado region hummingbird community is extremely diverse, with at least 36 species listed (most common include: *Amazilia fimbriata*, *Chlorostilbon aureoventris*, *Colibri serrirostris*, *Eupetomena macroura*, *Phaethornis pretrei*, and *Thalurania furcata*; Negret et al. 1984; Oliveira 1998). It is important to keep in mind that the elevated number of species listed to date is for the whole Cerrado Biome, an area of approximately 2 million km^2, while hummingbirds recorded at La Selva are within a 1,000 ha area. Nonetheless, the elevated number of hummingbirds in the region may reflect a generalistic and opportunistic foraging behavior: namely, the use of flowers that do not have the floral attributes commonly associated with hummingbird pollination (Oliveira 1998). Several of the flowers visited by hummingbirds (e.g., species of *Inga*, *Vochysia*, *Qualea*, *Bauhinia* and *Caryocar*) are regularly pollinated by other animals, ranging from moths and bees to bats (Oliveira 1998; chapter 17). Moreover, an experimental study with artificial feeders suggests that cerrado region hummingbirds may not have fixed preferences for certain floral characteristics, but instead a large adaptive capacity to exploit resources that are regionally advantageous (Carvalho and Macedo in prep.).

IMPACT OF FIRE

Fire may destroy or damage individuals, affect growth forms or reproduction, or in some way alter the environment, providing new opportunities for some organisms (Frost 1984; chapters 4, 9). The impact of fire upon bird communities has been studied primarily in temperate regions; fewer studies have occurred in the tropics, and these mostly in Africa (Frost 1984; Trollope 1984) and Australia (Luke and McArthur 1977).

In the cerrado region, both natural and intentional fires occur more frequently toward the end of the dry season. Cavalcanti and Alves (1997) conducted a study in a cerrado area to examine the impact of fire on the avian community and to test some concepts proposed by Catling and Newsome (1981): (1) the existence of fire specialists; (2) a prevalence of ecological generalists; and (3) an expected tendency toward low species diversity in areas subjected to frequent burning. They sampled the avifauna before and after burning occurred, and recorded parameters involving population changes, site fidelity of marked individuals, foraging behavior, and the identification of specialists. Their results were partly consistent with Catling and Newsome's (1981) arguments. They identified a fire specialist, *Charitospiza eucosma*, which was seen only sporadically before the fire but was captured freely after the fire. However, the authors also point out that the predictions regarding a fire-adapted avifauna do not exclude other explanations for the low species diversity. A study comparing burned and unburned open savanna sites during a five-year period showed similar species' numbers and abundances, which does not support the suggestion of fire-adapted avifaunas (Figueiredo 1991).

Although particular traits characterizing a fire-adapted avifauna may not be obvious, birds exhibit some typical behaviors during fires. For example, small birds may seek refuge in protected sites (termite mounds) or simply avoid the area by moving to unaffected areas. In various instances birds are attracted to fires to seek out prey trying to escape. Migrant species are often associated with fires that provide them with easy prey. Raptors, the most frequent fire-followers, include: *Heterospizias meridionalis*, *Polyborus plancus*, *Buteogallus urubitinga*, *Ictinia plumbea*, and *Cathartes aura* (Sick 1983). Among terrestrial birds, fires attract *Cariama cristata* and *Rhea americana*, and often other birds such as *Streptoprocne zonaris* and *Tyrannus melancholicus* (Sick 1965). The incidence of fire may also be associated with a populational increase in some species such as *Rhynchotus rufescens* and *Nothura maculosa* (Setubal 1991).

Most studies concerning the impact of fire upon the cerrado region avifauna have been restricted to birds of more open cerrado physiognomies (chapter 6). Community changes after a fire in a gallery forest have been described only in Marini and Cavalcanti (1996). They found that similarity coefficients for the avian communities differed little before and after the fire, considering the total community. However, considering groups of species related to the types of habitat most frequently used, they found that species more dependent upon gallery forests were more severely affected, with community structure differing between periods. The authors speculate that possibly cerrado species have evolved behavioral and perhaps physiological adaptations for fire (e.g., feeding on arthropods after the fire), whereas forest birds may not have the same responses. Therefore, fires may decrease bird populations in gallery forest while having very little impact on cerrado bird community structure.

BREEDING PATTERNS

Very few tropical birds have been studied in detail and over a long-term period; consequently, a vast number of questions concerning their reproduction remain unanswered. Additionally, because of the contrast between temperate and tropical regions in temperature extremes, availability of resources, predator pressure and other crucial factors, we can expect to find important differences in birds' breeding characteristics, such as the role of territories, investment in parental care, male strategies through singing and displays, and the evolution of social systems, to name a few. The literature is prolific in speculative concepts, sometimes based on studies of temperate birds and occasionally resulting from analyses of general data from a limited number of tropical birds. For instance, although Ricklefs (1976) advanced several hypotheses that might explain the slower growth rate of nestlings of tropical species despite higher nest predation rates, these remain largely unverified.

In this section I have highlighted some facets of the reproductive biology of a few species, ranging from very common species, adapted to disturbed habitats, to those requiring specific habitats. While a few studies exist concerning behavior and breeding biology of a limited number of species in the cerrado region, in-depth studies over long-term periods for even the most common birds are virtually nonexistent.

Columbina talpacoti, for example, a widely distributed and common bird, has been studied mostly in disturbed areas, although it also occurs in urban areas and undisturbed cerrado. In the region it reproduces all 12 months of the year (Couto 1985; Cintra and Cavalcanti 1997), peaking during the rainy period (Couto 1985). The ability to maintain reproduction even during the dry season may be due to the capacity of producing "pigeon milk" to nurture the young, as well as to the plasticity allowing occupation of disturbed areas. The main food source that probably allowed year-round reproduction in the study population was the natural grain from the surrounding cerrado as well as from neighboring farms, which included peanuts, sunflowers, wheat, rice, beans, corn, etc. Nests were found mostly in coffee bushes, and the state of the plants was important in determining reproductive activity: when the foliage decreased, so did the number of nests.

Another very common, though essentially unstudied bird in disturbed cerrado *sensu stricto* and *campo sujo* areas is *Volatinia jacarina*. This species is widely distributed, and its range includes areas other than the cerrado region. However, its reproductive behavior has been virtually ignored, although it presents interesting questions relevant to the understanding of the evolution of mating systems in general. In this species, males form loose aggregations and execute displays, in the form of vertical leaps from perches, that resemble traditional leks. Almeida and Macedo (2001) found that the number of displaying males declined throughout the season, although the intensity in displays showed no variation. Focal males observed had significantly different display rates and also defended territories of different sizes. The size of territories, ranging from 13.0 m² to 72.5 m², was not associated with the average display rate of their owners. Additionally, no association was found between the display of males and the vegetation structure of territories. Observations of parental care indicate that nests are commonly built within male territories, and that both sexes feed chicks. These results suggest that the mating system of this species, despite the aggregations of displaying males, does not fit the traditional lek system, as had been previously proposed in Murray's (1982) and Webber's (1985) brief observations. However, the underlying reasons for the aggregations of males and their territories are still obscure, as well as mating strategies for males and females.

The availability of nesting sites and specificity of their characteristics are important factors regulating breeding in many birds. Cavity nesting is common in the tropics, as it allows a certain degree of freedom from weather extremes and predation. However, natural cavities may constitute a limited resource. The macaw *Ara ararauna*, an obligatory cavity-nesting species distributed from southern Central America to the central region of Brazil (Sick 1997), is an obligate cavity-nesting species, with very specific habitat requirements. The breeding season for this species occurs between August and December, although courtship is initiated in May (Bianchi 1998). In the Brazilian cerrados these birds nest within cavities of the buriti-palm *Mauritia flexuosa* in the *vereda* formations (chapter 6). They lay from two to four eggs that hatch asynchronously, and nestlings develop during approximately 78 days. The availability of cavities found during the study was low (about 5% within a palm tree field), and their occupancy much reduced, ranging from 10–15%, that is, less than 1% of palms per field. In disturbed areas the availability of cavities was higher (43.5%), but the occupancy was even lower (5.1%). Analyses suggested that the important parameters that determine cavity choice are: (1) total trunk height; (2) cavity height; and (3) cavity depth.

Antilophia galeata is an atypical manakin of the dry and flooded gallery forests of central Brazil, in contrast to most others of its family, which inhabit the Amazonian and Atlantic forests (Marini 1992a, b; Sick 1997). Reproduction coincides with the rainy season. It resembles other manakins in that it is highly dichromatic and essentially frugivorous, but differs in that it establishes long term pairbonds, whereas other known members of the Pipridae are lekking species and polygynous (Marini 1992b). Nest building and nestling care appear to be performed only by females. Its nesting biology, in general, conforms to what is known about other manakins, including details concerning nest architecture, male gonadal development, and molting patterns (Marini 1992b).

SOCIALLY BREEDING SPECIES

Patterns of sociality have long interested biologists, and the number of studies of social animals has been increasing steadily (Brown 1987). Because the most common breeding pattern in temperate birds is that of pairs in territories, it was not until Davis (1940a,b, 1941) and Skutch (1959) described year-round social groups of birds in tropical America that social organization in birds became a topic of interest. These birds were dubbed cooperative breeders, and can be subdivided loosely into species where: (1) only a pair reproduces, but with the aid of several nonparental helpers; and (2) several co-breeders share breeding opportunities within the group, using one or several nests within the same territory (Brown 1987; Stacey and Koenig 1990). Cooperative breeding is relatively rare, and is known to occur in only about 220 of the roughly 9,000-plus species of birds (Brown 1987). However, because the geographic distribution of cooperative breeders is biased toward lower latitudes, reaching its greatest abundance in the neotropics (Brown 1987), it is likely that many more such species will be described in the future.

The *Guira guira* (Cuculidae) system has been under investigation for approximately 10 years, revealing a complex and intriguing social system briefly described here (Macedo 1992, 1994; Quinn et al. 1994; Macedo and Bianchi 1997a,b; Melo and Macedo 1997; Macedo and Melo 1999). Guira cuckoos breed primarily during the period from August to March, with groups renesting as many as five times. Groups average approximately six individuals, but membership may reach up to 15 birds occupying a single territory and using the same nest. Communal clutch size ranges from a couple of eggs to as many as 26 eggs laid by the various females. One of the most intriguing aspects of breeding in this species is the practice of egg ejection by group members, with nests averaging a loss of about four eggs. Although an average of 10 eggs are laid per nest, due to the ejection of eggs only about five survive to hatching. The brood that hatches is then further reduced through infanticide practiced by group members. The genetic relations among group members and chicks has been studied in a preliminary way, indicating that: (1) the mating system is polygynous as well as polyandrous; (2) several individuals contribute to the communal brood; (3) reproductive monopoly by a single pair does not occur; and (4) some individuals may be excluded from a breeding bout.

The social system and the helping behavior of the tanager *Neothraupis fasciata* have been described in Alves and Cavalcanti (1990) and Alves (1990). This species forms groups of up to six members, which actively participate in mixed-species flocks. Nesting occurs from September to at least January. The most common social structure recorded was that of a pair with nestlings and young from former nesting bouts. Groups were generally stable in composition and remained so between years. Pairs are socially monogamous and, while males helped care for nestlings, only the females were seen building nests. Average clutch size is 2.8 eggs, but ranges from two to five eggs, and incubation lasts for about 15 to 17 days. Nests are frequently parasitized by *Molothrus bonariensis*. Of the seven nests monitored in these studies, male and female helpers occurred in three, and assisted through feeding of nestlings, acting as sentinels and in territorial defense.

Although the study of these cooperative breeders provides some detailed information, additional research is needed to answer many remaining questions. For example, for *Guira guira*, the genetic identity of infanticidal adults, hierarchical organization within groups, and the distribution of reproductive opportunities among group members are promising lines of investigation. There are also studies comparing this species with the other crotophagine in the region, *Crotophaga ani*. Several research questions remain for *Neothraupis fasciata* as well; it would be of interest not only to determine the genetic identity of helpers, but also to quantify their help and compare the success rates of groups with and without auxiliaries, as well as dispersal patterns of the young.

RESEARCH POSSIBILITIES

In presenting a general overview of the information available for the cerrado region avifauna, I hope to have provided a catalyst for future research. Although descriptive work has provided meaningful background data, a thorough understanding of the Brazilian cerrados bird community is mostly lacking, and there are extensive opportunities for research in most areas. At the community level, bird abundance and distribution information remain scanty for the vast areas of uncensused cerrados of central and northern Brazil. Additionally, there is a serious gap in our knowledge of gallery forest birds, because they are difficult to observe and capture. Studies of habitat partitioning patterns, dependent upon resource abundance (which is different in each cerrado region habitat), would also provide important data on community structure. In the area of conservation it is likely that the study of natural patches of forest within the cerrado region (e.g., gallery forests) will provide important information concerning the likely consequences of anthropogenic fragmentation. Also, studies on seed dispersal and forest regeneration would

be pertinent to conservation issues. Autoecological and behavioral studies of cerrado region birds are almost nonexistent, and information is
needed on the social structure, mating system, and reproductive biology
of almost all species. Other topics of interest include: competition and predation, differential resistance to parasites, long-term impact of fire upon
bird communities, habitat fragmentation and its effects (e.g., expansion
in host species for *Molothrus bonariensis*; changes in community composition), timing of breeding as it relates to nest predation and/or weather
and food abundance, and social behavior (e.g., in breeding and flocking)
in relation to behavioral and morphological attributes of the species
involved. In short, the topics listed above, a far from comprehensive list,
include broad categories for which understanding is scanty at best.

ACKNOWLEDGMENTS

I thank J. M. C. Silva and two anonymous reviewers for their comments
and suggestions on previous drafts of the manuscript. I also thank the
Brazilian Research Council (CNPq) for a research grant during the time
this manuscript was produced.

REFERENCES

Almeida, J. B. and R. H. Macedo. 2001. Lek-like mating system of the monog-
    amous blue-black grassquit. *Auk* 118:404–411.
Alves, M. A. S. 1990. Social system and helping behavior in the white-banded
    tanager (*Neothraupis fasciata*). *Condor* 92:470–474.
Alves, M. A. S. 1991. Dieta e táticas de forrageamento de *Neothraupis fasci-
    ata* em cerrado no Distrito Federal, Brasil (Passeriformes: Emberizidae).
    Ararajuba 2:25–29.
Alves, M. A. S. and R. B. Cavalcanti. 1990. Ninhos, ovos e crescimento de
    filhotes de *Neothraupis fasciata*. *Ararajuba* 1:91–94.
Alves, M. A. S. and R. B. Cavalcanti. 1996. Sentinel behavior, seasonality, and
    the structure of bird flocks in a Brazilian savanna. *Ornitol. Neotropical*
    7:43–51.
Bagno, M. A. 1998. As aves da Estação Ecológica de Águas Emendadas. In
    J. Marinho-Filho, F. Rodrigues, and M. Guimarães, eds., *Vertebrados da
    Estação Ecológica de Águas Emendadas*, pp. 22–33. Brasília: Governo
    do Distrito Federal.
Bianchi, C. A. 1998. "Biologia Reprodutiva da Arara-canindé (Ara ararauna,
    Psittacidae) no Parque Nacional das Emas, Goiás." Master's thesis, Uni-
    versidade de Brasília, Brasília, Brazil.

Blake, J. G. and B. A. Loiselle. 1991. Variation in resource abundance affects capture rates of birds in three lowland habitats in Costa Rica. *Auk* 108:114–130.

Brown, J. H. and M. A. Bowers. 1985. Community organization in hummingbirds: Relationships between morphology and ecology. *Auk* 102:251–269.

Brown, J. L. 1987. *Helping and Communal Breeding in Birds.* Princeton, N.J.: Princeton University Press.

Carvalho, C. B. V. and R. H. Macedo. In prep. Memory, learning and floral characteristics in foraging patterns of neotropical hummingbirds.

Catling, P. C. and A. E. Newsome. 1981. Responses of the Australian vertebrate fauna to fire: An evolutionary approach. In A. M. Gill, R. H. Groves, and I. R. Noble, eds., *Fire and the Australian Biota*, pp. 273–310. Canberra: Australian Academy of Science.

Cavalcanti, R. B. 1992. The importance of forest edges in the ecology of open country cerrado birds. In P. A. Furley, J. Proctor, and J. A. Ratter, eds., *The Nature and Dynamics of Forest-Savanna Boundaries*, pp. 513–517. London: Chapman and Hall.

Cavalcanti, R. B. and M. A. S. Alves. 1997. Effects of fire on savanna birds in central Brazil. *Ornitol. Neotropical* 8:85–87.

Cintra, R. and R. B. Cavalcanti. 1997. Intrapopulational variation in growth rates of nestling ruddy ground-doves (*Columbina talpacoti*, Aves: Columbidae) in Brazil. *Rev. Bras. Ecol.* 1:10–14.

Couto, E. A. 1985. "O Efeito da Sazonalidade na População da Rolinha (Columbina talpacoti) no Distrito Federal." Master's thesis, Universidade de Brasília, Brasília, Brazil.

Davis, D. E. 1940a. Social nesting habits of the smooth-billed ani. *Auk* 57:179–218.

Davis, D. E. 1940b. Social nesting habits of *Guira guira*. *Auk* 57:472–484.

Davis, D. E. 1941. Social nesting habits of *Crotophaga major*. *Auk* 58:179–183.

Davis, D. E. 1946. A seasonal analysis of mixed flocks of birds in Brazil. *Ecology* 27:168–181.

Figueiredo, S. V. 1991. "Efeito do Fogo Sobre o Comportamento e Sobre a Estrutura da Avifauna de Cerrado." Master's thesis, Universidade de Brasília, Brasília, Brazil.

Fleming, T. H., R. Breitwisch, and G. H. Whitesides. 1987. Patterns of tropical vertebrate frugivore diversity. *Ann. Rev. Ecol. Syst.* 18:91–109.

Frost, P. G. H. 1984. The responses and survival of organisms in fire-prone environments. In P. de V. Booysen and N. M. Taiton, eds., *Ecological Effects of Fire in South African Ecosystems*, pp. 273–309. Berlin: Springer-Verlag.

Fry, C. H. 1970. Ecological distribution of birds in north-eastern Mato Grosso State, Brazil. *An. Acad. Bras. Ciênc.* 42:275–318.

Gottsberger, G. and I. Silberbauer-Gottsberger. 1983. Dispersal and distribution in the cerrado vegetation of Brazil. *Sonderbd. naturwiss. Ver. Hamburg* 7:315–352.

Haffer, J. 1969. Speciation in Amazonian forest birds. *Science* 165:131–137.

Haffer, J. 1974. *Avian Speciation in South America*. Cambridge: Publ. Nuttall Ornith. Club 14.

Herrera, C. M. 1981. Are tropical fruits more rewarding to dispersers than temperate ones? *Amer. Nat.* 118:896–907.

Howe, H. F. 1977. Bird activity and seed dispersal of a tropical wet forest tree. *Ecology* 58:539–550.

Kodric-Brown, A., J. H. Brown, G. S. Byers, and D. F. Gori. 1984. Organization of a tropical island community of hummingbirds and flowers. *Ecology* 65:1358–1368.

Lack, D. 1968. *Ecological Adaptations for Breeding in Birds*. London: Meuthuen.

Levey, D. J. 1988. Tropical wet forest treefall gaps and distributions of understory birds and plants. *Ecology* 69:1076–1089.

Levey, D. J. and F. G. Stiles. 1992. Evolutionary precursors of long-distance migration: Resource availability and movement patterns in neotropical landbirds. *Amer. Nat.* 140:447–476.

Loiselle, B. A. and J. G. Blake. 1991. Temporal variation in birds and fruits along an elevational gradient in Costa Rica. *Ecology* 72:180–193.

Luke, R. H. and A. G. McArthur. 1997. *Bushfires in Australia*. Canberra: Australian Government Publishing Service.

MacArthur, R. H. and J. W. MacArthur. 1961. On bird species diversity. *Ecology* 42:594–600.

Macedo, R. H. 1992. Reproductive patterns and social organization of the communal guira cuckoo (*Guira guira*) in central Brazil. *Auk* 109:786–799.

Macedo, R. H. 1994. Inequities in parental effort and costs of communal breeding in the guira cuckoo. *Ornitol. Neotropical* 5:79–90.

Macedo, R. H. and C. A. Bianchi. 1997a. Communal breeding in tropical guira cuckoos (*Guira guira*): Sociality in the absence of a saturated habitat. *J. Avian Biology* 3:207–215.

Macedo, R. H. and C. A. Bianchi. 1997b. When birds go bad: Circumstantial evidence for infanticide in the communal South American guira cuckoo. *Ethol. Ecol. and Evolution* 9:45–54.

Macedo, R. H. and C. Melo. 1999. Confirmation of infanticide in the communally-breeding guira cuckoo. *Auk* 116:847–851.

Marini, M. A. 1992a. Foraging behavior and diet of the helmeted manakin. *Condor* 94:151–158.

Marini, M. A. 1992b. Notes on the breeding and reproductive biology of the helmeted manakin. *Wilson Bull.* 104:168–173.

Marini, M. A. and R. B. Cavalcanti. 1993. Habitat and foraging substrate use of three *Basileuterus* warblers from central Brazil. *Ornitol. Neotropical* 4:69–76.

Marini, M. A. and R. B. Cavalcanti. 1996. Influência do fogo na avifauna do sub-bosque de uma mata de galeria do Brasil central. *Rev. Bras. Biol.* 56:749–754.

Martin, T. E. 1987. Food as a limit on breeding birds: A life-history perspective. *Ann. Rev. Ecol. Syst.* 18:453–487.

Melo, C. and R. H. Macedo. 1997. Mortalidade em ninhadas de *Guira guira* (Cuculidae): Competição por recursos. *Ararajuba* 5:49–56.

Moermond, T. C. and J. S. Denslow. 1985. Neotropical avian frugivores: Patterns of behavior, morphology, and nutrition, with consequences for fruit selection. *Ornithol. Monogr.* 36:865–897.

Müller, P. 1973. *The Dispersal Centers of Terrestrial Vertebrates in the Neotropical Realm: A Study in the Evolution of the Neotropical Biota and its Native Landscapes.* Haag: Dr. W. Junk.

Murray Jr., B. G. 1982. Territorial behaviour of the blue-black grassquit. *Condor* 84:119.

Negret, A. J. 1983. "Diversidade e Abundância da Avifauna da Reserva Ecológica do IBGE, Brasília—D.F." Master's thesis, Universidade de Brasília, Brasília, Brazil.

Negret, A. J. and R. A. Negret. 1981. *As Aves Migratórias do Distrito Federal.* Boletim Técnico No. 6. Brasília: Ministério da Agricultura, Instituto Brasileiro de Desenvolvimento Florestal.

Negret, A. J., J. Taylor, R. C. Soares, R. B. Cavalcanti and C. Johnson. 1984. Aves da Região Geopolítica do Distrito Federal. Brasília: Ministério do Interior, Secretaria Especial do Meio Ambiente.

Negret, R. A. 1978. "O Comportamento Alimentar Como Fator de Isolamento Ecológico em Oito Espécies de Tyrannidae (Aves) do Planalto Central, Brasil." Master's thesis, Universidade de Brasília, Brasília, Brazil.

Nores, M. 1992. Bird speciation in subtropical South America in relation to forest expansion and retraction. *Auk* 109:346–357.

Nores, M. 1994. Quaternary vegetational changes and bird differentiation in subtropical South America. *Auk* 111:499–503.

Oliveira, G. M. 1998. "Disponibilidade de Recursos Florais para Beija-flores no Cerrado de Uberlândia/MG." Master's thesis, Universidade de Brasília, Brasília, Brazil.

Orians, G. H. 1969. The number of bird species in some tropical forests. *Ecology* 50:783–801.

Paes, M. M. N. 1993. "Utilização de Frutos por Aves em uma Área de Cerrado do Distrito Federal." Master's thesis, Universidade de Brasília, Brasília, Brazil.

Pinheiro, F. 1999. "Síndromes de Dispersão de Sementes de Matas de Galeria do Distrito Federal." Master's thesis, Universidade de Brasília, Brasília, Brazil.

Quinn, J. S., R. H. Macedo, and B. White. 1994. Genetic relatedness of communally-breeding guira cuckoos. *Anim. Behav.* 47:515–529.

Ricklefs, R. E. 1976. Growth rates of birds in the humid New World tropics. *Ibis* 118:179–207.

Ricklefs, R. E. 1990. Speciation and diversity: The integration of local and regional processes. In D. Otte and J. A. Endler, eds., *Speciation and its Consequences*, pp. 599–622. Chicago: University of Chicago Press.

Ricklefs, R. E. and D. Schluter. 1993. Species diversity: Regional and historical influences. In R. E. Rcklefs and D. Schluter, eds., *Species Diversity in Ecological Communities*, pp. 350–363. Chicago: University of Chicago Press.

Rojas, R. and R. Ribon. 1997. Guilda de aves em *Bowdichia virgilioides* (Fabaceae: Faboideae) em área de cerrado de Furnas, Minas Gerais. *Ararajuba* 5:189–194.

Setubal, S. S. 1991. "Biologia e Ecologia dos Tinamídeos Rhynchotus rufescens (Temminck, 1815) e Nothura maculosa (Temminch, 1815) na Região do Distrito Federal, Brasil." Master's thesis, Pontifícia Universidade Católica do Rio Grande do Sul, Porto Alegre, Brazil.

Sick, H. 1955. O aspecto fitofisionômico da paisagem do médio Rio das Mortes, Mato Grosso e a avifauna da região. *Arq. Mus. Nac.* XLII 2:541–576.

Sick, H. 1965. A fauna do cerrado. *Arq. Zool.* 12:71–93.

Sick, H. 1966. As aves do cerrado como fauna arborícola. *An. Acad. Bras. Ciênc.* 38:355–363.

Sick, H. 1983. *Migrações de aves na América do Sul continental* (tradução: Walter A. Voss). Publicação Técnica No. 2. Brasília: CEMAVE—Centro de Estudos de Migrações de Aves.

Sick, H. 1997. *Ornitologia Brasileira*. Rio de Janeiro: Ed. Nova Fronteira.

Silberbauer-Gottsberger, I. and G. Gottsberger. 1988. A polinização de plantas do cerrado. *Rev. Bras. Biol.* 48:651–663.

Silva, E. M. D. 1980. "Composição e Comportamento de Grupos Heteroespecíficos de Aves em Área de Cerrado, no Distrito Federal." Master's thesis, Universidade de Brasília, Brasília, Brazil.

Silva, J. M. C. 1995a. Avian inventory of the cerrado region, South America: implications for biological conservation. *Bird. Conserv. Int.* 5:291–304.

Silva, J. M. C. 1995b. Biogeographic analysis of the South American cerrado avifauna. *Steenstrupia* 21:49–67.

Silva, J. M. C. 1995. Birds of the Cerrado Region, South America. *Steenstrupia* 21:69–92.

Silva, J. M. C. 1996. Distribution of Amazonian and Atlantic birds in gallery forests of the Cerrado Region, South America. *Ornitol. Neotropical* 7:1–18.

Silva, J. M. C. 1997. Endemic bird species and conservation in the Cerrado Region, South America. *Biodiv. and Conserv.* 6:435–450.

Skutch, A. F. 1959. Life history of the groove-billed ani. *Auk* 76:281–317.

Smythe, N. 1970. Relationships between fruiting seasons and seed dispersal methods in a neotropical forest. *Amer. Nat.* 104:25–35.

Soares, R. C. 1983. "Taxas de Crescimento de Filhotes de Rolinha, Columbina talpacoti (Aves: Columbidae), em Relação a Fatores Ecológicos e Populacionais no Planalto Central." Master's thesis, Universidade de Brasília, Brasília, Brazil.

Stacey, P. B. and W. D. Koenig (eds.). 1990. *Cooperative Breeding in Birds.* Cambridge: Cambridge University Press.

Stiles, F. G. 1983. Birds: introduction. In D. H. Janzen, ed., *Costa Rican Natural History*, pp. 502–530. Chicago: University of Chicago Press.

Stiles, F. G. 1985. Seasonal patterns and coevolution in the hummingbird-flower community of a Costa Rican subtropical forest. In P.A. Buckley, M. S. Foster, E. S. Morton, R. S. Ridgely, and F. G. Buckely, eds., *Neotropical Ornithology*, pp. 757–787. Lawrence: Allen Press.

Trollope, W. S. W. 1984. Fire in savanna. In P. de V. Booysen and N. M. Taiton, eds., *Ecological Effects of Fire in South African Ecosystems*, pp. 149–175. Berlin: Springer-Verlag.

Tubelius, D. P. 1997. "Estrutura de Comunidades de Aves em Habitats Preservados e Alterados de Cerrado, na Região do Distrito Federal." Master's thesis, Universidade de Brasília, Brasília, Brazil.

Villalobos, M. P. 1994. "Guilda de Frugívoros Associada com o Buriti (Mauritia flexuosa: Palmae) em uma Vereda no Brasil Central." Master's thesis, Universidade de Brasília, Brasília, Brazil.

Webber, T. 1985. Songs, displays, and other behaviour at a courtship gathering of blue-black grassquits. *Condor* 87:543–546.

Young, B. E. 1994. The effects of food, nest predation and weather on the timing of breeding in tropical house wrens. *Condor* 96:341–353.

Appendix Avifauna Classified by Breeding Status, Habitat, and Endemicity (as Reported in Silva 1995)

| Family | No. species | Breeding | | Migrants | Habitat | | | Endemics |
		In region	Un-known		Cerrado[a]	Forest	Both	
Rheidae	1	1	0	No	1	0	0	No
Tinamidae	16	16	0	No	5	11	0	Yes: 2
Podicipedidae	2	2	0	No	2	0	0	No
Phalacrocoracidae	1	1	0	No	1	0	0	No
Anhingidae	1	1	0	No	1	0	0	No
Ardeidae	16	16	0	No	13	1	2	No
Ciconiidae	3	3	0	No	3	0	0	No
Threskiornithidae	6	6	0	No	5	0	1	No
Anhimidae	2	2	0	No	2	0	0	No
Anatidae	11	9	2	No	10	0	1	No
Cathartidae	4	4	0	No	3	0	1	No
Pandionidae	1	0	0	1 NA	1	0	0	No
Accipitridae	33	29	2	2 NA	14	11	8	No
Falconidae	13	12	0	1 NA	5	5	3	No
Cracidae	11	11	0	No	0	9	2	Yes: 1
Aramidae	1	1	0	No	1	0	0	No
Rallidae	16	11	5	No	8	0	8	No
Heliornithidae	1	1	0	No	1	0	0	No
Eurypygidae	1	1	0	No	0	0	1	No
Cariamidae	1	1	0	No	1	0	0	No
Jacanidae	1	1	0	No	1	0	0	No
Recurvirostridae	1	1	0	No	1	0	0	No
Charadriidae	4	3	0	1 NA	4	0	0	No
Scolopacidae	13	2	0	11 NA	13	0	0	No
Laridae	3	2	1	No	3	0	0	No
Rhynchopidae	1	1	0	No	1	0	0	No
Columbidae	18	18	0	No	7	7	4	Yes: 2
Psittacidae	33	33	0	No	4	15	14	Yes: 2
Opisthocomidae	1	1	0	No	0	0	1	No
Cuculidae	14	13	0	1 NA	3	5	6	No
Tytonidae	1	1	0	No	1	0	0	No
Strigidae	14	14	0	No	3	7	4	No
Nyctibiidae	3	3	0	No	0	1	2	No
Caprimulgidae	15	13	1	1 NA	9	3	3	Yes: 1
Apodidae	9	4	5	No	4	1	4	No
Trochilidae	36	36	0	No	5	17	14	Yes: 1
Trogonidae	8	8	0	No	0	8	0	No
Alcedinidae	5	5	0	No	1	0	4	No
Momotidae	4	4	0	No	0	4	0	No
Galbulidae	5	5	0	No	0	4	1	No
Bucconidae	13	13	0	No	1	10	2	No
Capitonidae	1	1	0	No	0	1	0	No
Ramphastidae	10	10	0	No	0	9	1	No
Picidae	25	25	0	No	2	17	6	No
Dendrocolaptidae	23	23	0	No	1	20	2	No

Appendix (*continued*)

Family	No. species	In region	Un-known	Migrants	Cerrado[a]	Forest	Both	Endemics
Furnariidae	41	41	0	No	9	25	7	Yes: 5
Formicariidae	58	58	0	No	2	49	7	Yes: 2
Conopophagidae	1	1	0	No	0	1	0	No
Rhinocryptidae	2	2	0	No	1	1	0	Yes: 2
Tyrannidae	122	111	2	3 AB, 6 SA	31	65	26	Yes: 3
Pipridae	17	17	0	No	0	16	1	Yes: 1
Cotingidae	10	6	0	4 AB	0	9	1	No
Hirundinidae	14	8	1	5 NA	12	1	1	No
Motacillidae	2	2	0	No	2	0	0	No
Troglodytidae	9	9	0	No	3	4	2	No
Mimidae	2	2	0	No	2	0	0	No
Muscicapidae	11	7	0	1 AB, 1 NA, 2 SA	1	7	3	No
Emberizidae	103	87	12	4 SA	40	47	16	Yes: 9
Parulidae	12	11	0	1 NA	1	10	1	Yes: 1
Vireonidae	7	7	0	No	0	6	1	No
Icteridae	18	16	1	1 NA	11	3	4	No
Corvidae	4	4	0	No	1	1	2	Yes: 1
Total	837	758	33 (3.9)	46	258	411	168	32
(%)		(90.6)		(5.6)	(30.8)	(49.1)	(20.1)	(3.8)

Header note:

	Breeding				Habitat			

[a] Includes all open formations (e.g., cerrado *sensu stricto*, *campo cerrado, campo sujo, campo limpo, vereda*, and *campo rupestre*). See chapter 6 for description of vegetation physiognomies.

Abbreviations: NA and SA = long-distance migrants from North and South America, respectively; AB = altitudinal migrant within Brazil.

14

The Cerrado Mammals: Diversity, Ecology, and Natural History

Jader Marinho-Filho, Flávio H. G. Rodrigues,
and Keila M. Juarez

THE FIRST FORMAL RECORDS OF CERRADO MAMMALS WERE MADE by one of the first Brazilian scientists, Alexandre R. Ferreira, who from 1783 to 1792 explored the provinces of Grão-Pará, Rio Negro, Mato Grosso, and Cuiabá (Hershkovitz 1987). However, only in the second half of the 20th century have Brazilian zoologists made the transition from a merely taxonomic treatment of the fauna towards a more naturalistic and ecological approach. Herein we present a review and analysis of the available information on natural history and geographical ranges of species and groups of the cerrado mammalian fauna, delineating patterns, making comparisons with other tropical savannas, and indicating lacunas and lines of investigation remaining to be explored.

DATABASE OF CERRADO MAMMALS

We followed Wilson and Reeder (1993) as a guide for the taxonomic status and distribution of mammals. Since the limits of the distributional ranges of most Brazilian mammals are far from well defined, we established the database on the distribution of Brazilian mammals from a number of different sources, including comprehensive works such as Vieira (1942), Moojen (1952), Cabrera (1957, 1961), Alho (1982), Koopman (1982, 1993), Streilein (1982), Emmons and Feer (1990), and Eisenberg and Redford (1999); published compilations with analyses on local and/or regional faunas, such as Schaller (1983), Redford and Fonseca

(1986), Marinho-Filho and Reis (1989), Medellin and Redford (1992), Fonseca et al. (1996), Marinho-Filho and Sazima (1998), and references cited therein; technical reports and local/regional inventories such as Naturae (1996), PCBAP (1997), Marinho-Filho (1998); and unpublished original information from scientific collections at the Universidade de Brasília, the Museu de História Natural da Universidade Estadual de Campinas, Museu Nacional, and the authors' personal observations.

This database generated a checklist of mammalian species for the entire cerrado region (see table 14.1). We considered as endemics those species occurring only within this biome in Brazil. Those species whose entire known range is contained within a circular area of 300 km of diameter were considered as restricted. There are still many gaps to be filled, controversies about the status of some species, and many species remaining to be described. Genera such as *Nectomys, Dasyprocta, Oryzomys, Galea,* and *Cavia* are in urgent need of extensive revision. Even with these potential sources of error, the general picture presented here seems adequate for our purposes.

The limits of the Cerrado Biome are those presented by the Brazilian Institute of Geography and Statistics (IBGE 1993; chapter 6; and fig. 6.1), with minor modifications. The maps showing these limits were produced by the Biodiversity Conservation Data Center of the Fundação Biodiversitas (Belo Horizonte, Brazil). Our analysis considers all habitat types within the Cerrado biome, including more dense formations such as gallery and dry forests.

CHARACTERIZATION OF FAUNA: COMPOSITION, SPECIES RICHNESS, AND ABUNDANCE

In all, 194 mammalian species from 30 families and 9 orders are recognized from the Cerrado Biome (table 14.1), making this biome the third most speciose in Brazil, after the Amazon and the Atlantic Forest, and followed by the *caatinga* and the *pantanal* (see Fonseca et al. 1999 for a better understanding of the Brazilian mammalian fauna). The largest groups are bats and rodents, represented by 81 and 51 species, respectively, including the notably speciose families Phyllostomidae and Muridae. Likewise, carnivores, didelphimorph marsupials, and xenarthrans are rather diversified groups, the last two being distinctive elements of the neotropical mammalian fauna.

In general, this fauna is essentially composed of small-sized animals: 85% of the species have body masses no greater than 5 kg, and only five

Table 14.1 Checklist of Cerrado Mammals with Endemic Species, and Species Included in the Brazilian Official List of Species Threatened with Extinction (Thr)

Scientific Name	Endemic	Thr.	Abun-dance	Range	Habitat Open	Habitat Forest	Weight (g)	Diet
DIDELPHIMORPHIA								
Family Didelphidae								
Caluromys lanatus			R	w		x	310–410	om
Caluromys philander			R	w		x	140–270	om
Chironees minimus			R	w		x	600–700	om
Didelphis albiventris			A	w	x	x	500–2000	om
Didelphis marsupialis			A	w	x	x	500–2000	om
Gracilinanus agilis			A	w	x	x	20–30	om
Lutreolina crassicaudata			R	w	x	x	200–540	om
Marmosa murina			A	w	x	x	45–60	om
Metachirus nudicaudatus			R	w		x	300–450	om
Micoureus demerarae			R	w	x	x	80–150	om
Monodelphis americana			A	w	x	x	11–35	om
Monodelphis domestica			A	w		x	35–100	om
Monodelphis kunsi			R	w	x		8.5–14	om
Monodelphis rubida	x		R	r	x		45–46	om
Philander opossum			R	w		x	200–600	om
Thylamys pusilla			A	w	x		12–30	om
Thylamys velutinus			R	w	x		16–32	om
XENARTHRA								
Family Myrmecophagidae								
Myrmecophaga tridactyla	x		r	w	x	x	22000–40000	in
Tamandua tetradactyla			a	w	x	x	3500–8500	in
Family Bradypodidae								
Bradypus variegatus			r	w		x	2300–5500	fo
Family Dasypodidae								
Cabassous tatouay			r	w	x		3400–6400	in
Cabassous unicinctus			r	w	x	x	1500–5000	in
Dasypus novemcinctus			r	w	x	x	2500–6300	om
Dasypus septemcinctus			a	w	x	x	1500–2000	om
Euphractus sexcintus			a	w	x		3000–7000	om
Priodontes maximus	x		r	w	x		30000–60000	in
Tolypeutes matacus			r	w	x		1000–1150	in
Tolypeutes tricinctus	x		r	w	x	x	1000–1800	in

Table 14.1 (*continued*)

Scientific Name	Endemic	Thr.	Abun-dance	Range	Habitat Open	Habitat Forest	Weight (g)	Diet
CHIROPTERA								
Family Emballonuridae								
Centronycteris maximiliani			r	w	x	x		in
Peropteryx kappleri			r	w	x	x	6–11	in
Peropteryx macrotis			r	w	x	x	4–8	in
Rhinchonycteris naso			a	w	x	x	4–7	in
Saccopteryx bilineata			a	w	x	x	7–12	in
Saccopteryx leptura			r	w	x	x	4–7	in
Family Noctilionidae								
Noctilio albiventris			a	w	x	x	21–55	in
Noctilio leporinus			a	w	x	x	60–90	fi
Family Mormoopidae								
Pteronotus parnellii			a	w	x	x	11–28	in
Pteronotus personatus			r	w	x	x	6–9	in
Pteronotus gymnonotus			r	w	x	x	10–16	in
Family Phyllostomidae								
Anoura caudifer			a	w	x	x	10–13	ne
Anoura geoffroyi			a	w	x	x	13–19	ne
Artibeus cinereus			a	w	x	x	12–14	fr
Artibeus concolor			a	w	x	x	18–20	fr
Artibeus jamaicensis			a	w	x	x	50–65	fr
Artibeus lituratus			a	w	x	x	60–85	fr
Artibeus planirostris			a	w	x	x	50–65	fr
Carollia perspicillata			a	w	x	x	12–25	fr
Chiroderma trinitatum			r	w	x	x	13–15	fr
Chiroderma villosum			r	w	x	x	44–50	fr
Choeroniscus minor			r	w	x	x	10	ne
Chrotopterus auritus			r	w		x	60–95	ca
Desmodus rotundus			a	w	x	x	25–50	sa
Diaemus youngi			a	w	x	x	27–35	sa
Diphylla ecaudata			r	w	x	x	20–35	sa
Glossophaga soricina			a	w	x	x	10–14	ne
Lonchophylla bokermanni			a	r	x	x	9–12	ne
Lonchophylla de keyseri	x		a	a	x	x	10–12	ne

Table 14.1 (continued)

Scientific Name	Endemic	Thr.	Abun-dance	Range	Habitat Open	Habitat Forest	Weight (g)	Diet
Lonchorhina aurita			a	w		x	12–22	in
Macrophylum macrophylum			r	w	x	x	7–11	in
Micronycteris behni			r	r		x	<15	in/fr
Micronycteris megalotis			a	w		x	5–7	in/fr
Micronycteris minuta			r	w		x	7–9	in/fr
Mimon benetti			r	w		x	10–15	in/fr
Mimon crenulatum			a	w		x	10–12	in/fr
Phylloderma stenops			r	w		x	41–65	in/fr
Phyllostomus discolor			a	w	x	x	30–45	om
Phyllostomus elongatus			r	w		x	38–57	in/fr
Phyllostomus hastatus			a	w	x	x	90–140	om
Platyrrhinus helleri			a	w	x	x	12–16	fr
Platyrrhinus lineatus			a	w	x	x	20–27	fr
Rhinophylla pumilio			a	w	x	x	8–12	fr
Sturnira lilium			a	w		x	16–25	fr
Sturnira tildae			a	w		x	21–30	fr
Tonatia bidens			a	w		x	23–30	in
Tonatia brasiliense			a	w		x	9–12	in
Tonatia silvicola			a	w		x	20–36	in
Trachops cirrhosus			a	w		x	28–45	ca
Uroderma bilobatum			a	w		x	13–20	fr
Uroderma magnirostrum			r	w		x	16–21	fr
Vampyressa pusilla			r	w		x	7–8	fr
Family Natalidae								
Natalus stramineus			r	w	x	x	4–7	in
Family Furipteridae								
Furipterus horrens			r	w		x	4–5	in
Family Vespertilionidae								
Eptesicus brasiliensis			a	w	x	x	8–10	in
Eptesicus diminutus			r	w	x	x	5–7	in
Eptesicus furinalis			r	w	x	x	7–14	in
Histiotus velatus			r	w	x	x	12–14	in
Lasiurus borealis			a	w	x	x	8–14	in

Table 14.1 (*continued*)

Scientific Name	Endemic	Thr.	Abun-dance	Range	Habitat Open	Habitat Forest	Weight (g)	Diet
Lasiurus cinereus			r	w	x	x	20	in
Lasiurus ega			a	w	x	x	10–15	in
Myotis albescens			r	w	x	x	7–11	in
Myotis nigricans			a	w	x	x	4–8	in
Myotis riparius			r	w	x	x	4–7	in
Rhogeessa tumida			r	w	x	x	3–5	in
Family Molossidae								
Eumops auripendulus			a	w	x	x	62–66	in
Eumops bonariensis			r	w	x	x	11–20	in
Eumops glaucinus			r	w	x	x	22–28	in
Eumops hansae			r	w	x	x		in
Eumops perotis				w	x	x	60–76	in
Molossops abrasus			r	w	x	x	25–42	in
Molossops mattogrossensis				w	x	x	7–9	in
Molossops planirostris				w	x	x	5–9	in
Molossops temminckii			a	w	x	x	4–9	in
Molossus ater			a	w	x	x	21–43	in
Molossus molossus			a	w	x	x	12–28	in
Nyctinomops aurispinosus				w	x	x		in
Nyctinomops laticaudatus			a	w	x	x	8–16	in
Nyctinomops macrotis				w	x	x	16–20	in
Promops nasutus			r	w	x	x	14–25	in
Tadarida brasiliensis			r	w	x	x	9–19	in
PRIMATES								
Family Callithrichidae								
Callithrix jacchus			a	w	x	x	250–325	om
Callithrix melanura			r	w		x	380–500	om
Callithrix penicillata			a	w	x	x	250–350	om
Family Cebidae								
Alouatta caraya			a	w		x	3000–10000	fo/fr
Aotus infulatus			r	w		x	600–1000	fr
Cebus apella			a	w		x	1700–4500	in/fr
CARNIVORA								
Family Canidae								
Cerdocyon thous			a	w	x	x	4000–9000	om
Chrysocyon brachyurus	x		r	w	x		20,000–30,000	om

Table 14.1 (*continued*)

Scientific Name	Endemic	Thr.	Abun-dance	Range	Habitat Open	Habitat Forest	Weight (g)	Diet
Pseudalopex vetulus	x		r	w	x		3000–4500	in/fr
Speothos venaticus		x	r	w	x	x	5000–7000	ca
Family Procyonidae								
Nasua nasua			a	w	x	x	3000–7500	om
Potos flavus			r	w		x	2000–3500	fr
Procyon cancrivorus			a	w	x	x	3500–7500	om
Family Mustelidae								
Conepatus semistriatu			r	w	x		1500–3500	in
Eira barbara			r	w	x	x	2700–7000	om
Galictis cuja			a	w	x	x	1000–2500	om
Galictis vittata			a	w	x	x	1500–2500	om
Lontra longicaudis		x	r	w		x	5000–15,000	fi
Pteronura brasiliensis		x	r	w		x	24,000–34,000	fi
Family Felidae								
Herpailurus yaguaroundi			r	w	x	x	4000–9000	ca
Leopardus pardalis		x	r	w	x	x	8000–15,000	ca
Leopardus tigrinus		x	r	w	x	x	1300–3000	ca
Leopardus wiedii		x	r	w	x	x	3000–9000	ca
Oncifelis colocolo		x	r	w	x		1700–3650	ca
Panthera onca		x	r	w	x	x	30,000–150,000	ca
Puma concolor		x	r	w	x	x	30,000–120,000	ca
PERISSODACTYLA								
Family Tapiridae								
Tapirus terrestris			r	w	x	x	200,000–250,000	fo/fr
ARTIODACTYLA								
Family Tayassuidae								
Pecari tajacu			r	w	x	x	17,000–30,000	om
Tayassu pecari			r	w	x	x	25,000–40,000	om
Family Cervidae								
Blastoceros dichotomus		x	r	w	x	x	100,000–150,000	fo
Mazama americana			r	w		x	24,000–50,000	fo/fr
Mazama gouazoupira			a	w	x	x	13,000–23,000	fo/fr
Ozotoceros bezoarticus		x	r	w	x		28,000–35,000	fo
RODENTIA								
Family Muridae								
Akodon cursor			a	w		x	20–65	fr/gr/in
Akodon montensis			a	w		x	20–65	fr/gr/in

Table 14.1 (continued)

Scientific Name	Endemic	Thr.	Abundance	Range	Habitat Open	Habitat Forest	Weight (g)	Diet
Akodon lindberghi	x		r	r	x		16–19	fr/gr/in
Bibimys labiosus	x		r	r		x	20–35	fr/gr/in
Bolomys lasiurus			a	w	x		20–58	fr/gr/in
Calomys callosus			a	w	x		22–30	fr/gr/in
Calomys laucha			r	w	x		30–38	fr/gr/in
Calomys tener	x		a	w	x		15–23	fr/gr/in
Holochilus brasiliensis			r	w		x	275–455	fo
Holochilus sciureus			r	w		x	144–177	fo
Juscelinomys candango	x	x	r	r	x			fr/gr/in
Kunsia fronto			r	r	x		110–400	fr/gr/in
Kunsia tomentosus		x	r	r	x		200–600	fr/gr/in
Microakodontomys transitorius	x		r	r	x	x		
Nectomys rattus			a	w		x	200–450	om
Oecomys bicolor			r	w		x	21–41	fr/gra
Oecomys cleberi	x		r	?		x	19–25	fr/gra
Oecomys concolor [a]			r	w		x	45–96	fr/gra
Oligoryzomys chacoensis			r	r		x	14–25	fr/gr/in
Oligoryzomys eliurus	x		r	w	x	x	10–25	fr/gr/in
Oligoryzomys nigripes			r	w		x	14–25	fr/gr/in
Oryzomys megacephalus			a	w		x	30–65	fr/gr/in
Oryzomys lamia	x		r	r		x		fr/gr/in
Oryzomys ratticeps			r	w			120–157	fr/gr/in
Oryzomys subflavus [b]			a	w	x	x	60–140	fr/gr/in
Oxymycterus delator	x		a	r	x		40–105	fr/gr/in
Oxymycterus roberti	x		a	w	x		40–105	fr/gr/in
Pseudoryzomys simplex	x		r	w	x	x	30–56	fr/gr/in
Rhipidomys emiliae				w		x	40–100	om
Rhipidomys macrurus			a	w		x	40–100	om
Thalpomys cerradensis	x		r	r	x			fr/gr/in
Thalpomys lasiotis	x		r	r	x			fr/gr/in
Wiedomys pyrrhorhinos			r	w	x	x	24–32	fr/gr/in

Table 14.1 (continued)

Scientific Name	Endemic	Thr.	Abun- dance	Range	Habitat Open	Habitat Forest	Weight (g)	Diet
Family Erethizontidae								
Coendou prehensilis			r	w		x	3200–5300	fr/fo
Family Caviidae								
Cavia aperea			a	w	x		500–1000	fo
Galea spixii			a	w	x		300–600	fo
Kerodon rupestris			r	w	x		900–1000	fo
Family Hydrochaeridae								
Hydrochaeris hydrochaeris			a	w	x	x	35000–65000	fo
Family Agoutidae								
Agouti paca			r	w		x	5000–13000	fr/fo
Family Dasyproctidae								
Dasyprocta leporina			a	w	x	x		fr/gr
Dasyprocta azarae			a	w	x	x	2500–3200	fr/gr
Family Ctenomyidae								
Ctenomys brasiliensis	x		r	r	x			fo
Family Echimyidae								
Carterodon sulcidens	x		r	r	x			fo
Clyomys bishopi			a	r	x		100–300	fr/gr/in
Clyomys laticeps			r	w	x		100–300	fr/gr/in
Dactylomys dactylinus			r	w		x	600–700	fo
Echimys braziliensis [c]			r	r		x		fr/gr
Proechimys roberti			a	w		x	160–500	fr/gr/in
Proechimys longicaudatus			a	w		x	160–500	fr/gr/in
Trinomys moojeni	x		r	r		x		fr/gr
Thrichomys apereoides			a	w	x	x	247–500	fr/gr/in
LAGOMORPHA								
Family Leporidae								
Sylvilagus brasiliensis			a	w	x	x	450–1200	fo
TOTAL: 194 species	19	17						

species weigh more than 50 kg (table 14.1), strongly contrasting with the mammalian fauna of African savannas, where large mammals abound. There are also differences concerning species composition, number of species, and biomass of mammals between African savannas and other savannas in the world. In Africa there are practically as many bovid species as there are murid rodents (Sinclair, 1983). Almost 100 species of ungulates inhabit Africa (Sinclair 1983; Ojasti 1983), compared to a little over 20 species in South America. The average number of ungulates for eight African savanna locations is 14.5 ± 3.7 (range = 11–20), whereas in the savannas of southern Asia, this number is only 7.3 ± 1.8 (range = 6–10; n = 6; see data compilation in Bourlière 1983a). In South America, this number may be even smaller: only two ungulate species inhabit the Llanos in Masaguaral, Venezuela (Einsenberg et al. 1979). In the Brazilian cerrados, the number of ungulates in a given location varies between 6 and 7 species, depending on the presence of the marsh deer, *Blastocerus dichotomus*. There are no native ungulates in Australia, and the niche of the large herbivores is filled by kangaroos (Freeland 1991). Only six species of large Macropodidae occur in the Australian savannas, and only 3 to 4 may be found per location (Freeland 1991). Another difference between the ungulates of the African savannas and those of the Brazilian cerrados is feeding behavior. The great herbivores of the cerrados are browsers (Rodrigues and Monteiro-Filho 1999; Tomas and Salis 2000), unlike their African counterparts, the majority of which are grazers. In the cerrados, the role of the grazers is carried out by rodents, like the capybara (*Hydrochaeris hydrochaeris*), cavies (Caviidae), and the lagomorph *Sylvilagus brasiliensis* (see table 14.1). Aside from the ungulates, carnivores represent another prominent group in African savannas. Between 8 and 27 (average 14.7 ± 10.7; n = 3) species of carnivores can be found at

---

Table Note: Categories of abundance: r = rare, a = abundant; distributional range: w = widely distributed, r = restricted distribution. Categories of feeding habit: om = omnivore, in = insectivore, fo = folivore, fi = fish specialist, ne = nectarivore, fr = frugivore, as = sanguinivore, ca= carnivore, gr = grainivore.

[a]Musser and Carleton (1993) state that *Oecomys concolor* is restricted to localities north of the Amazon. However, this specific name has been frequently used in the literature referring to large-bodied *Oecomys* forms in the cerrado range, and that is why we decided to keep it here.

[b]*Oryzomys subflavus* has been split into a number of species, four of which occur in the cerrado domain (Percequillo 1998). However, these species are not decribed yet and were not considered herein.

[c]Emmons and Feer (1990) recognize *Echimys braziliensis* as *Nelomys* sp., stating that the correct specific name is not clear. This species has also been called *Phylomys braziliensis*, which Cabrera (1961:540) considers a *nomem nudum*.

a given location in the African savannas (Bourlière 1983b). The number of carnivores per location in the cerrado is not much less (13.6 ± 1.8; range 12–16), considering five sampling areas: Emas National Park (GO) (Silveira, 1999), the Distrito Federal (DF) (Fonseca and Redford, 1984; Marinho-Filho et al., 1998, and records of the Brasília Zoo), Grande Sertão Veredas National Park (MG), Serra da Mesa (GO), and the Jatobá Ranch (BA) (unpublished data). Actually this average may be even greater, because some species may be very difficult to sample, especially the small felines, which may go undetected even when present in an area.

The great majority of cerrado mammalian species have wide distributions, and, although the total number of individuals for a given species may be considered high throughout the entire range of the biome, most species tend to be locally rare. A comparative analysis of communities of small, non-flying mammals from 11 cerrado areas of central Brazil shows great variation among areas for the abundance of 39 species of marsupials and rodents (Marinho-Filho et al. 1994). Approximately one third of the individuals captured in all study areas were *Bolomys lasiurus*. Though the dominant species in most areas, it represented only 2.2% of the captures in one area and was absent from two additional areas. Similar patterns were found for other species, high in number at a given site, but rare or even absent in another. Thus, another third of the total number of captured individuals was represented by five species, and the remaining 33 species corresponded approximately to 30% of the total number of individuals of small mammals in the sampling areas. (Marinho-Filho et al. 1994).

In the same study, the beta diversity and distances between each of the 11 areas in relation to all others were calculated. For the 55 location pairs examined, Marinho-Filho et al. (1994) found a high mean beta diversity (mean = 0.58, SD = 0.13; range = 0.29–0.80), but there was no strong association between the distance between areas (5 km to 1,300 km) and beta diversity.

HABITAT UTILIZATION AND ENDEMICITY

The mammalian fauna of the cerrado region consists mainly of elements inhabiting a great variety of environments (table 14.1). About 54% of the mammalian species occupy forest environments as much as open areas, whereas 16.5% are exclusive to open areas and 29% exclusive to forests. The mammalian fauna of the cerrado appears to be derived primarily from a set of forest species (Redford and Fonseca 1986; Marinho-Filho and Sazima 1998). Gallery forests appear to play an important role as mesic corridors that allow for the establishment of elements not adapted to the conditions found in dry, open cerrado areas (Mares et al. 1985; Redford and Fonseca 1986). This results in the low endemism found in the cerrado: only 18 species (9.3%) may be considered exclusive to this biome.

Endemism rates for plants are considered high (see chapters 6, 7). In contrast, the cerrado fauna shares many elements with other open formations in tropical South America and is strongly influenced by two adjacent forest biomes, the Atlantic forest and the Amazonian rainforest. For vertebrates, the degree of endemism is low (Vanzolini 1963; Sick 1965; Silva 1995a, 1995b; see also chapters 11–13), and these animals exhibit no specific adaptations for life in the cerrado.

Most (56%) of the endemic mammalian species of the cerrado inhabit exclusively open areas. Of the remaining 44%, four species are forest inhabitants, and four occur in forests and open areas (table 14.1). Of the open area species, four (*Ctenomys brasiliensis, Carterodon sulcidens, Juscelinomys candango*, and *Oxymycterus roberti*) are semi-fossorial and thus avoid the environmental extremes that savanna inhabitants must confront.

Considering the 18 endemic species, five are restricted to a type locality: *Bibimys labiosus, Juscelinomys candango, Microakodontomys transitorius, Oecomys cleberi*, and *Carterodon sulcidens*. Of these, *J. candango, M. transitorius*, and *O. cleberi* were described from the Distrito Federal at the core of the cerrado region. The other two are known from the Lagoa Santa area (state of Minas Gerais), in the southeastern portion of the cerrado. Of the remaining 13 species, five are distributed in the central and central-southeastern portion of the cerrado; four are restricted to the southern and two to the western-southwestern portion; one species is found in all the Cerrado range (*Pseudalopex vetulus*); and the only endemic bat is found in the central-northern region. There are no known endemics restricted to the northern portion, but this picture may only represent the greater concentration of studies in the south-central region, which is more accessible and closer to important scientific centers.

The analysis of habitat utilization by the cerrado mammals (fig. 14.1A) confirms the predominance of generalists over specialists, except for the primates, which are predominantly forest specialists, and rodents, which have as many specialist species for forests as for open areas. Xenarthra is the only taxon with a predominance of open-area species.

DIETS

The mammals of the cerrados are grouped here according to feeding habits resulting in 12 diet categories. The insectivorous feeding habit

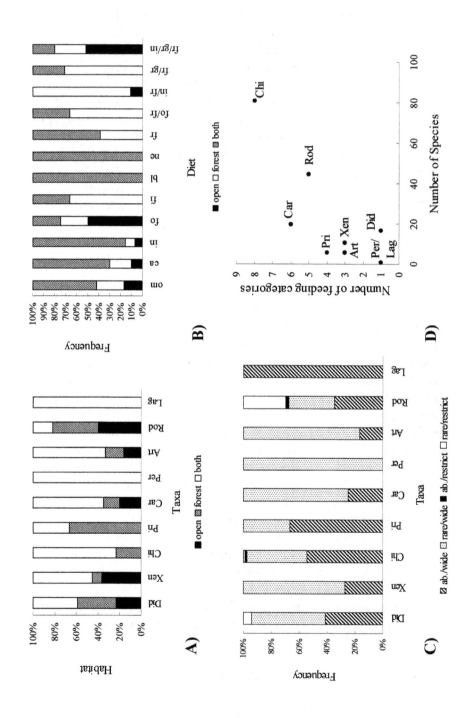

A)

Habitat

open ▨ forest ☐ both

Taxa: Did, Xen, Chi, Pri, Car, Per, Art, Rod, Lag

B)

Frequency

Diet

■ open ☐ forest ▨ both

Diet categories: om, ca, in, fo, fi, bi, ne, fr, fo/fr, in/fr, fr/gr, fr/gr/in

C)

Frequency

Taxa: Did, Xen, Chi, Pri, Car, Per, Art, Rod, Lag

▨ ab./wide ▨ rare/wide ■ ab./restrict ☐ rare/restrict

D)

Number of feeding categories

Number of Species

•Chi

•Rod

•Car

•Pri •Xen •Per/ •Did
      Art Lag

ranks as the most frequent, including around 27% of the species. Among the mammals that basically feed on insects, the orders Chiroptera and Xenarthra stand out, along with a single member of the Carnivora (table 14.1). More than 80% of the species with insectivorous feeding habits exploit open areas as much as forests (fig. 14.1B). The second most frequent group, comprising about 18% of the species, consists of omnivorous mammals, which feed on items of both animal and plant origin. Despite representing a rather diverse group, including species from several mammalian orders, 49% are from the order Didelphimorphia. The great majority are small-sized animals (54% of the omnivorous species weight less than 500 g), and use open areas as much as forests (fig. 14.1B). Fruits, representing a highly important food resource for cerrado mammals, are consumed by 55% of the species, ranging in size from small to large in several mammalian orders. Primarily frugivorous mammals include many bat species, one primate and one carnivore, representing about 9% of the species total.

Diets represented by only one food category are relatively frequent among cerrado mammals. In contrast to common observations in relation to habitat, most of the mammalian fauna consists of dietary specialists. Carnivores, frugivores, insectivores, folivores, piscivores, sanguivores, and nectarivores account for 54% of the total number of species. Orders with a greater number of species also tend to have more feeding categories ($r^2 = 0.58$; $P = .017$, fig. 14.1D), reflecting a possible niche segregation among species, as observed for canids, which present very little overlap among diets (Juarez 1997).

Figure 14.1 (*preceding page*) (A) Pattern of habitat utilization by cerrado mammalian orders. Did = Didelphimorpha; Xen = Xenarthra; Chi = Chiroptera; Pri = Primates; Car = Carnivora; Per = Perissodactyla; Art = Artiodactyla; Rod = Rodentia; Lag = Lagomorpha. (B) Feeding habits of cerrado mammals associated with habitat types; om = omnivorous; ca = carnivorous; in = insectivorous; fo = folivorous; fi = fish; bl = blood; ne = nectarivorous; fr = frugivorous; fo/fr = folivorous/frugivorous; in/fr = insectivorous/frugivorous; fr/gr = frugivorous/granivorous; fr/gr/in = frugivorous/granivorous/insectivorous. (C) Relative frequency of Brazilian cerrado mammalian species in the four categories of rarity: locally abundant and widespread; locally rare and widespread; locally abundant with restricted distribution; and locally rare with restricted distribution. (D) Number of feeding categories in relation to the number of species in each mammal order in the Brazilian cerrados. Twelve feeding categories are recognized, as indicated in (B).

CONSERVATION STATUS OF THE
CERRADO MAMMALIAN FAUNA

Based on the Official Brazilian List of Species Threatened with Extinction (Bernardes et al. 1989), 17 species that are confirmed to occur in the Cerrado Biome are threatened (table 14.1). Following Arita (1993), who made a similar analysis of Central American bats, we classified the cerrado mammalian species into four categories, considering their distributional ranges and relative abundances from data in the literature and our own experience: species (a) locally abundant with restricted distribution; (b) locally abundant and widespread; (c) locally scarce and widespread; and (d) locally scarce with restricted distribution. Even considering the variation in numbers of species with wide distributions, this classification seems adequate to delineate a general pattern. Species with broad distributions are less threatened than restricted species, and locally scarce species tend to be more vulnerable than locally abundant ones. Rare species with narrow distributions face the highest risks.

Of the species here analyzed (see table 14.1), 47.6% possess wide distributions and are locally rare; 42.7% are locally abundant and widely distributed; 1.1% are locally abundant but have restricted distributions; and 8.6% are locally rare and have a restricted distribution. The latter categories are principally found in the order Rodentia, along with a representative from the order Chiroptera and another from Didelphimorphia (fig. 14.1C). Most of the cerrado species that are considered threatened with extinction (83%; see Bernardes et al. 1989) fall in the category of locally rare with widespread distribution, and only two species are in the higher risk category (locally rare and restricted distribution).

This discrepancy may reflect the lack of knowledge regarding mammals of the cerrado region (which can be expanded to all of South America). Few data are available for the true status of many species, especially small, rare, and geographically restricted species. The larger species, which tend to have a greater emotional appeal, are more frequently listed, as are those with wider distribution. In fact, 80% of the cerrado species weighing over 50 kg are listed as threatened; whereas 36.4% of the species between 5 and 50 kg, 17.4% of the species between 0.5 and 5 kg, and just 0.8% of the species less than 0.5 kg are also considered at risk. One should also consider that certain abundant or widely distributed species, theoretically not at high risk for extinction, are nonetheless threatened by factors extrinsic to their biology, related instead to anthropogenic pressures (chapters 5, 18). The pattern currently recognized here is probably the sum of both situations.

ACKNOWLEDGMENTS

We thank Marina Anciães, Nana Rocha, Ludmilla Aguiar, and Patrícia S. de Oliveira for helping with the data compilation and elaboration of the mammal list. Ricardo "Pacheco" Machado produced the species distribution maps used in our analysis. We also thank the Fundação Biodiversitas and Conservation International for the GIS treatment and analysis. Marc Johnson helped with the English version. The Brazilian Research Council (CNPq) provided financial support to JMF (Proc. 300591/86-1).

REFERENCES

Alho, C. J. R. 1982. Brazilian rodents: Their habitats and habits. In M. A. Mares and H. H. Genoways, eds., *Mammalian Biology in South America,* pp. 143–166. Pittsburgh: Pymatuning Laboratory of Ecology, University of Pittsburgh.

Arita, H. T. 1993. Rarity in Neotropical bats: Correlations with phylogeny, diet, and body mass. *Ecol. Appl.* 3:500–517.

Bernardes, A. T., A. B. M. Machado, and A. B. Rylands. 1989. *Fauna Brasileira Ameaçada de Extinção*. Belo Horizonte: Fundação Biodiversitas.

Bourlière, F. 1983a. The Savanna mammals: Introduction. In F. Bourlière, ed., *Tropical Savannas*, pp. 359–361. Amsterdam: Elsevier.

Bourlière, F. 1983b. Mammals as secondary consumers in savanna ecosystems. In F. Bourlière ed., *Tropical Savannas*, pp. 463–475. Amsterdam: Elsevier.

Cabrera, A. 1957. Catalogo de los Mamiferos de America del Sur (Metatheria-Unguiculata-Carnivora). *Rev. Mus Arg. Cienc. Nat. Bernardino Rivadavia* 4:1–307.

Cabrera, A. 1961. Catálogo de los Mamíferos de America del Sur (Sirenia-Perissodactyla-Artiodactyla-Lagomorpha-Rodentia-Cetacea). *Rev. Mus Arg. Cienc. Nat. Bernardino Rivadavia* 4:1–732.

Eisenberg, J. F., M. A. O'Connel, and P. V. August. 1979. Density, productivity, and distribution of mammals in two Venezuelan habitats. In J. F. Eisenberg, ed., *Vertebrate Ecology in the Northern Neotropics*, pp. 187–207. Washington, D.C., Smithsonian Institution Press.

Eisenberg, J. F. and K. H. Redford, 1999. *Mammals of the Neotropics: The Central Neotropics, Ecuador, Peru, Bolivia, Brazil*. Chicago: University of Chicago Press.

Emmons, L. H. and F. Feer. 1990. *Neotropical Rainforest Mammals: A Field Guide*. Chicago: University of Chicago Press.

Fonseca, G. A. B. and K. H. Redford. 1984. The mammals of IBGE's Ecological Reserve, Brasília, and an analysis of the role of gallery forests in increasing diversity. *Rev. Bras. Biol.* 44:517–523.

Fonseca, G. A. B., G. Herrmann, Y. L. R. Leite, R. A. Mittermeier, A. B. Rylands, and J. L. Patton. 1996. Lista anotada dos mamíferos do Brasil. *Occas. Pap. Cons. Biol.* 4:1–38.

Fonseca, G. A. B., G. Herrmann, and Y. L. R.Leite. 1999. Macrogeography of Brazilian mammals. In J. Eisenberg and K. H. Redford, eds., *Mammals of the Neotropics: The Central Neotropics, Ecuador, Peru, Bolivia, Brazil* , pp. 549–563. Chicago: University of Chicago Press.

Freeland, W. J. 1991. Large herbivorous mammals: Exotic species in northern Australia. In: P. A. Werner, ed., *Savanna Ecology and Management: Australian Perspectives and Intercontinental Comparisons*, pp. 101–105. Oxford: Blackwell Scientific Publications.

Hershkovitz, P. 1987. A history of the recent mammalogy of the Neotropical Region from 1492 to 1850. In B. D. Patterson and R. M. Timm, eds., *Studies in Neotropical Mammalogy: Essays in Honor of Philip Hershkovitz. Fieldiana: Zoology* 39:11–98

IBGE. 1993. *Mapa de Vegetação do Brasil.* Brasília: Fundação Instituto Brasileiro de Geografia e Estatística.

Juarez, K. M. 1997. "Dieta, Uso de Habitat e Atividade de Três Especies de Canídeos Simpátricas do Cerrado." Master's thesis, Universidade de Brasília, Brasília, Brazil.

Koopman, K. 1982. Biogeography of bats of South America. In M. A. Mares and H.H. Genoways (eds). *Mammalian Biology in South America*, pp. 273–302. Pittsburgh: Pymatuning Laboratory of Ecology, University of Pittsburgh.

Koopman, K. F. 1993. Order Chiroptera. In D. E. Wilson and D. M. Reeder, eds., *Mammal Species of the World: A Taxonomic and Geographic Reference*, pp. 137–241. Whashington, D.C.: Smithsonian Institution Press.

Mares, M. A., M. R. Willig, and T. E. Lacher Jr. 1985. The Brazilian caatinga in South American zoogeography: Tropical mammals in a dry region. *J. Biogeogr.* 12:57–69.

Marinho-Filho, J. 1998. Informações prévias para o grupo temático "Mastozoologia". In *Ações Prioritárias para a Conservação da Biodiversidade do Cerrado e Pantanal.* Brasília: Conservation International, Funatura, Fundação Biodiversitas, and Universidade de Brasília.

Marinho-Filho, J. and M. L. Reis. 1989. A fauna de mamíferos associada as matas de galeria. In L. M. Barbosa, ed., *Simpósio sobre Mata Ciliar, Anais*, pp. 43–60. Campinas: Fundação Cargill.

Marinho-Filho, J., M. L. Reis, P. S. Oliveira, E. M. Oliveira, and M. N. Paes, 1994. Diversity standards, small mammal numbers and the conservation of the cerrado biodiversity. *An. Acad. Bras. Ciênc.* 66 (supp.):149–157.

Marinho-Filho, J., F. H. G. Rodrigues, M. M. Guimarães, and M. L. Reis. 1998. Os mamíferos da Estação Ecológica de Águas Emendadas, Planaltina, DF. In J. Marinho-Filho, F.H.G. Rodrigues and M. Guimarães, eds., *Vertebrados da Estação Ecológica de Águas Emendadas: História Natural e Ecologia em um Fragmento de Cerrado do*

Brasil Central. pp. 34–63. Brasília: Secretaria de Meio Ambiente e Tecnologia e Instituto de Ecologia e Meio Ambiente.

Marinho-Filho, J. and I. Sazima 1998. Brazilian bats and conservation biology: A first survey. In T. H. Kunz and P. A. Racey, eds., *Bat Biology and Conservation*, pp. 282–294. Washington D.C.: Smithsonian Institution Press.

Medellin, R. A. and K. H. Redford. 1992. The role of mammals in forest-savanna boundaries. In P. A. Furley, J. Proctor, and J. Ratter, eds., *Nature and Dynamics of Savanna Boundaries*, pp. 519–548. London: Chapman and Hall.

Moojen, J. 1952. *Os Roedores do Brasil.* Rio de Janeiro, Instituto Nacional do Livro.

Musser, G. G. and M. D. Carleton 1993. Family Muridae. In D. E. Wilson and D. M. Reeder, eds., *Mammal Species of the World: A Taxonomic and Geographic Reference.* 2nd. ed. Washington, D.C.: Smithsonian Institution Press.

Naturae, 1996. *Inventário da Fauna da Área sob Influência da UHE Serra da Mesa.* Relatório Técnico. Rio de Janeiro: Furnas Centrais Elétricas.

Ojasti, J. 1983. Ungulates and large rodents of South America. In F. Bourlière, ed., *Tropical Savannas*, pp. 427–439. Amsterdam: Elsevier.

PCBAP, 1997. *Plano de Conservação da Bacia do Alto Paraguai (Pantanal).* Vol II, Tomo III. Brasília: Ministério do Meio Ambiente Recursos Hídricos e da Amazônia Legal e Secretaria de Coordenação dos Assuntos de Meio Ambiente.

Percequillo, A. R. 1998. "Sistemática de *Oryzomys* Baird 1858 do Leste de Brasil (Muroidea, Sigmodontinae)." Master's thesis, Universidade de São Paulo, São Paulo, Brazil.

Redford, K. H., and G. A. B. Fonseca. 1986. The role of gallery forests in the zoogeography of the Cerrado's non-volant mammalian fauna. *Biotropica* 18:126–135.

Rodrigues, F. H. G. and E. L. A. Monteiro-Filho. 1999. Feeding behaviour of the pampas deer: A grazer or a browser? *IUCN Deer Specialist Group News* 15:12–13.

Sarmiento, G. 1983. The savannas of Tropical America. In F. Bourlière, ed., *Tropical Savannas*, pp. 245–288. Amsterdam: Elsevier.

Schaller, G. B. 1983. Mammals and their biomass on a Brazilian ranch. *Arq. Zool.* 33:1–36.

Sick, H., 1965. A fauna do cerrado. *Arq. Zool.* 12:71–93.

Silva, J. M. C. 1995a. Biogeographic analysis of the South American Cerrado avifauna. *Steenstrupia* 21:47–67.

Silva, J. M. C. 1995b. Birds of the Cerrado region, South America. *Steenstrupia* 21:69–92.

Silveira, L. 1999. "Ecologia e Conservação dos Mamíferos Carnívoros do Parque Nacional das Emas, Goiás." Master's thesis, Universidade Federal de Goiás, Goiás, Brazil.

Sinclair, A. R. E. 1983. The adaptations of African ungulates and their effects on community function. In F. Bourlière ed., *Tropical Savannas*, pp. 401–426. Amsterdam: Elsevier.

Streilein, K. E. 1982. Behavior, Ecology and distribution of the South American marsupials. In M. .A. Mares and H. H. Genoways eds., *Mammalian Biology in South America*. pp. 231–250. Pittsburgh: Pymatuning Laboratory of Ecology, University of Pittsburgh.

Tomas, W. M. and S. M. Salis. 2000. Diet of the marsh deer (*Blastocerus dichotomus*) in the Pantanal wetland, Brazil. *Stud. Neotrop. Fauna and Environ.* 35:165–172.

Vanzolini, P. E. 1963. Problemas faunísticos do cerrado. In M. G. Ferri, ed., *Simpósio Sobre o Cerrado*, pp. 307–320. São Paulo: Editora da Universidade de São Paulo.

Vieira, C. C. 1942. Ensaio monográfico sobre os quirópteros do Brasil. *Arq. Zool. São Paulo*, 3:219–471.

Wilson, D. E. and D. M. Reeder. (eds). 1993. *Mammal Species of the World: A Taxonomic and Geographic Reference.* 2nd. ed. Washington, D.C.: Smithsonian Institution Press.

Part IV

Insect-Plant Interactions

15

Ant Foraging on Plant Foliage: Contrasting Effects on the Behavioral Ecology of Insect Herbivores

Paulo S. Oliveira, André V. L. Freitas,
and Kleber Del-Claro

ANTS ARE DOMINANT ORGANISMS WHOSE INDIVIDUAL COLONIES may contain several million workers. Their numerical dominance in terrestrial habitats is combined with a broad taxonomic diversity and a widespread distribution throughout the Globe (Hölldobler and Wilson 1990). The ecological success of ants is attributed to their eusocial mode of life, local abundance, and diversity of adaptations, among other things (Wilson 1987). Such traits result in a wide variety of feeding habits and foraging strategies, including the use of plant foliage as a foraging substrate (Carroll and Janzen 1973). Intense foraging on vegetation appears to have set the scenario for a multitude of interactions with many plant species worldwide, ranging from facultative to obligate ant-plant associations (reviewed by Davidson and McKey 1993; Bronstein 1998). Incidentally, by frequently foraging on the plant surface, ants often affect the life of a particular trophic group: the herbivores.

Why are ants so common on foliage? First, ants may nest in plant structures, and therefore the plant itself is part of the colony's immediate patrolled area (Janzen 1967). Second, ground-nesting ants may extend their foraging areas by climbing on plants to search for food (Carroll and Janzen 1973). A predictable food source can reinforce ant visitation to a particular plant location, and plant-derived food products such as extrafloral nectar and/or food bodies are known to promote ant activity on foliage (Bentley 1977; Koptur 1992). Additionally, some insect herbivores may also produce food secretions that are highly attractive to a variety of ant species (Way 1963; Malicky 1970; DeVries and Baker 1989).

Whatever the factor promoting their activity on plants, ants may affect the life of insect herbivores in different ways, resulting in positive, negative, or neutral consequences (Bronstein 1994). Most studies on ant-plant interactions, however, have focused on the deterrence of insect herbivores by ants and the possible influence of such activity on plant fitness (Bronstein 1998, and included references). Rarely has this interface been studied from the herbivore's standpoint (Heads and Lawton 1985). In this chapter we illustrate how intense ant activity on plant foliage can strongly affect the behavioral ecology of insect herbivores in the cerrado. We first present the factors that likely promote ant foraging on cerrado plants, and then describe two case studies that demonstrate a close link between the behavior of insect herbivores and their encounters with ants on the plant surface.

ANT FORAGING ON CERRADO PLANTS

Several factors contribute to the ubiquity of ants on cerrado plant foliage. First, the stems of many plants are hollowed out by boring beetles, and the galleries are then used as nesting sites by numerous arboreal ant species. Morais (1980) recorded a total of 204 arboreal ant colonies in 1,075 m² of *campo cerrado* (scattered shrubs and trees; see chapter 6), and within this area 136 live woody individuals and 17 dead standing trunks were found to house stem-nesting ants. Such a high occurrence of ant nests in the vegetation likely results in intensive foraging on cerrado foliage (Morais 1980; Morais and Benson 1988) and rivals similar censuses undertaken in tropical forests (Carroll 1979). Second, plants bearing extrafloral nectaries are abundant among local woody floras (Oliveira and Leitão-Filho 1987; Oliveira and Oliveira-Filho 1991; Oliveira and Pie 1998), and such glands have been shown to be important promoters of ant activity on the cerrado foliage (see fig. 15.1A, B; see also Oliveira et al. 1987, Oliveira and Brandão 1991; Costa et al. 1992; Oliveira et al. 1995). Third, insect herbivores that produce food secretions play a key role in attracting ants to leaves, and both honeydew-producing homopterans and lycaenid butterfly larvae are known to induce ant foraging on cerrado plants (fig. 15.1C, D; Dansa and Rocha 1992; Lopes 1995; Del-Claro and Oliveira 1996, 1999; Diniz and Morais 1997).

Given that ants are dominant components of the insect fauna found on the cerrado foliage, experimental investigation of ant-herbivore inter-

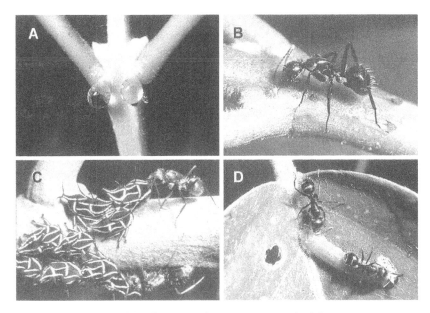

Figure 15.1 Liquid food sources for ants on cerrado foliage. (A) Accumulated extrafloral nectar in *Qualea grandiflora* (Vochysiaceae). (B) *Camponotus* sp. collecting extrafloral nectar at *Q. grandiflora*. (C) *Camponotus blandus* collecting honeydew from a *Guayaquila xiphias* treehopper. (D) *Synargis* (Riodininae) caterpillar being tended by *Camponotus* ants.

actions in this vegetation type should be particularly profitable for testing hypotheses concerning the impact of ants on herbivore survival and behavior. Recent experimental work with two distinct systems has provided strong evidence that the occurrence of ant-herbivore encounters on the host plant can be largely mediated by behavioral patterns of the herbivore. Results show that, depending on the nature of the impact of the ants (negative or positive), herbivore behavior can promote either the breakage or the reinforcement of the relationship, thereby decreasing or increasing the chance of encountering an ant on the host plant (see fig. 15.2).

ANT-BUTTERFLY INTERACTIONS

Ant effects on butterfly biology and behavior have been investigated for decades, with most studies focusing on myrmecophilous lycaenid species whose larvae are protected against natural enemies by tending ants (Malicky 1970; Pierce and Mead 1981; Pierce and Elgar 1985; DeVries 1984, 1991). By living in close proximity to ants, however, butterfly larvae risk

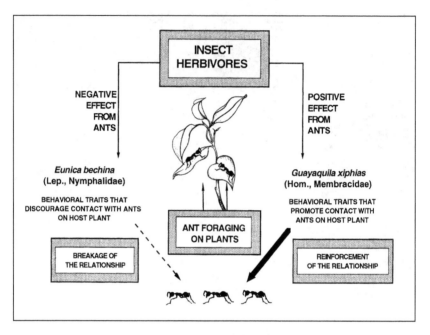

Figure 15.2 Diagram illustrating how behavioral traits of insect herbivores can mediate contact with ants on the host plant. Depending on the nature of the impact from foraging ants—negative or positive—herbivores can mediate the interaction either by avoiding (*Eunica* butterflies) or promoting (*Guayaquila* treehoppers) contact with ants on the host plant.

attack by the latter (but see DeVries 1991). This risk is minimized/avoided by lycaenid larvae via traits that decrease physical damage from ant attacks, reduce ant aggressiveness, and/or incite tending behavior. Such traits include a protective thick cuticle, the production of sweet appeasing substances, ant-mimicking vibration calls, and reduction of the beat reflex upon disturbance (Malicky 1970; DeVries 1990, 1991). The energetic costs to lycaenid larvae of feeding associated ants may include a prolongation of larval development (Robbins 1991) and a sex-related loss of pupal weight (Fiedler and Hölldobler 1992). However, no measurable cost to larvae has been found in other ant-lycaenid systems (DeVries and Baker 1989; Cushman et al. 1994; Wagner and Martinez del Rio, 1997).

On the other hand, larvae of non-myrmecophilous butterflies are frequently preyed upon or removed from host plants by foraging ants (Jones 1987; Freitas and Oliveira 1992, 1996; Freitas 1999). Caterpillars of many lepidopteran species have evolved traits to escape ant predation, especially on ant-visited plants (reviewed by Heads and Lawton 1985; Salazar and Whitman 2001). Few studies, however, have been conducted on these systems, and most were with *Heliconius* butterflies (Benson et al. 1976; Smiley 1985, 1986). To date only one ant-butterfly system has been documented in greater detail in the cerrado (Freitas and Oliveira 1992, 1996; Oliveira 1997). We here summarize the negative effects of ants on a non-myrmecophilous butterfly, *Eunica bechina*, and show that both the larval and adult stages possess traits that result in decreased contact with ants on a highly ant-visited host plant (fig. 15.2).

Negative Impact of Ants: Eunica Butterflies vs. Ants on a Nectary Plant

Eunica bechina (Nymphalidae) is a non-myrmecophilous butterfly whose larvae feed on *Caryocar brasiliense* (Caryocaraceae). This host plant bears extrafloral nectaries on the sepals and leaf buds, and is visited day and night by 34 species of nectar-gathering ants in an area of cerrado *sensu stricto* (see chapter 6) near Itirapina, SE Brazil (Oliveira and Brandão 1991). Controlled ant-exclusion experiments revealed that visiting ants decrease the infestation levels of three common herbivores of *Caryocar*, including eggs and larvae of *E. bechina* (see Oliveira 1997).

Females lay eggs singly on young leaves, on which the caterpillars preferably feed (Oliveira and Freitas 1991). As also recorded for a number of other nymphalids (DeVries 1987), *Eunica* larvae rest on stick-like frass chains constructed at leaf margins (see fig. 15.3B). A series of field observations and experiments on the system involving *Eunica* and ants (Freitas and Oliveira 1992, 1996) has demonstrated that the behavioral biology of the butterfly is closely linked with ant activity on the host plant, and can be summarized as follows.

Ants and butterfly eggs. Although ants are known to prey on or remove insect eggs from plants (Letourneau 1983), they do not consume or remove *Eunica bechina* eggs from the host plant (Freitas and Oliveira 1996). Field observations indicated that foraging ants frequently walk in the vicinity of *Eunica* eggs but ignore them. On plants other than *Caryocar*, we have observed *Pheidole* ants preying on eggs of the nymphalid butterflies *Actinote pellenea* and *Dione juno*, whereas *Crematogaster* ants prey on eggs of *Placidula euryanassa*. Both *Pheidole* and *Crematogaster* also consume eggs of *Anaea otrere* (J. M. Queiroz and P. S. Oliveira, unpublished data). Such butterfly eggs (all non-euryteline Nymphalidae) consumed by foraging ants have a soft chorion and are weakly attached to the host plants. Features like toughness and firm attachment to leaves possibly account for the lack of attractiveness of *Eunica* eggs to the ants that forage on *Caryocar*.

Ant activity and caterpillars. Foraging ants frequently found and

Figure 15.3 Interaction between *Eunica bechina* and ants. (A) *Camponotus* sp. attacking a third-instar caterpillar. (B) Second-instar caterpillar resting motionless on the tip of its stick-like frass chain, as a *Camponotus* ant forages nearby. Note a previously used frass chain at upper left. (C) Rubber ants and (D) control rubber circles used in field experiments to test whether adult *Eunica* visually avoid ovipositing on ant-occupied plant locations. See also fig. 15.4.

attacked *Eunica* caterpillars on the host plant (fig. 15.3A), and field experiments revealed that larval mortality is affected by the rate of ant visitation to the host plant (see fig. 15.4C). Larval vulnerability to ant predation, however, varies with the ant species and size of the caterpillar (Freitas and Oliveira 1992, 1996). If touched by ants, larvae usually display the beat reflex (curling and wriggling; see Malicky 1970) and/or also jump off the leaf and hang by a silken thread. When an ant bites a caterpillar, the latter vigorously bends its body towards the ant and frequently regurgitates, eventually inhibiting further ant attacks. Moreover, field experiments have demonstrated that the stick-like frass chains built by caterpillars at leaf margins (fig. 15.3B) constitute a safe refuge against ant predation on the host plant (fig. 15.4D). Although frass chains have long been described by naturalists, and their function has been assumed to be predator avoidance (DeVries 1987, and included references), the field experiment on ant-*Eunica* interactions demonstrated their relevance for larval survival on a host plant with high rates of ant visitation.

Ant activity and ovipositing females. Female butterflies avoid

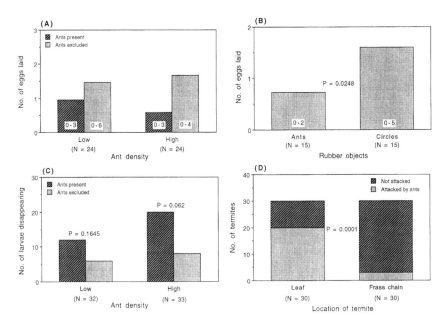

Figure 15.4   Field experiments on the interaction between *Eunica bechina* and ants on shrubs of *Caryocar brasiliense*. (A) Oviposition by *Eunica* females on egg-free experimental branch pairs during a 24 h period. Ant presence negatively affects butterfly oviposition, but the effect is significant only under high rates of ant visitation (mean > 0.5 ants per branch in six previous censuses). (Mann-Whitney *U*-tests; ranges are given inside bars). (B) In 2-choice experiments using egg-free plants, butterflies laid more eggs (after 24 hours) on plant branches bearing rubber circles than on neighboring branches with rubber ants (Mann-Whitney *U*-tests; ranges are given inside bars). (C) Ant foraging negatively affects caterpillar survival on the host plant, but mortality after 24 hours is significant only on branches with high ant density (*G* tests). (D) During 10-min trials, foraging ants attack live termites in significantly greater numbers on *Caryocar* leaves than on the frass chains constructed by *Eunica* caterpillars (*G* tests). (After Freitas and Oliveira 1996). See also fig. 15.3.

ovipositing on plant locations with high ant densities (fig. 15.4A). Although chasing by ants can have an inhibitory effect on the oviposition behavior of female insects (Janzen 1967; Schemske 1980), this was not detected in our observations of *E. bechina*. Since ants do not chase egg-laying *Eunica*, and the oviposition event lasts only 1–3 seconds, the differential occurrence of butterfly eggs on ant-visited and ant-excluded *Caryocar* plants (Oliveira 1997) presumably results from the discriminating abilities of the ovipositing female. This hypothesis was tested by simultaneously placing artificial rubber ants and rubber circles at neighboring branches of the host plant (fig. 15.3C, D). The results unequivocally indicate that branches with rubber ants were less infested than those with rubber circles (fig. 15.4B) and that visual cues (i.e., ants) likely mediated egg-laying decisions by the butterfly. Although ant presence *per se* was shown to produce an avoidance response by *E. bechina* females, ant behavior and/or chemical cues could also potentially affect female oviposition.

In conclusion, *E. bechina*, in both immature and adult life stages, possesses traits that facilitate living in an ant-rich environment. Although such traits are probably more clear-cut in the larvae (i.e., jumping off the leaf, construction of frass chains) than in the adults (selection of plants with low ant densities), the correct decision of the egg-laying female can be crucial for the survival chances of her offspring.

Ants can inhibit herbivore occupation of host plants and have been thought to provide a consistent defense system relatively immune to evolutionary changes by the herbivore (Schemske, 1980). One may expect that lepidopteran larvae bearing ant-avoiding traits would have an advantage in the cerrado ant-rich environment. Even if larvae-constructed frass chains did not evolve as a direct response to the risk of ant predation, they may have initially facilitated the use of ant-visited plants by increasing larval safety against ant attacks. Data from field experiments strongly suggest that such stick-like structures at leaf margins provide protection against walking predators (Freitas and Oliveira 1996; Machado and Freitas 2001).

Butterflies are known to use visual cues prior to oviposition to evaluate both plant quality and the presence of conspecific competitors (Rausher 1978; Williams and Gilbert 1981; Shapiro 1981). The field study of *Eunica bechina* demonstrated that visual detection of ant presence can also mediate egg-laying decision by female butterflies (Freitas and Oliveira 1996). Although the influence of ants on oviposition decisions of butterflies has been documented in species with myrmecophilous larvae (Pierce and Elgar 1985), the precise cues eliciting the oviposition response have never been determined. Although our work has shown that visual detection of ant presence can inhibit butterfly oviposition, there is likely an array of ant-avoiding traits still to be discovered.

ANT-HOMOPTERA INTERACTIONS

The honeydew produced by phloem-feeding Homoptera (primarily aphids, membracids, and scales) is an ant attractant consisting of a mixture of sugars, amino acids, amides, and proteins (Auclair 1963). Associations between ants and such homopteran groups have been commonly considered mutualistic (Way 1963). Tending ants may harvest the energy-rich fluid around the clock (fig. 15.1C) and in turn provide a range of benefits to the homopterans, including protection from predators and parasitoids, and increased fecundity (Bristow 1983; Buckley 1987). Honeydew can be a relevant item in the diet of many ant species (Tobin 1994; Del-Claro and Oliveira 1999), and intra- and interspecific competition among homopteran aggregations for the services of ants can negatively affect homopteran fitness through reduced tending levels (Cushman and Whitham 1991). Ant-derived benefits to honeydew-producing Homoptera can also vary with factors such as the species of ant partner, size of homopteran group, developmental stage of homopterans, frequency of ant attendance, and predator abundance (Cushman and Whitham 1989; Breton and Addicott 1992; Del-Claro and Oliveira 2000; Queiroz and Oliveira 2000). Therefore, the outcomes of ant-homopteran associations are strongly dependent upon the ecological conditions in which they occur (Cushman and Addicott 1991; Bronstein 1994).

Although experimental research on ant-plant-homopteran interactions has increased markedly over the past two decades, most studies come from temperate areas (e.g., Bristow 1983, 1984; Buckley 1987; Cushman and Whitham 1989, 1991). Only recently have these associations been studied in tropical habitats, including the Brazilian cerrados (Dansa and Rocha 1992; Del-Claro and Oliveira 1999, 2000). We report here on the system involving the treehopper *Guayaquila xiphias* (Membracidae) and ants, and show that ant-tending can positively affect both homopteran survival and fecundity, and that the homopterans' capacity to attract ants early in life is a crucial behavioral trait reinforcing this relationship (fig. 15.2).

Positive Impact of Ants: Guayaquila *Treehoppers and Honeydew-Gathering Ants*

The honeydew-producing treehopper *Guayaquila xiphias* feeds on shrubs of *Didymopanax vinosum* (Araliaceae) in the cerrado vegetation (*sensu stricto*, see chapter 6) near Mogi-Guaçu, SE Brazil, and occurs in aggregations of 1 to 212 individuals near the flowers or the apical meristem (see fig. 15.5A, C; Del-Claro and Oliveira 1999). *Guayaquila* females exhibit parental care and guard both the egg mass and young nymphs (fig. 15.5A, B). Nymphs develop into adults in 20–23 days, and then disperse from the natal aggregations. Treehopper aggregations are tended day and

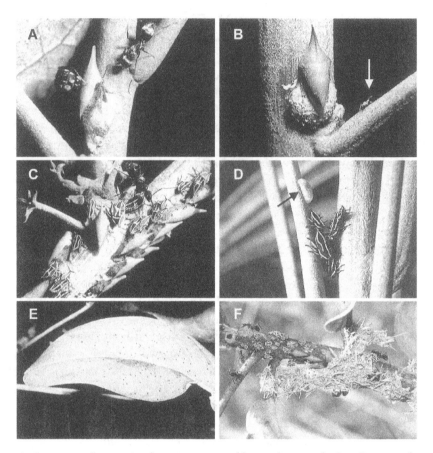

Figure 15.5 Interaction between ants and honeydew-producing *Guayaquila xiphias* treehoppers. (A) Brood-guarding *Guayaquila* female being tended by *Camponotus blandus* ants. (B) *Gonatocerus* parasitoid wasp (arrow) near an untended brood-guarding female. (C) *Camponotus rufipes* tending a *Guayaquila* aggregation. (D) Larvae of predatory *Ocyptamus* syrphid fly (arrow) near untended treehopper nymphs. (E) Scattered droplets of flicked honeydew on leaves beneath an untended *Guayaquila* aggregation. (F) Ant-constructed shelter for *Guayaquila*. See also fig. 15.6.

night by an assemblage of 21 honeydew-gathering ant species, which may construct shelters as satellite nests to house the homopterans (fig. 15.5F; Del-Claro and Oliveira 1999). The attractiveness of *Guayaquila*'s honeydew to ants is high enough to maintain tending activities unchanged, even after the ants have discovered an alternate sugar source on the host plant (Del-Claro and Oliveira 1993).

A series of field observations and controlled experiments has revealed that the treehoppers can receive a range of benefits from ant-tending and that their behavior can promote contact with ants on the host plant (Del-Claro and Oliveira 1996, 2000). The ecology of the system can be summarized as follows.

Ant effects on Guayaquila*'s natural enemies.* Due to continuous honeydew-gathering activity, ant density at any given time is higher near the treehoppers than at other plant locations, and this can markedly affect the spatial distribution and foraging behavior of *Guayaquila*'s natural enemies, such as parasitoid wasps, salticid spiders, and syrphid flies (fig. 15.5B, D). For instance, parasitoid distribution on the plant was shown to be significantly affected by increased ant activity near brood-guarding *Guayaquila*, and parasitization of treehopper ovipositions was more successful in the absence of ants (fig. 15.5A, B; Del-Claro and Oliveira 2000). Aggressive toward intruding predators and parasitoids, tending ants not only ward off such enemies from the vicinity of the treehoppers, but may also attack and kill the intruders. Controlled ant-exclusion experiments revealed that ant presence decreases the abundance of *Guayaquila*'s natural enemies on the host plant (see fig. 15.6A).

Ant-derived benefits to Guayaquila xiphias. Ant-exclusion experiments have demonstrated that tending ants can have a positive impact on treehopper survival (fig. 15.6 B). Moreover, ants can also confer a direct reproductive benefit to *Guayaquila* (see also Wood 1977; Bristow 1983). By transferring parental care to ants, ant-tended brood-guarding females (fig. 15.5A) have a higher chance of producing an additional clutch than untended females (91% *vs.* 54% of the cases; $P = 0.018$, $\chi^2 = 5.61$; $N = 22$ females in each experimental group). Two years of experimental manipulations, however, have shown that ant-derived benefits related to protection and fecundity can vary with time and/or with the species of tending ant (Del-Claro and Oliveira 2000). Several other studies have also shown that species of ants may differ greatly in the protection they afford to homopterans, and this may depend on the ants species-specific traits such as size, promptness to attack intruders, morphological and chemical weapons, as well as recruitment behavior (e.g., Addicott 1979; Messina 1981; Buckley 1987; Buckley and Gullan 1991).

Attraction of ants through honeydew flicking. Ant-tending unequivocally plays a crucial role in the survival of developing brood of *Guayaquila xiphias* in the cerrado, as also shown for other temperate ant-membracid systems (e.g., Bristow 1983; Cushman and Whitham 1989). It is therefore reasonable to predict that any behavior promoting early contact with ants would be advantageous for ant-tended treehoppers (see also DeVries 1990; DeVries and Baker 1989, on ant-tended caterpillars). *Guayaquila xiphias* females, as well as developing nymphs, frequently flick away the accumulated honeydew if it is not promptly collected by

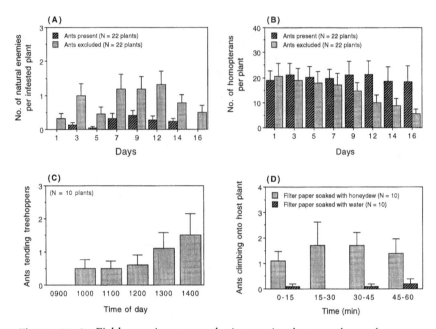

Figure 15.6 Field experiments on the interaction between honeydew-producing *Guayaquila xiphias* and tending ants on shrubs of *Didymopanax vinosum*. (A) Ant presence significantly reduces the number of *Guayaquila*'s natural enemies (spiders, syrphid flies, and parasitoid wasps) on the host plant (Treatment: $F = 11.54$, $df = 1$, $P = .0015$). (B) Ant-tending positively affects treehopper survival through time (Treatment × Time: $F = 4.33$, $df = 7$, $P = .0001$). (C) After finding scattered droplets of flicked honeydew on the ground beneath untended treehoppers, the number of ants involved with tending activities increases with time due to recruitment behavior ($F = 2.44$, $df = 5$, $P = .04$). (D) Pieces of honeydew-soaked filter paper placed beneath treehopper-free plants induce significantly more ground-dwelling ants to climb onto the plant than control papers with water (Treatment: $F = 15.89$, $df = 1$, $P = .001$). All tests performed with repeated-measures ANOVA. Data from (A) and (B) after Del-Claro and Oliveira (2000); (C) and (D) after Del-Claro and Oliveira (1996). See also fig. 15.5.

tending ants; this results in the occurrence of scattered honeydew droplets below untended or poorly tended treehopper aggregations (fig. 15.5E). Field experiments have shown that honeydew flicking by untended *Guayaquila* can provide cues to ground-dwelling ants, which climb onto the plant and start tending activities (fig. 15.6C, D; Del-Claro and Oliveira 1996). Groups of untended *Guayaquila* nymphs start secreting honeydew soon after introduction on previously unoccupied host plants.

Upon encountering the droplets on the ground, alerted ants climb onto the plant and eventually find the homopterans. The number of ants engaged in tending activities increases with time due to recruitment to the newly discovered food source (fig. 15.6C). Honeydew-soaked filter papers placed beneath unoccupied host plants further confirmed that flicked honeydew provides cues to ants and induces them to climb onto the plant (fig. 15.6D).

Attraction of Ants by Ant-Tended Insects

The presence of honeydew on lower foliage or on the ground beneath untended homopterans is well documented (Buckley 1987; Hölldobler and Wilson 1990). Douglas and Sudd (1980) discounted the possibility that scattered aphid honeydew attracted *Formica* ants since they had seen these ants ignoring fallen droplets. In the *Guayaquila*-ant association, however, we have shown that flicking accumulated honeydew can mediate this ant-homopteran system by promoting contact between potentially interacting species. Honeydew accumulated on the bodies or in the vicinity of untended homopterans may result in increased mortality due to fungal infections (Buckley 1987). It is therefore possible that ant attraction through honeydew flicking has evolved as a by-product of a primarily defensive behavior against fungi-induced damage.

Ant-tending may also confer a range of benefits to butterfly larvae in the family Lycaenidae (Pierce and Mead 1981; DeVries 1991). Some adult butterflies promote contact with ants by choosing ant-occupied plant individuals (Pierce and Elgar 1985). Myrmecophilous butterfly larvae and pupae produce substrate-borne vibrational calls, which have been demonstrated to attract nearby ants (DeVries 1990, 1992; Travassos and Pierce 2000). Therefore, for myrmecophilous butterflies, contact with tending ants can be promoted by both adults and immatures. Cocroft (1999) has recently shown that substrate-borne vibration calls are used in offspring-parent communication by *Umbonia* treehoppers. DeVries (1991b) has speculated that vibrational communication by ant-tended membracids as well as by other myrmecophilous insects could be used to maintain ant association.

CONCLUSIONS AND RESEARCH DIRECTIONS

Ant-plant-herbivore interactions offer numerous promising avenues for future research in the cerrado, with ramifications for different areas of experimental field biology and applied ecology. The uniqueness of the cerrado for this type of research relies on the prevalence of ants on the plant substrate, and on the abundance of predictable liquid food sources in the form of extrafloral nectar and insect-derived secretions. Moreover, arboreal ants commonly nest inside hollowed-out stems of cerrado plants (Morais 1980), and this per se promotes intense ant patrolling activity on leaves, regardless of the presence of liquid food rewards on the plant. The prevalence of ants on foliage makes ant-herbivore-plant interactions especially pervasive in the cerrado, as revealed by the high abundance of extrafloral nectary-mediated interactions (Oliveira and Oliveira-Filho 1991; Oliveira 1997), as well as the large number of ant-tended treehoppers (Lopes 1995) and lycaenids (Brown 1972) occurring in this biome. The data summarized in this chapter illustrate how foraging by ants on cerrado plants can affect herbivore biology in contrasting ways, and at the same time point to a number of facets in ant-herbivore systems that have not yet been investigated. For instance, although it is clear that both butterfly adults and larvae can either avoid (as in *Eunica*) or promote (as in ant-tended lycaenids) encounters with ants on the host plant, we are only beginning to understand the mechanisms through which such interactions can be behaviorally mediated by the herbivore. Although visual stimuli play an important role for *Eunica* females to avoid ants, the cues used by lycaenids to lay eggs on ant-occupied plants are still unknown. Similarly, we know virtually nothing of the decision mechanisms used by ant-tended treehoppers in selecting individual host plants. Is ant presence somehow perceived by treehopper females, and can this mediate oviposition? Can ant-tended treehoppers use vibrational communication to attract ant partners? Moreover, since the negative/positive impact of ants on a given herbivore species can vary among different ant species, can the herbivore tell ants apart and behave/respond differently to them depending on the intensity of their harmful/beneficial effects? Finally, the cerrado savanna is unique for the study of ant-plant-herbivore systems because in most cases the researcher can have full visual access to the foliage. Field work under this situation permits not only a more accurate description of the behavioral traits mediating the interactions, but also the development of controlled field experiments to identify the selective forces operating within such multitrophic systems.

ACKNOWLEDGMENTS

We thank R.J. Marquis, P.J. DeVries, S. Koptur, T.K. Wood, R.K. Robbins, G. Machado, T. Quental, and H. Dutra for helpful suggestions on the manuscript. P.J. DeVries also provided information on butterflies. Our studies in the cerrado were supported by Brazilian Federal agencies (CNPq, CAPES), and by research grants from the Universidade Estadual de Campinas (FAEP).

REFERENCES

Addicott, J. F. 1979. A multispecies aphid-ant association: Density dependence and species-specific effects. *Can. J. Zool.* 57:558–569.

Auclair, J. L. 1963. Aphid feeding and nutrition. *Ann. Rev. Entomol.* 8:439–490.

Benson, W. W., K. S. Brown, and L. E. Gilbert. 1976. Coevolution of plants and herbivores: Passion flower butterflies. *Evolution* 29:659–680.

Bentley, B. L. 1977. Extrafloral nectaries and protection by pugnacious bodyguards. *Ann. Rev. Ecol. Syst.* 8:407–428.

Breton, L. M. and J. F. Addicott. 1992. Density-dependent mutualism in an aphid-ant interaction. *Ecology* 73:2175–2180.

Bristow, C. M. 1983. Treehoppers transfer parental care to ants: A new benefit of mutualism. *Science* 220:532–533.

Bronstein, J. L. 1994. Conditional outcomes in mutualistic interactions. *Trends Ecol. Evol.* 9:214–217.

Bronstein, J. L. 1998. The contribution of ant-plant protection studies to our understanding of mutualism. *Biotropica* 30:150–161.

Brown Jr., K. S. 1972. Maximizing daily butterfly counts. *J. Lep. Soc.* 26:183–196.

Buckley, R. C. 1987. Interactions involving plants, Homoptera, and ants. *Ann. Rev. Ecol. Syst.* 18:111–138.

Buckley, R. C. and P. Gullan. 1991. More aggressive ant species (Hymenoptera: Formicidae) provide better protection for soft scales and mealybugs (Homoptera: Coccidae, Pseudococcidae). *Biotropica* 23:282–286.

Carroll, C. R. 1979. A comparative study of two ant faunas: The stem-nesting ant communities of Liberia, West Africa, and Costa Rica, Central America. *Amer. Nat.* 113:551–561.

Carroll, C. R. and D. H. Janzen. 1973. Ecology of foraging by ants. *Ann. Rev. Ecol. Syst.* 4:231–257.

Cocroft, R. B. 1999. Parent-offspring communication in response to predators in a subsocial treehopper (Hemiptera: Membracidae: *Umbonia crassicornis*). *Ethology* 105:553–568.

Costa, F. M. C. B., A. T. Oliveira-Filho, and P. S. Oliveira. 1992. The role of extrafloral nectaries in *Qualea grandiflora* (Vochysiaceae) in limiting herbivory: An experiment of ant protection in cerrado vegetation. *Ecol. Entomol.* 17:362–365.

Cushman, J. H. and J. F. Addicott. 1991. Conditional interactions in antplant-herbivore mutualisms. In C. R. Huxley and D. F. Cutler, eds., *Ant-Plant Interactions*, pp. 92–103. Oxford: Oxford University Press.

Cushman, J. H., V. K. Rashbrook and A. J. Beattie. 1994. Assessing benefits to both participants in a lycaenid-ant association. *Ecology* 75:1031–1041.

Cushman, J. H. and T. G. Whitham. 1989. Conditional mutualism in a membracid-ant association: Temporal, age-specific, and density-dependent effects. *Ecology* 70:1040–1047.

Cushman, J. H. and T. G. Whitham. 1991. Competition mediating the outcome of a mutualism: Protective services of ants as a limiting resource for membracids. *Amer. Nat.* 138:851–865

Dansa, C. V. A. and C. F. D. Rocha. 1992. An ant-membracid-plant interaction in a cerrado area of Brazil. *J. Trop. Ecol.* 8:339–348.

Davidson, D. W. and D. McKey. 1993. The evolutionary ecology of symbiotic ant-plant relationships. *J. Hym. Res.* 2:13–83.

Del-Claro, K. and P. S. Oliveira. 1993. Ant-homoptera interaction: Do alternative sugar sources distract tending ants? *Oikos* 68:202–206.

Del-Claro, K. and P. S. Oliveira. 1996. Honeydew flicking by treehoppers provides cues to potential tending ants. *Anim. Behav.* 51:1071–1075.

Del-Claro, K. and P. S. Oliveira. 1999. Ant-Homoptera interactions in a neotropical savanna: The honeydew-producing treehopper *Guayaquila xiphias* (Membracidae) and its associated ant fauna on *Didymopanax vinosum* (Araliaceae). *Biotropica* 31:135–144.

Del-Claro, K. and P. S. Oliveira. 2000. Conditional outcomes in a neotropical treehopper-ant association: Temporal and species-specific variation in ant protection and homopteran fecundity. *Oecologia* 124:156–165.

DeVries, P. J. 1984. Of crazy-ants and Curetinae: Are *Curetis* butterflies tended by ants? *Zool. J. Linn. Soc.* 79:59–66.

DeVries, P. J. 1987. *The Butterflies of Costa Rica and Their Natural History.* Princeton: Princeton University Press.

DeVries, P. J. 1990. Enhancement of symbiosis between butterfly caterpillars and ants by vibrational communication. *Science* 248:1104–1106.

DeVries, P. J. 1991a. Mutualism between *Thisbe irenea* butterflies and ants, and the role of ant ecology in the evolution of larval-ant associations. *Biol. J. Linn. Soc.* 43:179–195.

DeVries, P. J. 1991b. Call production by myrmecophilous riodinid and lycaenid butterfly caterpillars (Lepidoptera): Morphological, acoustical, functional, and evolutionary patterns. *Amer. Mus. Nov.* 26:1–23.

DeVries, P. J. 1992. Singing caterpillars, ants and symbiosis. *Scient. Amer.* 267:76–82.

DeVries, P. J. and I. Baker. 1989. Butterfly exploitation of a plant-ant mutualism: Adding insult to herbivory. *J. N. Y. Entomol. Soc.* 97:332–340.

Diniz, I. R. and H. C. Morais. 1997. Lepidopteran caterpillar fauna of cerrado host plants. *Biodiv. and Conserv.* 6:817–836.

Douglas, J. M. and J. H. Sudd. 1980. Behavioral coordination between an aphid and the ant that tends it: An ethological analysis. *Anim. Behav.* 28:1127–1139.

Fiedler K. and B. Hölldobler. 1992. Ants and *Polyommatus icarus* immatures (Lycaenidae): Sex-related developmental benefits and costs of ant attendance. *Oecologia* 91:468–473.

Freitas, A. V. L. 1999. An anti-predator behavior in larvae of *Libytheana carinenta* (Nymphalidae, Libytheinae). *J. Lep. Soc.* 53:130–131.

Freitas, A. V. L. and P. S. Oliveira. 1992. Biology and behavior of *Eunica bechina* (Lepidoptera: Nymphalidae) with special reference to larval defense against ant predation. *J. Res. Lepid.* 31:1–11.

Freitas, A. V. L. and P. S. Oliveira. 1996. Ants as selective agents on herbivore biology: effects on the behaviour of a non-myrmecophilous butterfly. *J. Anim. Ecol.* 65:205–210.

Heads, P. A. and J. H. Lawton. 1985. Bracken, ants and extrafloral nectaries: III. How insect herbivores avoid predation. *Ecol. Entomol.* 10:29–42.

Hölldobler, B. and E. O. Wilson. 1990. *The Ants.* Cambridge, Mass.: The Belknap Press of Harvard University Press.

Janzen, D. H. 1967. Interaction of the bull's horn acacia (*Acacia cornigera* L.) with an ant inhabitant (*Pseudomyrmex ferruginea* F. Smith) in Eastern Mexico. *Univ. Kansas Sci. Bull.* 47:315–558.

Jones, R. E. 1987. Ants, parasitoids, and the cabbage butterfly *Pieris rapae*. *J. Anim. Ecol.* 56:739–749.

Koptur, S. 1992. Extrafloral nectary-mediated interactions between insects and plants. In E. Bernays, ed., *Insect-Plant Interactions*, Vol. 4, pp. 81–129. Boca Raton: CRC Press.

Letourneau, D. K. 1983. Passive aggression: An alternative hypothesis for the *Piper-Pheidole* association. *Oecologia* 60:122–126.

Lopes, B. C. 1995. Treehoppers (Homoptera: Membracidae) in the Southeast Brazil: Use of host plants. *Rev. Bras. Zool.* 12:595–608.

Machado, G. and A. V. L. Freitas 2001. Larval defence against ant predation in the butterfly *Smyrna blomfildia*. *Ecol. Entomol.* 26:436–439.

Malicky, H. 1970. New aspects of the association between lycaenid larvae (Lycaenidae) and ants (Formicidae, Hymenoptera). *J. Lep. Soc.* 24:190–202.

Messina, F. J. 1981. Plant protection as a consequence of ant-membracid mutualism: Interactions on Goldenrod (*Solidago* sp.). *Ecology* 62:1433–1440

Morais, H. C. 1980. "Estrutura de uma comunidade de formigas arborícolas em vegetação de campo cerrado." Master's thesis, Universidade Estadual de Campinas, Campinas, Brazil.

Morais, H. C. and W. W. Benson 1988. Recolonização de vegetação de cerrado após queimada, por formigas arborícolas. *Rev. Bras. Biol.* 48:459–466.

Oliveira, P. S. 1997. The ecological function of extrafloral nectaries: Herbivore deterrence by visiting ants and reproductive output in *Caryocar brasiliense* (Caryocaraceae). *Funct. Ecol.* 11:323–330.

Oliveira, P. S. and C. R. F. Brandão. 1991. The ant community associated with extrafloral nectaries in Brazilian cerrados. In C. R. Huxley and D. F. Cutler, eds., *Ant-Plant Interactions*, pp. 198–212. Oxford: Oxford University Press.

Oliveira, P. S. and A. V. L. Freitas. 1991. Hostplant record for *Eunica bechina magnipunctata* (Nymphalidae) and observations on oviposition sites and immature biology. *J. Res. Lepid.* 30:140–141.

Oliveira, P. S., C. Klitzke, and E. Vieira. 1995. The ant fauna associated with the extrafloral nectaries of *Ouratea hexasperma* (Ochnaceae) in an area of cerrado vegetation in Central Brazil. *Ent. Mo. Mag.* 131:77–82.

Oliveira, P. S. and H. F. Leitão-Filho. 1987. Extrafloral nectaries: Their taxonomic distribution and abundance in the woody flora of cerrado vegetation in southeast Brazil. *Biotropica* 19:140–148.

Oliveira, P. S. and A. T. Oliveira-Filho. 1991. Distribution of extrafloral nectaries in the woody flora of tropical communities in western Brazil. In P. W. Price, T. M. Lewinsohn, G. W. Fernandes, and W. W. Benson, eds., *Plant-Animal Interactions: Evolutionary Ecology in Tropical and Temperate Regions*, pp. 163–175. New York: John Wiley and Sons.

Oliveira, P. S. and M. R. Pie. 1998. Interaction between ants and plants bearing extrafloral nectaries in cerrado vegetation. *An. Soc. Entomol. Brasil* 27:161–176.

Oliveira, P. S., A. F. da Silva, and A. B. Martins. 1987. Ant foraging on extrafloral nectaries of *Qualea grandiflora* (Vochysiaceae) in cerrado vegetation: Ants as potential antiherbivore agents. *Oecologia* 74:228–230.

Pierce, N. E. and M. A. Elgar. 1985. The influence of ants on host plant selection by *Jalmenus evagora*, a myrmecophilous lycaenid butterfly. *Behav. Ecol. Sociobiol.* 16:209–222.

Pierce, N. E. and P. S. Mead. 1981. Parasitoids as selective agents in the symbiosis between lycaenid butterfly larvae and ants. *Science* 211:1185–1187.

Queiroz, J. M. and P. S. Oliveira. 2000. Tending ants protect honeydew-producing whiteflies (Homoptera: Aleyrodidae). *Environ. Entomol.* 30:295–297.

Rausher, M. D. 1978. Search image for leaf shape in a butterfly. *Science* 200:1071–1073.

Robbins, R. K. Cost and evolution of a facultative mutualism between ants and lycaenid larvae (Lepidoptera). *Oikos* 62:363–369.

Salazar, B. A. and D. W. Whitman. 2001. Defensive tactics of caterpillars against predators and parasitoids, pp. 161–207. In T. N. Ananthakrishnan (ed.), *Insect and Plant Defense Dynamics*. Enfield: Science Publisher Inc.

Schemske, D. W. 1980. The evolutionary significance of extrafloral nectar production by *Costus woodsonii* (Zingiberaceae): An experimental analysis of ant protection. *J. Ecol.* 68:959–967.

Shapiro, A. M. 1981. The pierid red-egg syndrome. *Amer. Nat.* 117:276–294.

Smiley, J. T. 1985. *Heliconius* caterpillars mortality during establishment on plants with and without attending ants. *Ecology* 66:845–849.

Smiley, J. T. 1986. Ant constancy at *Passiflora* extrafloral nectaries: Effects on caterpillar survival. *Ecology* 67:516–521.

Tobin, J. E. 1994. Ants as primary consumers: Diet and abundance in the Formicidae. In J. H. Hunt and C. A. Nalepa, eds., *Nourishment and Evolution in Insect Societies*, pp. 279–308. Oxford: Westview Press.

Travassos, M. A. and N. E. Pierce. 2000. Acoustics, context and function of vibrational signalling in a lycaenid butterfly-ant mutualism. *Anim. Behav.* 60:13–36.

Wagner, D. and C. Martinez del Rio. 1997. Experimental tests of the mechanism for ant-enhanced growth in an ant-tended lycaenid butterfly. *Oecologia* 112:424–429.

Way, M. J. 1963. Mutualism between ants and honeydew-producing Homoptera. *Ann. Rev. Entomol.* 8:307–344

Williams, K. S. and L. E. Gilbert. 1981. Insects as selective agents on plant vegetative morphology: Egg mimicry reduces egg laying by butterflies. *Science* 212:467–469.

Wilson E. O. 1987. Causes of ecological success: The case of the ants. *J. Anim. Ecol.* 56:1–9.

Wood, T. K. 1977. Role of parent females and attendant ants in maturation of the treehopper, *Entylia bactriana* (Homoptera: Membracidae). *Sociobiology* 2:257–272.

16

Interactions Among Cerrado Plants and Their Herbivores: Unique or Typical?

Robert J. Marquis, Helena C. Morais,
and Ivone R. Diniz

A LONG DRY SEASON, FREQUENT FIRES, AND VERY LOW SOIL nutrient quality are stress factors that make life difficult for cerrado plants (see chapters 2, 4, 9). Perhaps as a result, ecological studies of plant adaptation to cerrado environments have emphasized the role of abiotic factors in shaping plant adaptation and plant distribution in these environments (Lewinsohn et al. 1991). Investigation of biotic factors such as herbivory, and of plant-animal interactions in general, have lagged behind such efforts.

Initial studies of plant-herbivore interactions in cerrado have been descriptive, focusing on basic information. Questions addressed by such studies include: how many herbivore species are involved, and what are their abundances and diets? What is the relationship between herbivorous insect abundance and plant phenology? How much damage do herbivores cause, and which plant traits account for interspecific differences in damage level? Here we provide at least partial answers to these questions, highlighting our own work at the Fazenda Água Limpa (FAL), a reserve in central Brazil (Brasília, DF), while drawing on studies conducted at other sites when available.

When tackling an understudied ecosystem, it is useful to ask how different that system is from others already studied. Besides describing our current understanding of cerrado plant-herbivore interactions, we make comparisons when appropriate with data for other tropical savanna systems and tropical forests in general.

COMPOSITION OF A LOCAL LEPIDOPTERA FAUNA

Over the last 10 years, HCM and IRD and their students have focused on the surface-feeding caterpillars found on leaves of cerrado plants at FAL. Rainfall at FAL is approximately 1,400 mm annually, with a strong dry season from May to September. Average annual temperature is 22°C.

Fifteen individuals of each of 40 plant species, representing 21 families, were censused weekly for at least one year per plant species. Collected caterpillars were then reared in the laboratory on leaves of the host species on which they were encountered. A total of 3,347 individuals of 415 species were successfully reared. A smaller sample of caterpillars (147 total reared from 33 plant species of 17 families) were collected from flowers and developing fruits (see also Clark and Martins 1987; Del Claro et al. 1997 for studies of beetle and thrips attack of flowers, respectively).

These rearing efforts demonstrated that most cerrado folivorous species were rare. Though 38 families were represented in the collection, 50% of all reared species were from just five families: Elachistidae (formerly the Oecophoridae), Gelechiidae, Pyralidae, Geometridae, and the Arctiidae. For 42% of these reared species, a single individual was reared, and in 51% of the total species, only 2–5 individuals were reared. Only 4% of the species were represented by more than 50 individuals (see Diniz and Morais 1997).

Not only were most caterpillar species rare, but caterpillars as a whole were rare. Occurrence by one or more leaf-feeding caterpillars on plants censused (= occupancy; a total of 30,000 individual plants were censused over the 10-year period) was only 10%, varying from 0.7% to 34% per plant species. Similar low rates of occupancy by caterpillars have been reported for tropical wet forest plants, including *Piper arieianum* shrubs in Costa Rican wet forest (occupancy equaled 0.07%, Marquis 1991), seedlings of various tree species in French Guiana wet forest (Basset 1999), and from sweep samples in the understory of tropical wet forest (Elton 1973; Boinski and Fowler 1989).

In a detailed study at FAL, the cumulative number of species yielded no asymptotic level of richness for three species of *Erythroxylum* (*E. deciduum*, *E. suberosum*, and *E. tortuosum*) over three time periods of 6 mo, 7 mo, and 23 mo (Price et al. 1995). The caterpillar fauna of these three tree species was strikingly different from that of a sample of woody plants from a temperate savanna (Arizona, U.S.) at the same altitude. Estimated richness per plant species was 2–3 times higher in cerrado, abundance per plant was 11-fold higher in the temperate site, and the number of plants with at least one caterpillar was 12% in cerrado versus 49% in the temperate savanna (Price et al. 1995).

Caterpillar species averaged 19 per plant species but varied greatly from four to 53 species ($N = 40$ plant species). Previous studies have shown that interspecific variation in local species richness of herbivorous insects is related to plant size, taxonomic isolation, and both local and regional estimates of abundance (reviewed by Strong et al. 1984; Marquis 1991). We used detrended correspondence analysis (DCA) to determine the plant traits that might account for this interspecific variation in cerrado Lepidoptera faunal diversity per plant species (Morais et al. in prep.). Analysis was based on a matrix of 40 species and 7 variables: growth form (herb, shrub, or tree), foliar phenology (after Morais et al. 1995), the presence of latex and extrafloral nectaries, leaf pubescence, the number of species per family in the Distrito Federal (an estimate of taxonomic isolation), and the number of censuses, an estimate of sampling intensity. The first axis explained 33.6% of the interspecific variation in lepidopteran species richness, while the cumulative variance explained by the two first axes was 61.8% (eigenvalues for axis 1 and 2 were 0.095 and 0.079, respectively). Richness was negatively related to the presence of latex, extrafloral nectaries, and leaf pubescence, with the first two variables accounting for 60% of the separation among host plant species (see fig. 16.1).

Local species richness of another guild of herbivorous insects, galling species, likewise does not show a relationship with host plant geographic range for cerrado. Rather, galling insect richness is related to plant growth form and habitat quality, with the number of galling species greatest on shrubs rather than on trees, contrary to the usual pattern. The most important factor for galling species richness in cerrado seems to be hygrothermic stress, as richness is higher in xeric than in mesic sites and declines with altitude (Fernandes and Price 1991).

Contrary to expectation, 29% of all caterpillar species (55 species from 12 families) reared from reproductive structures had been found and reared previously on leaves (Diniz and Morais 1995; Diniz and Morais submitted). The separate and combined impacts on plant fitness of herbivores feeding on both leaves and reproductive parts has not been considered (Strauss 1997), perhaps because rarely are insect herbivores recognized to feed on both plant part types. Three families represented 71% of all species eating flowers and young fruits: Gelechiidae (19 species), Tortricidae (12 species), and Pyralidae (eight species).

The reasons for the observed rarity of caterpillars in cerrado are not well understood. Presumably low abundance is due to some combination of the effect of low relative atmospheric humidity, especially at the end of

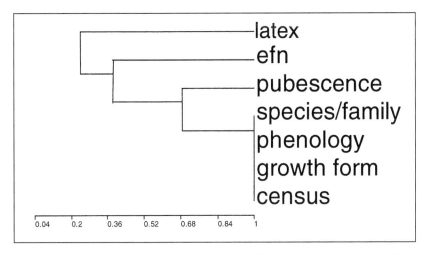

Figure 16.1   Similarity analysis (Sorensen) showing groupings, and plant traits associated with those groupings, of plant species in terms of the number of associated caterpillar species. Sixty percent of the variability is accounted for by the first two groupings, which in turn were associated with the presence of latex in the leaves and extrafloral nectaries (efn).

dry season; low nutrient content of plant material (see below); and predation pressure from natural enemies. Elton (1973; see also Coley and Barone 1997) hypothesized that high predation and parasitism were the factors that maintained low abundance of herbivorous insects at the Brazilian and Panamanian wet forest sites that he sampled. High parasitism levels (see below) would suggest that parasitoids may have an important influence on the abundance of cerrado caterpillars.

SEASONALITY AND INSECT HERBIVORE ABUNDANCE

Previous research throughout the tropics, both in the Old and New World, demonstrates in general that insect herbivore abundance becomes more seasonal as the distribution of rainfall becomes more seasonal. With a more seasonal distribution of rainfall, and increasing deciduousness of the foliage, abundance and activity of herbivores is limited to a smaller portion of the year: namely, the narrowing wet season. Further, a peak in herbivore abundance is associated with the flushing of new leaves with the coming of the wet season. In contrast, in the wettest and most aseasonal of forests, abundance changes little over the year (e.g., Barlow and

Woiwod 1990) and may sometimes be highest in the drier portion of the year (e.g., Wong 1984).

Rainfall is highly seasonal in distribution in cerrado of central Brazil, and as a result leaf flushing at the community level peaks just prior to the beginning of the wet season (Morais et al. 1995). Flowering and fruiting occur year-round, with a peak in flowering activity at the end of dry season (August to September) and a peak in fruiting at the beginning of the wet season (October to November) (Oliveira 1998). In contrast to other sites with highly seasonal rainfall, cerrado plants retain leaves throughout the dry season. Plant species show one of three patterns of leaf flush and leaf loss (Morais et al. 1995): (1) all leaves are abscised just before the major flush of new leaves (deciduous species); (2) leaves are steadily abscised beginning at the end of the dry season, throughout leaf flush, and on into the wet season (semideciduous species); or (3) leaves are produced and abscised all year round (evergreen species) and live approximately one year (most evergreen species) or, rarely, up to three years (a few evergreen species).

Herbivorous insect abundance does not fit well into the expected pattern based on seasonal distribution of rainfall alone. Based on one year of data from window, pitfall, and malaise trapping by IRD and her students, the abundance of immature and adult Homoptera, immature and adult Orthoptera, and adult Lepidoptera were not strongly seasonal, occurring in relatively high numbers throughout the year, although the peak in abundance of Homoptera and Lepidoptera adults did coincide with the beginning of the wet season (see fig. 16.2A). Distributions for all were not different from random across the year (circular statistic S_0 = 96.9, 98.0, and 88.4, r = 0.24, 0.23, 0.30, for Homoptera, adult Lepidotera, and Orthoptera, respectively) (Pinheiro et al. submitted). Most strikingly, in contrast to what has been reported for many other seasonal forest sites in which insect herbivore abundance is greatest at leaf flush (e.g., Frith and Frith 1985), or increases as the rainy season progresses (e.g., Wolda 1978; Murali and Sukumar 1993), caterpillar abundance FAL peaked 7–8 months after the main leaf flush period, coinciding instead with the early dry season (Price et al. 1995; Pinheiro et al. 1997; Morais et al. 1999; fig. 16.2B). Only Coleoptera peaked in abundance with the beginning of the wet season and just after the time of highest new leaf availability (Pinheiro et al. submitted; fig. 16.2A). This general lack of seasonality in some taxa clearly shows that herbivorous insects can persist during a harsh dry season if their food plants are available.

Morais et al. (1999) suggest escape from parasitism as a possible reason for the end of the wet season peak in caterpillar abundance. Parasitism at FAL peaked in September–October, just after the time of leaf flush (fig. 16.2B). If escape from natural enemies has been the selective

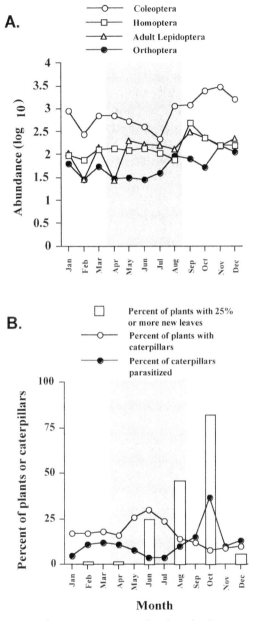

Figure 16.2 Seasonal variation in cerrado plant-herbivore-parasitoid interactions. (A) Monthly number of individuals to four herbivore orders estimated by three collection methods (pitfall, window, and Malaise traps) from May 1997 to April 1998 in cerrado *sensu stricto*, Distrito Federal, Brazil. The dry season during the period of study is indicated by the shading. (B) Seasonal pattern of new leaf production, caterpillar abundance, and parasitism of reared caterpillars in an area of cerrado in the Distrito Federal of Brazil (after Morais et al. 1995, 1999). The dry season during the period of study is indicated by the shading.

factor resulting in caterpillars feeding on mature leaves, then this suggests that selection to avoid mortality by the third trophic level has outweighed the cost of feeding on less nutritious, mature leaves.

Within this general overall pattern of caterpillar abundance are idiosyncratic patterns that do not conform to the overall trends. For example, some species can be found throughout the entire year on their host plants, such as *Cerconota achatina*, *Stenoma cathosiota* (Elachistidae: Stenomatinae), and *Fregela semiluna* (Arctiidae), whereas others are found principally during the rainy season, such as *Siderone marthesia* (Nymphalidae) and *Aucula munroei* (Noctuidae) (Andrade et al. 1995; Morais et al. 1996; Morais et al. 1999; Diniz et al. 2000; Morais and Diniz 2001).

The described seasonal patterns for central Brazil in particular may not apply to all of Brazilian cerrado. For example, there is a marked reduction in herbivorous insect abundance in general (Cytrynowicz 1991), and for Membracidae (Homoptera) in particular (Lopes 1995), during the dry and cold period between May and September in southeastern Brazil. These data suggest that herbivorous insects cannot withstand the combined stresses of low temperatures and low humidity experienced in that region.

HOST SPECIFICITY OF LEPIDOPTERAN CATERPILLARS

Herbivorous insects show a wide range of host specificity, from those that specialize on an individual plant part of only one plant species, to those that include literally hundreds of host species in their diet list. Most, however, are specialists, and initial studies in the tropics suggested that a high degree of specificity is characteristic of tropical herbivorous insects. Median host range for 12 taxa (weevils, flea beetles, hispine beetles, butterflies, and one subfamily of Geometridae) ranged from 1–4 host plant species per insect species (see review in Marquis 1991).

Since this initial list was compiled, a number of other studies have been published on the host specificity of herbivore faunas in general. In cerrado, HCM and IRD successfully reared 3,347 caterpillars from the leaves of 86 host plant species, representing 40 plant families at FAL. Of those reared, 174 species were reared only once and are not considered for this analysis. For the remainder, results show that the majority of reared caterpillar species were host-plant specialists (155 species or 64% feed within one host plant family) (see table 16.1). However, this level of specificity was lower than that reported for a Panamanian forest and a Costa Rican forest (see table 16.2), but not as low as that for a set of wet

Table 16.1   Host Specificity of Lepidopteran
Caterpillars in the Cerrado *Sensu Stricto*
of Fazenda Água Limpa, Brasília

No. of caterpillar species	No. of host plant families	No. of adults reared
155	1	1766
45	2	604
11	3	203
8	4	55
6	5	90
3	6	29
2	7	42
4	8	89
2	9	140
1	11	109
1	12	20
2	13	138
1	23	62
Total 241	38	3347

forest trees in Papua New Guinea (Bassett 1996) and one tree species in a Queensland rain forest of Australia (Bassett 1992; table 16.2). Within the cerrado sample, families of Lepidoptera vary greatly in their degree of specialization, from approximately 83% restricted to a single plant family (Pyralidae) to none (Megalopygidae) (Diniz and Morais in prep.; see table 16.3). Cerrado Membracidae (sapsucking treehoppers), like other tropical Membracidae (Wood and Olmstead 1984), demonstrate a high degree of polyphagy (Lopes 1995; table 16.2). Together, these data support the view that tropical herbivorous insect species are greatly variable in their degree of specificity, both within taxa across sites, and across taxa within sites.

The degree of specialization of the herbivore fauna of a particular plant is important from the plant's point of view. Plants attacked mainly by generalists may be able to escape those herbivores in evolutionary time by evolving chemically novel deterrents as toxins to distinguish themselves from their neighbors, with whom they share herbivores. In contrast, an herbivore fauna consisting of mostly specialists may be much more difficult to evade. For 17 species of plants the degree of herbivore specialization varied from 36% to 90%. Such a variable degree of specialization among herbivores of individual host plant species was also found for a sample of 10 Panamanian tree species (Barone 1998). The proportion of

Table 16.2 Host Specificity of Insect Herbivores in Tropical Regions

STUDY SITE	Australia	Papua New Guinea	BCI Panama	Santa Rosa Costa Rica	Cerrado Distrito Federal	Cerrado São Paulo
Reference	Basset 1992	Basset 1996	Barone 1998	Janzen 1988	Diniz and Morais, this paper	Lopes 1995
Insect groups	Chewing and sucking herbivores	Chewing herbivores	Chewing herbivores	Lepidoptera	Lepidoptera	Membracidae
Methodology	Field and literature observations	Feeding trial	Feeding trial	Rearing to maturity	Rearing to maturity	Field observations
No. insect species	283	340	46	400	241	26
No. of host plant species/ families	—	10/10	10/6	725[a]	86/40	40/20
Feeding on just one plant family	11%	54%	85%	90%	64%	31%

[a]725 = number of species of vascular plants in Santa Rosa.

specialists in the fauna was not related to the number of species involved for either the cerrado or the Panamanian sample (Barone 1998). The next step should be to assess the amount of damage to a given host plant species caused by generalists versus specialists, as Barone (1998) has done.

INTERSPECIFIC VARIATION IN LEAF DAMAGE AND LEAF QUALITY TRAITS

One approach for understanding the traits that influence the interactions between plants and their herbivores is a comparative one (i.e., to measure a number of putative defensive traits on various plant species in the same habitat, and then determine if the variation in traits among plant species accounts for differences in levels of herbivore attack among those plant species). Classic work by Coley (1983) used this approach to explain variation in leaf chewing damage for a set of 42 tree species in Panama. Similar studies involved interspecific variation in leaf chewing damage in Mexico (Filip et al. 1995) and in Australia (Lowman and Box 1983), and attack by sucking insects in Indonesia (Hodkinson and Casson 1987). Results show that young leaves are often the most vulnerable to attack

Table 16.3   Host Specificity of Families of 114
Lepidopteran Species Reared to Maturity

Family	No. of species	% of polyphagous species
Arctiidae	8	87.5
Geometridae	12	50.0
Hesperiidae	9	22.2
Limacodidae	7	87.5
Megalopygidae	6	100.0
Mimallonidae	11	18.8
Oecophoridae	8	75.0
Pyralidae	29	17.2
Riodinidae	8	87.5
Saturniidae	9	44.0
Tortricidae	7	28.6

Note: A polyphagous species is one that feeds on a host species from more than one plant family.

(Coley and Kursar 1996), and that depending on the study, leaf toughness, and nitrogen and phenolic content, can explain some of the observed interspecific variation. An added benefit of these types of studies is that they begin to indicate how rates of herbivore attack vary among locations, and which factors are responsible for such variation.

Marquis et al. (2001) undertook a study at FAL to determine the plant traits that contribute to variation among plant species in damage by insects and pathogens. For the 25 species studied (10 trees, 10 shrubs, and 5 herbs), they found that damage by herbivorous insects at the end of the leaf life (one year old at the end of the dry season) ranged between 0.5% and 14.3% per plant species (mean = 6.8%). Pathogen attack was much higher, ranging from 2.0% to 52.8% (mean = 17.3%). Damage by insects peaked at two points in the life of the leaves: during the leaf expansion period at the beginning of the wet season, and again sometime in the second half of the wet season. These two peaks in abundance coincided with known peaks in abundance of herbivorous insects in this system: the annual peak in abundance of leaf-chewing Coleoptera during the study year occurred during the leaf expansion period, while peaks in abundance of leaf-chewing Lepidoptera larvae in the years 1991–1993 occurred at the beginning of the dry season (Morais et al. 1999). The seasonal pattern of attack by pathogens was quite different, in that almost no damage occurred during the leaf expansion period, but fully expanded leaves continued to accrue damage throughout their lives.

Various traits were measured to test their impact on interspecific variation in insect and pathogen attack (leaf toughness, total phenolics, protein-binding capacity, nitrogen and water content, leaf pubescence, time to full leaf expansion, the presence of extrafloral nectaries and latex, plant size, and local species abundance). Of these, only protein-binding capacity was significantly negatively related to the amount of insect damage. Although cerrado leaves are very tough compared to those of more mesic sites (see table 16.4), and toughness was the most important predictor of interspecific variation in attack on Panamanian trees (Coley 1983), it was not a significant predictor for this set of cerrado plants. Further, leaf pubescence was found to provide protection for flushing leaves in a deciduous forest of Ghana (Lieberman and Lieberman 1984), while there was no influence of leaf pubescence on herbivore attack at the Brazil site. In fact, at least two lepidopteran herbivores of *Byrsonima crassa* and *B. verbascifolia* (Malpighiaceae) use leaf hairs to build shelters (Diniz and Morais 1997; Andrade et al. 1999).

Protein availability (nitrogen content/protein-binding capacity) and plant height were significantly positively correlated with pathogen attack. It may be that taller plant species are more susceptible to pathogen colonization because they are on average more likely to intercept windborne pathogen spores. Rate of leaf expansion was negatively correlated with pathogen attack. This result is consistent with the following scenario. Young leaves are first colonized and their tissues invaded during the leaf expansion period. The presence of the pathogen only becomes obvious after the fungi have further developed in the wet season. Plant species whose leaves rapidly pass through the vulnerable stage (period of leaf expansion) are less susceptible to attack.

The amount of pathogen and insect attack were uncorrelated for any time period or leaf age group, suggesting that insect herbivores do not contribute to infection levels, either as carriers of spores or by creating sites for infection, both activities shown to occur in other systems (e.g., de Nooij 1988). The lack of a negative correlation between herbivore and pathogen attack also suggests that herbivores were not avoiding pathogen-damaged leaves.

CROSS-SITE COMPARISONS IN INSECT DAMAGE AND LEAF QUALITY TRAITS

As a point of comparison for cerrado sites with other tropical terrestrial ecosystems, we suggest a unimodal relationship between total rainfall and length of the dry season, on the one hand, and rates of leaf area loss to

Table 16.4 Comparison of Leaf Quality Factors for Woody Plants of Three Different Neotropical Sites

Site	N	Percent nitrogen	Percent water	Total phenolics (mg/g)	Pubescence (bottom leaf surface only)	Toughness (g/mm²)	Days to full expansion
BCI Pioneers	22	3.2[a]/2.4[a]	74[a]/70[a]	127.0[a]/80.2[a]	5.9[a]/5.2[a]	11.4[a]/20.8[a]	35.1[a]
BCI Persistents	24	3.3[a]/2.2[b]	76[a]/62[b]	173.4[a]/95.6[a]	1.4[a]/0.5[a]	41.6[b]/33.8[b]	45.1[b]
Chamela	16	3.3[a]/2.7[a]	80[a]/71[a]	NA	10.1[a]/7.6[a]	8.4[a]/12.9[a]	NA
Brasília	20	1.7[b]/1.5[c]	60[b]/52[c]	289.8[b]/261.4[b]	159.1[c]/149.1[c]	91.2[c]/130.8[c]	37.6[a] ·

Note: First and second values are for young and mature leaves, respectively. N = number of species. Different superscript letters indicate significantly different means by ANOVA. BCI = Coley 1983, Chamela = Filip et al. 1995, Brasília = Marquis et al. 2001. NA = not available.

insect folivores on the other. Available data from closed canopy habitats are consistent with a pattern of decreasing leaf area loss to insect herbivores with increasing rainfall, within the rainfall range of 750–5000 mm/year (see fig. 16.3). Lower damage levels in wet evergreen forests are probably due to a combination of factors, including low host plant density, low seasonality in new leaf production, high natural enemy abundance, and the negative effects of high rainfall both on flight times of adults and feeding and survival of larvae. At sites with less rainfall, the physical effects of rainfall are ameliorated. Further, lower annual rainfall in the tropics inevitably means a greater seasonal distribution of that rain-

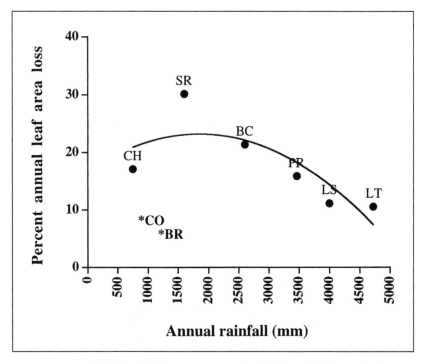

Figure 16.3 Relationship between annual rainfall and folivory by insects across various Neotropical sites. Solid circles are for relatively high soil nutrient locations, and stars are for low soil nutrient sites. BC = Barro Colorado Island (persistent trees species only (Coley 1982); CH = Chamela, Mexico (Filip et al. 1995); SR = Santa Rosa, Costa Rica (Stanton 1984); PR = El Verde, Puerto Rico (Angulo-Sandoval and Aide 2000); LS = La Selva, Costa Rica (Hartshorn et al. in Marquis and Braker 1994); LT = Los Tuxtlas, Mexico (de la Cruz and Dirzo 1987); CO = Corumbatai, São Paulo, Brazil (Fowler and Duarte 1991); BR = Fazenda Agua Limpa, Brasília, Brazil (Marquis et al. 2001). Fitted regression line: $y = -0.0000018x^2 + 0.007x + 16.732$; $r^2 = 0.66$.

fall. As a result, leaf production tends to be more synchronous not only within a plant species, but across many plant species, and timed to occur just before or at the onset of rains following the dry season. The availability of rains as a cue for herbivorous insects to emerge from diapause may allow for greater synchrony with sudden flushes of new leaves (Tauber et al. 1986). Even more so, this emerging herbivore fauna has the possibility of escaping control by their natural enemies, whose populations have also been reduced in activity during the dry season (Janzen 1993). Together, lower negative impact of rainfall and natural enemies, and greater synchrony (and predictability) of new leaf production, may lead to higher leaf area loss in dry deciduous (typhoon) forests than in wet evergreen forests.

Two estimates of end of the dry season leaf area loss by herbivores are available for cerrado (fig. 16.3; Fowler and Duarte 1991; Marquis et al. 2001). Both are approximately one-third that predicted by the relationship between rainfall and insect-caused leaf area loss described above. Differences could be due to site differences in seasonality, plant quality factors, or natural enemies. The cerrado sites are no more seasonal in their distribution of rainfall, as the length of dry season at both sites (Corumbatai = 6 mo, Brasília = 5–6 mo; Marquis et al. 2001) falls within the range found at Santa Rosa in Costa Rica (6 mo; Janzen 1993) and Chamela in Mexico (6–7 mo; Bullock and Solís-Magallanes 1990), where damage levels are much higher. Furthermore, damage levels are lower in cerrado compared to other seasonal deciduous forests, such as Santa Rosa and Chamela, despite the fact that leaves are maintained on the plant essentially year-round and thus are potentially exposed to attack for a longer time period. One other low-nutrient site in the neotropics also has correspondingly low leaf damage (Vasconcelos 1999).

One important aspect in which cerrado plants are unique compared to other sites is their phenology of leaf production. Cerrado contains an unusually high number of species that flush leaves before the rains begin (fig. 16.2B), so that the most vulnerable stage of development (the new leaf stage: Coley and Kursar 1996) has already passed by the time the herbivorous insect activity begins (at least for Coleoptera, Homoptera, and Orthoptera: fig. 16.2A). When comparisons have been made for other systems between species that flush leaves before the start of the rainy season and after, those species producing leaves prior to rains suffer less damage (Aide 1993; Murali and Sukumar 1993).

In addition to the effect of phenology of leaf flushing, initial data on plant quality factors show that leaves of cerrado plants at Brasília are tougher, and lower in nitrogen and water, than those at both Barro Colorado Island (BCI), Panama (Coley 1983) and Chamela (Filip et al. 1995; see table 16.4). Also, central Brazil plants are higher in phenolics than those on BCI (no comparative data are available from Chamela). Thus, differences in leaf quality may account for lower than expected (based on rainfall) levels of herbivory at the cerrado sites.

Finally, there are few comparative data on natural enemy composition or attack levels among savanna or tropical forest sites. Available data suggest that parasitism levels are highest sometime after the onset of the rainy season in seasonal forests, and that rates of parasitism are quite high. At the central Brazil site, parasitism of leaf-feeding Lepidoptera peaks at about 35% during the onset of the rainy season (Morais et al. 1999). A similar peak in caterpillar-feeding ichneumonids also occurs after the return of the rainy season in Kampala, Uganda (Owen and Chanter 1970). Olson (1994) found that attack by parasitoids on caterpillars of the saturniid moth *Rothschildia lebeau* was greatest (82%) during the middle rainy season, compared to early season (56%) values, at Santa Rosa in Costa Rica. In either case, rates of attack were very high. Rates of parasitism for the semideciduous forest of BCI, Panama are 15% to 25% (P. D. Coley and T. Kursar, pers. comm.).

The third trophic level also has an impact in the form of predatory ants visiting extrafloral nectaries. Specifically, ant activity at EFNs has been shown to reduce herbivory in two cerrado plants, *Qualea grandiflora* and *Caryocar brasiliense* (Costa et al. 1992; Oliveira 1997). Percentage of species and plants with EFNs at eight different cerrado sites varied from 15% to 26% and 8% to 31%, respectively (Oliveira and Leitão-Filho 1987; Oliveira and Oliveira-Filho 1991). These values all fall within the range reported for various other tropical forested habitats (5% to 80%; Coley and Aide 1991), so it does not appear that an unusually high abundance of EFNs in cerrado can explain the low rate of herbivory found there.

PATHOGEN ATTACK

A reasonable expectation for the relationship between pathogen attack and rainfall across sites is that higher pathogen attack would be found at sites with higher rainfall. The high levels of pathogen attack observed at FAL (mean = 17.3%, range = 2%–52.8%) would appear to contradict this result. In fact, leaf area loss to pathogen attack at the wet forest site of Los Tuxtlas in Mexico was on average only about 1% (G. Garcia-Guzman and R. Dirzo, pers. comm.). At the semideciduous forest on Barro

Colorado Island, Panama, pathogen damage to young leaves was higher than that in Mexico and more in accord with levels in cerrado (mean = 11.3%; range = 0 to 40.5%) (Barone 1998).

CONCLUSIONS AND FUTURE RESEARCH DIRECTIONS

In sharp contrast to neotropical savannas, the savannas of Africa, India, and Southeast Asia are (or until recently) dominated by ungulate herbivores. Studies in Africa have shown that these megaherbivores can have major impacts on plant species composition, particularly on the abundance and composition of woody biomass. This is not to say that vertebrates, both as browsers and seed predators, do not have impacts on vegetation structure in cerrado. Studies of the impacts of vertebrates as browsers, grazers, and seed predators in cerrado have not gone beyond diet studies and descriptions of community composition (Rodrigues and Monteiro-Filho 1999; chapter 14). Studies of seed predation by cerrado mammals and insects are still solely descriptive (Macedo et al. 1992; Lombardi and Toledo 1997; but see Scariot 1998).

Unfortunately, comparative data for plant-herbivorous insect interactions in Old World savannas do not appear to be available (Andersen and Lonsdale 1990; Werner 1991). This leaves us to make comparisons with other tropical forested sites. Such comparisons suggest that cerrado ecosystems differ from forests in a number of important ways. First, lepidopteran larval abundance is greatest at the end of the wet season and beginning of the dry season, a pattern unlike that reported for any other tropical site. However, as for other sites in the Old World, abundance of Coleoptera coincides with the onset of the rainy season, diet breadth of Lepidoptera is larger than once expected, and abundance of individual caterpillar species and caterpillars in general is low. Second, damage levels by herbivorous insects are much lower than expected based on our tentative model relating rainfall and annual leaf area loss to insect herbivores. Damage levels fall within the range found at low nutrient sites; however, both phenological escape (many plants produce new leaves before the increase in abundance of herbivorous insects) and low-quality tissues may account for relatively low damage levels in cerrado. Third, pathogen damage levels, at least at one cerrado site, are much higher than predicted a priori, given the level of rainfall that occurs there. How these data on pathogen attack conform with what happens at other locations, both in cerrado and out, awaits data collection from other sites.

We expect that herbivory has an important impact in cerrado, not only through its influence per se on plant growth and reproduction, but also through its potential interacting effect with abiotic factors. Although abiotic factors have great influence in cerrado, studies from other systems suggest that the impact of herbivory often modifies the influence of soil nutrient quality and water availability (Franco 1998) and interacts with fire frequency to determine plant population dynamics. Cerrado vegetation provides an excellent opportunity for future experimental work to determine the main and interactive effects of fire and herbivory on individual plant fitness, plant population dynamics, and plant succession.

Previous work in seasonally deciduous forest in Panama has shown that leafing phenology at the population level is an important determinant of early attack by chewing insects (Aide 1992, 1993). These results, together with our observation that leaf expansion rate influences attack by pathogens, suggest that a very profitable study would be an investigation of the potential role of intraspecific variation in leafing phenology on early attack by both chewing insects and pathogens in cerrado. This information then could be linked to correlated changes in plant physiology that allow plants to anticipate the coming of the rains, producing new leaves before the wet season begins. Such links between plant physiology and herbivore attack remain virtually unexplored in tropical habitats. The situation is further complicated in that the timing of the wet season is extremely variable from year to year. The consequences of such variability on plant phenology, plant reserves that allow leafing, and herbivorous insect population dynamics are unknown.

Studies of factors that control herbivory levels in tropical systems are beginning to examine influences beyond plant traits alone, including the potential role of the third trophic level. Initial data from central Brazil suggest that the third trophic level, in the form of parasitoids, may have a strong influence on the timing of lepidopteran life histories. The degree to which natural enemies maintain low population levels of caterpillars should be explored. The relative importance of top-down vs. bottom-up forces in influencing plant-insect herbivore interactions in this system is unknown.

Our focus thus far in the study of plant-insect herbivore interactions has been on cerrado vegetation itself. However, the cerrado landscape is one of savanna intermixed with gallery forests associated with streams and rivers. Because herbivores and their natural enemies may move from cerrado to gallery forests and back (e.g., Camargo and Becker 1999), the interactions that we have studied may depend on the surrounding landscape context. Gallery forests may provide shelter and alternative food sources for adults (both of herbivores and their natural enemies), and alternative food plants for larvae of herbivores. Distance to gallery forests would seem to be an important factor to consider in such a study.

In summary, there are a plethora of unanswered questions, many very basic to our understanding of cerrado plant-herbivore interactions. More detailed studies of basic natural history, including identification of the herbivores and their diet breadths, are essential. But sufficient information is now available to suggest feasible experiments that would reveal the impact of the herbivores on plant community structure and plant population dynamics in the context of the important abiotic factors. Similarly, manipulation of plants and the third trophic level can begin to reveal the factors that structure herbivore communities and drive insect population dynamics in cerrado ecosystems.

ACKNOWLEDGMENTS

We thank the many students of the project "Herbivores and Herbivory in Cerrado," who assisted in the collection and rearing of the caterpillars. This study was supported by FAPDF, FINATEC, CNPq, which also provided support for scientific initiation (PIBI-CNPq-UnB). We thank Katerina Aldás, Karina Boege, Phyllis Coley, Rebecca Forkner, Nels Holmberg, Damond Kyllo, John Lill, Eric Olson, and an anonymous reviewer for valuable comments on earlier versions.

REFERENCES

Aide, T. M. 1992. Dry season leaf production and escape from herbivory. *Biotropica* 24:532–537.

Aide, T. M. 1993. Patterns of leaf development and herbivory in a tropical understory community. *Ecology* 74:455–466.

Andersen, A. N. and W. M. Lonsdale. 1990. Herbivory by insects in Australian tropical savannahs: A review. *J. Biogeogr.* 17:433–444.

Andrade, I., I. R. Diniz, and H. C. Morais 1995. A lagarta de *Cerconota achatina* (Oecophoridae: Stenomatinae): Biologia e ocorrência em plantas hospedeiras do gênero *Byrsonima* (Malpighiaceae). *Rev. Bras. Zool.* 12:735–741.

Andrade, I., H. C. Morais, I. R. Diniz, and C. van den Berg. 1999. Richness and abundance of caterpillars on *Byrsonima* (Malpighiaceae) species in an area of cerrado vegetation in Central Brazil. *Rev. Biol. Trop.* 47:691–695.

Angulo-Sandoval, P. and T. M. Aide. 2000. Leaf phenology and leaf damage of saplings in the Luquillo Experimental Forest, Puerto Rico. *Biotropica* 32:425–422.

Barlow, H. S. and I. P. Woiwod. 1990. Seasonality and diversity of Macrolepidoptera in two lowland sites in the Dumoga-Bone National Park, Sulawesi Utara. In W. J. Knight and J. D. Holloway, eds., *Insects of the Rain Forests of South East Asia (Wallacea)*, pp. 167–172. London: Royal Entomological Society of London.

Barone, J. A. 1998. Host-specificity of folivorous insects in a moist tropical forest. *J. Anim. Ecol.* 67:400–409.

Basset, Y. 1992. Host specificity of arboreal and free-living insect herbivores in rain forests. *Biol. J. Linn. Soc.* 47:115–133.

Basset, Y. 1996. Local communities of arboreal herbivores in Papua New Guinea: Predictors of insect variables. *Ecology* 77:1906–1919.

Basset, Y. 1999. Diversity and abundance of insect herbivores foraging on seedlings in a rainforest in Guyana. *Ecol. Entomol.* 24:245–259.

Boinski, S. and N. L. Fowler. 1989. Seasonal patterns in a tropical lowland forest. *Biotropica* 21:223–233.

Bullock, S. H. and J. A. Solís-Magallanes. 1990. Phenology of canopy trees of a tropical deciduous forest in Mexico. *Biotropica* 22:22–35.

Camargo, A. J. A. and V. O. Becker. 1999. Saturniidae (Lepidoptera) from the Brazilian cerrado: Composition and biogeographic relationships. *Biotropica* 31:696–705.

Clark, W. E. and R. P. Martins. 1987. *Anthonomus biplagiatus* Redtenbacher (Coleoptera: Curculionidae): A Brazilian weevil associated with *Kielmeyera* (Guttiferae). *Coleopt. Bull.* 41:157–164.

Coley, P. D. 1982. Rates of herbivory on different tropical trees. In E. G. Leigh, Jr., A. S. Rand, and D. M. Windsor, eds., *The Ecology of a Tropical Forest: Seasonal Rhythms and Long-Term Changes*, pp. 123–132. Washington, D.C.: Smithsonian Institution Press.

Coley, P. D. 1983. Herbivory and defensive characteristics of tree species in a lowland tropical forest. *Ecol. Monogr.* 53:209–233.

Coley, P. D. and T. M. Aide. 1991. Comparison of herbivory and plant defenses in temperate and tropical broad-leaved forests. In P. W. Price, T. M. Lewinsohn, G. W. Fernandes, and W. W. Benson, eds., *Plant-Animal Interactions: Evolutionary Ecology in Tropical and Temperate Regions*, pp. 25–50. New York: John Wiley.

Coley, P. D. and J. A. Barone. 1996. Herbivory and plant defenses in tropical forests. *Ann. Rev. Ecol. Syst.* 27:305–335.

Coley, P. D. and T. A. Kursar. 1996. Anti-herbivore defenses of young tropical leaves: Physiological constraints and ecological tradeoffs. In S. S. Mulkey, R. L. Chazdon, and A. P. Smith, eds., *Tropical Forest Plant Ecophysiology*, pp. 305–336. New York: Chapman and Hall.

Costa, F. M. C. B., A. T. Oliveira-Filho, and P. S. Oliveira. 1992. The role of extrafloral nectaries in *Qualea grandiflora* (Vochysiaceae) in limiting herbivory: An experiment of ant protection in cerrado vegetation. *Ecol. Entomol.* 17:363–365.

Cytrynowicz, M. 1991. Resource size and predictability, and local herbivore richness in a subtropical Brazilian cerrado community. In P. W. Price, T. M. Lewinsohn, G. W. Fernandes, and W. W. Benson, eds., *Plant-Animal Interactions: Evolutionary Ecology in Tropical and Temperate Regions*, pp. 561–590. New York: John Wiley.

De la Cruz, M. and R. Dirzo. 1987. A survey of the standing levels of herbivory in seedlings from a Mexican rain forest. *Biotropica* 19:98–106.

Del Claro, K., R. Marullo, and L. A. Mound. 1997. A new Brazilian species of *Heteropthrips* (Insecta: Thysanoptera) co-existing with ants in the flower of *Peixotoa tomentosa* (Malpighiaceae). *J. Nat. Hist.* 31:1307–1312.

De Nooij, M. P. 1988. The role of weevils in the infection process of the fungus *Phomopsis subordinaria* in *Plantago lanceolata*. *Oikos* 52:51–58.

Diniz, I. R. and H. C. Morais. 1995. Larvas de Lepidoptera e suas plantas hospedeiras em um cerrado de Brasília, DF. *Rev. Bras. Ent.* 39:755–770.

Diniz, I. R. and H. C. Morais. 1997. Lepidopteran caterpillar fauna on cerrado host plants. *Biodiv. and Conserv.* 6:817–836.

Diniz, I. R. and H. C. Morais. Submitted. Lepidopteran larvae associated with reproductive structures of cerrado plants. *An. Soc. Entomol. Brasil.*

Diniz, I. R. and H. C. Morais. In prep. Local diet breadth of lepidopteran larvae in cerrado vegetation.

Diniz, I. R., H. C. Morais, S. Scherrer, and E. O. Emery. 2000. The polyphagous caterpillar *Fregela semiluna* (Lepidoptera: Arctiidae): Occurrence on plants in the Central Brazilian cerrado. *Bol. Herb. Ezechias Paulo Heringer* 5:103–112.

Elton, C. S. 1973. The structure of invertebrate populations inside neotropical rain forest. *J. Anim. Ecol.* 42:55–104.

Fernandes, G. W. and P. W. Price. 1991. Comparison of tropical and temperate galling species richness: The roles of environmental harshness and plant nutrient status. In P. W. Price, T. M. Lewinsohn, G. W. Fernandes, and W. W. Benson, eds., *Plant-Animal Interactions: Evolutionary Ecology in Tropical and Temperate Regions*, pp. 91–115. New York: John Wiley.

Filip, V., R. Dirzo, J. M. Maass, and J. Sarukhan. 1995. Within- and among-year variation in the levels of herbivory on the foliage of trees from a Mexican tropical deciduous forest. *Biotropica* 27:78–86.

Fowler, H. G. and L. C. Duarte. 1991. Herbivore pressure in a Brazilian cerrado. *Naturalia* 16:99–102.

Franco, A. C. 1998. Seasonal patterns of gas exchange, water relations and growth of *Roupala montana*, an evergreen savanna species. *Plant Ecol.* 136:69–76.

Frith, C. B. and D. W. Frith. 1985. Seasonality of insect abundance in an Australian upland tropical rainforest. *Austr. J. Ecol.* 10:237–248.

Hodkinson, I. D. and D. S. Casson. 1987. A survey of food-plant utilization by Hemiptera (Insecta) in the understorey of primary lowland rain forest in Salawesi, Indonesia. *J. Trop. Ecol.* 3:75–85.

Janzen, D. H. 1988. Ecological characterization of a Costa Rican dry forest caterpillar fauna. *Biotropica* 20:120–135.

Janzen, D. H. 1993. Caterpillar seasonality in a Costa Rican dry forest. In N. E. Stamp and T. M. Casey, eds., *Caterpillars: Ecological and Evolutionary Constraints on Foraging*, pp. 448–477. New York: Chapman and Hall.

Lewinsohn, T. M., G. M. Fernandes, W. W. Benson, and P. W. Price. 1991. Introduction: Historical roots and current issues in tropical evolutionary ecology. In P. W. Price, T. M. Lewinsohn, G. W. Fernandes, and W. W. Benson, eds., *Plant-Animal Interactions: Evolutionary Ecology in Tropical and Temperate Regions*, pp. 1–22. New York: John Wiley.

Lieberman, D. and M. Lieberman. 1984. The causes and consequences of synchronous flushing in a dry tropical forest. *Biotropica* 16:193–201.

Lombardi, J. A. and F. R. N. Toledo. 1997. *Davilla grandiflora* St.Hill. et Tul. (Dilleniaceae) as a host of larvae of *Mandarellus* sp. (Coleoptera: Curculionidae) in Minas Gerais, Brazil. *Ciênc. Cult.* 49:211–212.

Lopes, B. C. 1995. Treehoppers (Homoptera: Membracidae) in southeastern Brazil: Use of host plants. *Rev. Bras. Zool.* 12:595–608.

Lowman, M. D. and J. D. Box. 1983. Variation in leaf toughness and phenolic content among five Australian rainforest trees. *Austr. J. Ecol.* 8:17–25.

Macedo, M. V., T. M. Lewinsohn, and J. M. Kingsolver. 1992. New host records of some bruchid species in Brazil with the description of a new species of *Caryedes* (Coleoptera: Bruchidae). *Coleopt. Bull.* 46:330–336.

Marquis, R. J. 1991. Herbivore fauna of *Piper* (Piperaceae) in a Costa Rican wet forest: Diversity, specificity, and impact. In P. W. Price, T. M. Lewinsohn, G. W. Fernandes, and W. W. Benson, eds., *Plant-Animal Interactions: Evolutionary Ecology in Tropical and Temperate Regions*, pp. 179–208. New York: John Wiley.

Marquis, R. J. and H. E. Braker. 1994. Plant-herbivore interactions: Diversity, specificity, and impact. L. A. McDade, K. S. Bawa, H. A. Hespenheide, and G. S. Hartshorn, eds., *La Selva: Ecology and Natural History of a Neotropical Rain Forest*, pp. 261–281. Chicago: University of Chicago Press.

Marquis, R. J., I. R. Diniz, and H. C. Morais. 2001. Patterns and correlates of interspecific variation in foliar insect herbivory and pathogen attack in Brazilian cerrado. *J. Trop. Ecol.* 17:1–23.

Morais, H. C. and I. R. Diniz. 2001. Immature stage and host plant of the Brazilian cerrado moth *Aucula munroei* (Lepidoptera: Noctuidae: Agarastinae). *Trop. Lepidop.* (in press).

Morais, H. C., I. R. Diniz, and L. C. Baumgarten. 1995. Padrões de produção de folhas e sua utilização por larvas de Lepidoptera em um cerrado de Brasília, DF. *Rev. Bras. Bot.* 13:351–356.

Morais, H. C., I. R. Diniz, B. C. Cabral, and J. Mangabeira. Submitted. *Stenoma cathosiota* (Lepidoptera: Elachistidae) in the cerrado: Temporal and spatial variation of caterpillars abundance. *Ciênc. Cult.*

Morais, H. C., I. R. Diniz, and J. M. F. Maia. In prep. Richness and abundance of lepidopteran caterpillars associated with cerrado plants.

Morais, H. C., I. R. Diniz, and J. R. Silva. 1996. Larvas de *Siderone marthesia nemesis* (Lepidoptera: Nymphalidae, Charaxinae) em um cerrado de Brasília. *Rev. Bras. Zool.* 13:351–356.

Morais, H. C., I. R. Diniz, and D. M. S. Silva. 1999. Caterpillar seasonality in a central Brazilian cerrado. *Rev. Biol. Trop.* 47:1025–1033.

Murali, K. S. and R. Sukumar. 1993. Leaf flushing phenology and herbivory in a tropical dry deciduous forest, southern India. *Oecologia* 94:114–119.

Oliveira, P. E. 1998. Fenologia e biologia reprodutiva das espécies de cerrado. In S. M. Sano and S. P. Almeida, eds., *Cerrado: Ambiente e Flora*, pp. 169–192. Planaltina: Empresa Brasileira de Pesquisa Agropecuária.

Oliveira, P. S. 1997. The ecological function of extrafloral nectaries: Herbivore deterrence by visiting ants and reproductive output in *Caryocar brasiliense* (Caryocaraceae). *Funct. Ecol.* 11:323–330.

Oliveira, P. S. and H. F. Leitão-Filho. 1987. Extrafloral nectaries: Their taxonomic distribution and abundance in the woody flora of cerrado vegetation in southeast Brazil. *Biotropica* 19:140–148.

Oliveira, P. S. and A. T. Oliveira-Filho. 1991. Distribution of extrafloral nectaries in the woody flora of tropical communities in Western Brazil. In P. W. Price, T. M. Lewinsohn, G. W. Fernandes, and W. W. Benson, eds., *Plant-Animal Interactions: Evolutionary Ecology in Tropical and Temperate Regions*, pp. 163–175. New York: John Wiley.

Olson, E. J. 1994. "Dietary Ecology of a Tropical Moth Caterpillar, Rothschildia lebeau (Lepidoptera: Saturniidae)." Ph.D. thesis, University of Pennsylvania, Philadelphia, Pennsylvania, U.S.

Owen, D. F. and D. O. Chanter. 1970. Species diversity and seasonal abundance in tropical Ichneumonidae. *Oikos* 21:142–144.

Pinheiro, F., I. R. Diniz, D. Coelho, and M. P. S. Bandeira. Submitted. Temporal distribution abundance in the Brazilian cerrado: Effect of climatic variation. *Austral Ecol.*

Pinheiro, F., H. C. Morais, and I. R. Diniz. 1997. Composição de herbívoros em plantas hospedeiras com látex: Lepidoptera em *Kielmeyera* spp. (Guttiferae). In L. L. Leite and C. H. Saito, eds, *Contribuição ao Conhecimento Ecológico do Cerrado*, pp. 101–106. Brasília: ECL/Universidade de Brasília.

Price, P. W., I. R. Diniz, H. C. Morais, and E. S. A. Marques. 1995. The abundance of insect herbivore species in the tropics: High local richness of rare species. *Biotropica* 27:468–478.

Rodrigues, F. H. G. and E. Monteiro-Filho. 1999. Feeding behaviour of the pampas deer: A grazer or a browser? *Deer Specialist Group News* 15:12–13.

Scariot, A. 1998. Seed dispersal and predation of the palm *Acrocomia aculeata*. *Principes* 42:5–8.

Stanton, N. 1975. Herbivore pressure in two types of tropical forests. *Biotropica* 7:8–11.

Strauss, S. Y. 1997. Floral characters link herbivores, pollinators, and plant fitness. *Ecology* 78:1640–1645.

Strong, D. R. Jr., J. H. Lawton, and T. R. E. Southwood. 1984. *Insects on Plants: Community Patterns and Mechanisms*. Oxford: Blackwell Scientific.

Tauber, M. J., C. A. Tauber, and S. Masaki (eds). 1986. *Seasonal Adaptation of Insects*. Cambridge, UK: Oxford University Press.

Vasconcelos, H. L. 1999. Levels of leaf herbivory in Amazonian trees from different stages in forest regeneration. *Acta Amazonica* 29:615–623.

Werner, P. A. (ed.). 1991. *Savanna Ecology and Management*. Oxford, UK: Blackwell Scientific.

Wolda, H. 1978. Fluctuations in abundance of tropical insects. *Amer. Nat.* 112:1017–1045.

Wong, M. 1984. Understory foliage arthropods in the virgin and regenerating habitats of Pasoh Forest Reserve, West Malaysia. *Malaysian Forester* 47:43–69.

Wood, T. K. and K. L. Olmstead. 1984. Latitudinal effects on treehopper species richness (Homoptera, Membracidae). *Ecol. Entomol.* 9:109–115.

17

Pollination and Reproductive Biology in Cerrado Plant Communities

Paulo E. Oliveira and Peter E. Gibbs

STUDIES ON THE POLLINATION BIOLOGY AND BREEDING SYSTEMS of extensive samples of species in plant communities are valuable since they allow various conceptual issues to be addressed. With such data, comparisons can be made between different communities (e.g., moist versus seasonal forest), or between different components of the vertical strata (e.g., canopy versus understory taxa). In this way we can determine the role of diverse pollinators in different kinds of woodland, or analyze the frequency of obligate outbreeding versus self-compatible taxa in different communities or subunits of the same community. And with estimations of actual mating systems for particular species, usually using molecular genetic markers, we can attempt to determine whether self-incompatibility mechanisms do indeed result in outbreeding, or whether, due to small effective population size, consanguineous mating prevails. In this review we will focus on comparisons between community studies for the Brazilian cerrados and other neotropical woodlands.

Most community surveys of the reproductive biology of neotropical woodlands have been undertaken in Central America: Costa Rica, Panama (Barro Colorado Island) and Mexico. Preeminent among such studies are those for moist forest areas at La Selva (Costa Rica) which include flowering phenology, pollination biology, and breeding systems (Kress and Beach 1994). Other studies in Costa Rica include those at a variety of habitats ranging from moist to seasonal woodlands at localities near Cañas (Guanacaste province) (see Bawa 1990 for a review). These and other studies on Barro Colorado Island (Panama) and seasonal forests in Mexico were reviewed by Bullock (1995).

In South America, studies are available for Venezuelan secondary deciduous forest (Ruiz and Arroyo 1978), montane cloud forest in the Cordillera de la Costa at 1,700 m (Sobrevilla and Arroyo 1982), and savanna areas (e.g., Ramírez and Brito 1990). For Brazil, no community reproductive biology survey is available for Amazonia, although a number of individual species have been studied (e.g., Gribel et al. 1999). Some community studies of hummingbird and bat pollination systems have been published recently for the Atlantic coastal forest (Sazima et al. 1996, 1999).

The cerrado has a very rich flora, comprising more than 800 species of trees with perhaps three times that number of herbs and hemixyle, subshrubby species (Furley and Ratter 1988; Mendonça et al. 1998; see also chapters 6, 7). Moreover, the cerrados interdigitate with gallery forest and have extensive boundaries with mesophytic and moist forest, with its woody flora linked to neighboring tropical rainforest groups (Rizzini 1963; Sarmiento 1983; Prance 1992; Oliveira-Filho and Ratter 1995). A basic question, therefore, is whether the ecological differences between savanna and forest areas in the cerrado region involve different pollination and breeding systems.

REPRODUCTIVE BIOLOGY STUDIES IN CERRADO

Most reproductive biology studies of cerrado vegetation have been undertaken in the state of São Paulo, in the various "islands" of cerrado *sensu lato* within the SE fringe of the main cerrado area (chapter 6). Following in the tradition of Warming (1908) and his pioneer study of a cerrado community at Lagoa Santa, Mantovani and Martins (1988) studied the flowering and fruiting phenology of an area of cerrado *sensu stricto* at Mogi Guaçu by means of monthly censuses. Silberbauer-Gottsberger and Gottsberger (1988) studied the pollination biology of 279 woody and herbaceous species in an area of cerrado *sensu stricto* at Botucatu. These authors defined in general terms the floral visitors and aspects of pollination syndromes for the community, but without providing a list of species and their visitors. Saraiva et al. (1996) investigated the breeding systems of selected species in an area of cerrado *sensu stricto* at Corumbataí. They defined the sexual systems for 135 species and studied the breeding systems of a subsample of 21 species using controlled hand pollination.

In a cerrado *sensu stricto* area of 40 ha near Brasília, central Brazil, Oliveira and Gibbs (2000) studied the phenology, pollination biology and sexual systems of 59 woody species, with a subsample of 30 species studied for breeding system by means of hand pollination and observations on postpollination events in the pistil. In a similar survey, Barbosa (1997) investigated these reproductive parameters for 204 herbaceous and shrubby species in an area of ca. 3 ha of cerrado (*campo sujo*) near Uberlândia (state of Minas Gerais). Breeding systems for a subsample of 84 species were undertaken using hand pollinations, but details of the site of incompatibility were not studied.

These community surveys, together with other studies on reproductive biology of individual cerrado species (see Oliveira and Gibbs 2000 for a review), groups of species (e.g., Gottsberger 1986; Oliveira 1998b) or other reproductive aspects (e.g., dioecy, Oliveira 1996), provide a data base for pollination biology which includes around 113 species of the woody flora (estimated 800 species), around 200 of the understory flora (another 2,500 species at least), and breeding systems for some 70 woody species and 200 shrubs and herbs. Although the number of studied species in relation to the total flora is fairly modest, it is necessary to bear in mind that the cerrado flora as a whole is very heterogeneous. Ratter et al. (1996) encountered 534 woody and large shrub species in a survey of 98 sites throughout the cerrado region, of which 158 species (30%) occurred at only one site. Most studied sites will show less than 100 woody species. If we consider the 94 species which were found at 20 sites or more, and which Ratter et al. (1996) considered "a working list of the commonest tree species of the Cerrado," then our data for breeding systems represents 52% of these "core" cerrado species, while our data for pollination biology would include most of these species (71 spp., or 76%). Furthermore, data accumulated so far for sexual systems comprises 327 cerrado woody species (Oliveira 1996). Therefore, generalizations based on the data currently available are likely to give a reasonable picture of the reproductive biology of woody cerrado species. Certainly the cerrado data base in this respect is comparable with or better than that available for humid forest sites.

POLLINATION BIOLOGY

Studies of the reproductive biology of cerrado plants have shown a great diversity of pollination systems similar to those encountered in other neotropical forests (see fig. 17.1). In the cerrado *sensu lato*, bees are the main pollinators, as noted by Silberbauer-Gottsberger and Gottsberger (1988), but other pollinators have a role both for the woody and herbaceous cerrado flora (see table 17.1). They include pollinators characteristic

Figure 17.1 Diversity of pollination systems in cerrado vegetation: (A) Small bee (*Paratrigona* sp.) visiting a flower of *Casearia grandiflora*, a species pollinated mainly by flies. (B) Dynastidae beetles in a flower of *Annona coriacea*. (C) Hummingbird (*Amazilia* sp.) visiting flowers of *Palicourea rigida* (D) Noctuidae moth visiting *Aspidosperma macrocarpum* flowers. (E) Large-bee (*Centris violascens*) visiting a flower of *Eriotheca pubescens*. (F) Nectarivorous bat (*Glossophaga soricina*) visiting *Hymenaea stigonocarpa*. Sources: Original photographs by P. E. Oliveira, and (F) from Gibbs et al. 1999.

Table 17.1   Pollination Systems in Cerrado Woody Species
and Other Tropical Plant Formations

Gallery forest, Uberlândia (MG), P. E. Oliveira (unpublished)		Cerrado sensu stricto, Brasília (DF) (Oliveira and Gibbs 2000)		La Selva rainforest, Costa Rica (Kress and Beach 1994)	
System	% spp. (no.)	System	% spp. (no.)	System	% spp. (no.)
Wind	1 (1)	Wind	0	Wind	2.5 (7)
Very small insects	7 (7)	Very small insects	5 (3)	Small diverse insects	11.2 (31)
Small insects (small bees/ flies/wasps)	45 (46)	Small insects (small bees/flies/ wasps)	44 (26)	Flies	1.8 (5)
				Wasps	2.5 (7)
				Small bees	14.1 (39)
Large bees	23 (23)	Large bees	32 (19)	Large bees	24.3 (67)
Butterflies	2 (2)	Butterflies	0	Butterflies	4.3 (12)
Moths	12 (12)	Moths	12 (7)	Moths	8 (22)
Nonflying mammals	1 (1)	Nonflying mammals	0	Nonflying mammals	0
Bats	4 (4)	Bats	3 (2)	Bats	3.6 (10)
Humming- birds	0	Hummingbirds	2 (1)	Hummingbirds	14.9 (41)
Beetles	6 (6)	Beetles	2 (1)	Beetles	12.7 (35)

Note: Percentage of species in each pollination system for a gallery forest in the cerrado region, a cerrado *sensu stricto* area in Brasília, central Brazil, and at La Selva rainforest reserve in Costa Rica. Woody species only were included in the first two areas, whereas La Selva data included the complete flora. Small insects in the Brazilian data included fly, wasp, and small bee pollinated species. Very small insects or small diverse insects included small, unidentified insects, usually smaller than 5 mm.

of moist tropical forest, such as beetles and bats. As in other tropical communities (Bawa 1990), plant-pollinator relationships in the cerrado seem to involve guilds of pollinators associated with a given plant or group of plants (Oliveira and Gibbs 2000). There are few one-to-one plant-pollinator relationships, although many taxa are still poorly studied. Some cerrado plants are dependent on a restricted group of specialized vectors, as *Byrsonima* spp. (Malpighiaceae) with Centridinae bees (Anthophoridae), and *Annona* spp. with Dynastidae beetles (Gottsberger 1986). But most species rely on a broader spectrum of pollinators defined more by their size and foraging requirements than by specific interaction. Many species have small, apparently generalist flowers pollinated by a range of insects of different groups. Such species may be visited by flies, bees, and wasps, and although in some cases the main pollinators are evident, such as wasps in *Erythroxylum* (Barros 1998), or flies in *Casearia sylvestris* (Barbosa 1997), in most cases the main pollinators can be defined only on a local basis and in quantitative terms. In such cases, it may be better to group these less specialized systems as being pollinated by small generalist insects (as in Oliveira and Gibbs 2000).

The diversity of flower-visiting bees observed in different cerrado *sensu lato* studies, some 114–196 species, is similar to that at other tropical sites studied so far (Carvalho and Bego 1996). Small to medium-sized Apidae bees are the most common flower visitors in different cerrado areas (Carvalho and Bego 1996; Oliveira and Gibbs 2000). Eusocial bees such as the exotic *Apis mellifera scutellata* and the almost omnipresent *Trigona spinipes* are very common in the region. Although certainly pollinators of many plants, these bee species have also been noted for their destructive activity and pollen theft in buds and flowers (Roubik 1989). Other Meliponinae, Anthophoridae and Halictidae are small to medium-sized bees well represented in cerrado. Halictid bees have been observed as common pollinators of understory trees in gallery forests (personal observation), and their diversity seems to be higher in forests than in open habitats.

The most conspicuous and diverse group in the cerrado are the large anthophorid bee genera, such as *Centris* and *Xylocopa* (Carvalho and Bego 1996; Oliveira and Gibbs 2000). Some large bee genera of other families, such as *Eulaema*, *Bombus*, *Melipona* (Apideae), and *Oxaea* (Oxaeidae), are also common pollinators of cerrado plants (Carvalho and Bego 1996; Oliveira and Gibbs 2000), and a guild of large-bee flowers may be delimited for the cerrado areas. Many large-bee genera observed in cerrado are also cited for wet and dry forests in Costa Rica (Frankie et al., 1983; Bawa, Bullock, Perry et al. 1985; Kress and Beach 1994).

Other pollination systems common in tropical areas are also present in cerrado. Beetles are pollinators of cerrado species of Annonaceae (Gottsberger 1986, 1989). Hawkmoths are pollinators of some important species including *Qualea grandiflora* (Silberbauer-Gottsberger and Gottsberger 1975), probably the most widespread cerrado tree species (Ratter et al. 1996). Bat pollination is present in some tree species (Sazima and Sazima 1975; Gribel and Hay 1993; Gibbs et al. 1999) and also in small shrub species of *Bauhinia* (Barbosa 1997).

Some pollinators are absent or poorly represented in the cerrado *sensu lato*. Butterflies seem to be important pollinators of herbaceous Asteraceae (Silberbauer-Gottsberger and Gottsberger 1988), but in the

study by Oliveira and Gibbs (2000) no woody species was effectively pollinated by these vectors. At the same site, hummingbirds were pollinators
of only one species, although they were common opportunistic visitors of
more than 30% of the surveyed woody species. However, hummingbirds
were observed as true pollinators of some shrubs and herbs in open cerrado, *campo sujo*, areas (Barbosa 1997; chapter 6). It is notable that wind
pollination, which has been associated with seasonally dry areas (Bullock
1994), is rare among cerrado trees and shrubs, and occurs commonly only
in the grasses. Vertical stratification of pollination systems observed in
tropical forests (Bawa, Bullock, Perry et al. 1985; Kress and Beach 1994)
occurs also in cerrado. Moths and bats are pollinators mostly of trees,
while wind and hummingbird pollination appear mostly in the herbaceous layer. Dominance of animal pollination and virtual absence of wind
pollination may reflect the rainforest origin of the cerrado *sensu lato*
woody elements, whereas the herbaceous layer includes cosmopolitan
groups of different origins (Rizzini 1963; Sarmiento 1983).

BREEDING SYSTEMS

Studies on the breeding systems of Central American forest communities
indicate that most tree species are obligatory outbreeders (Bawa 1974;
Bawa, Perry, and Beach 1985; Bullock 1985, 1994). Studying at La Selva,
Kress and Beach (1994) estimated that 88% of the upper stratum species
were obligatory outbreeders, and this contrasted with predominant
(66%) self-compatibility in species of the understory stratum. The Brazilian cerrados conform to the Central American pattern in that most
woody species have obligatory outbreeding mechanisms (see tables 17.2,
17.3; Oliveira and Gibbs 2000), while self-compatibility is much more
common in the herbaceous and hemixyle taxa of understory (Saraiva et
al. 1996) or open cerrado (*campo sujo*; Barbosa 1997) communities (see
fig. 17.2).

Outbreeding cerrado woody species mostly have hermaphrodite
flowers and self-incompatibility, while dioecy has an incidence of only
10–15% in this community (Oliveira 1996). This contrasts with the 24%
dioecious species in the canopy of Central American woodlands (Kress
and Beach 1994). In fact, in typical cerrado *sensu stricto*, the incidence of
dioecy is around half of the 20% or so reported for seasonally dry forests
by Bullock (1995), and the rather higher frequency of dioecy in dense cerrado woodlands (*cerradão*) seems to be due to the occurrence of some
moist forest species in these areas (Oliveira 1996). As suggested by Bawa

Table 17.2   Sexual and Breeding Systems of Woody Species
in Cerrado and in Other Tropical Communities

	Cerrado sensu stricto	Dry forest, Costa Rica	Secondary forest, Venezuela	Cloud forest, Venezuela	Dry forest, Mexico	Rainforest, Costa Rica
Dioecy	15	22	24	31	20	23
Monoecy	5	10	0	3	10	11
SI	66	54	64	26	53	53
Inbreeding	14	14	12	43	17	13
Outbreeding	81	76	88	57	73	75
Tested species	30	34	13	36	33	28

Note: Percentage of species with each feature in cerrado *sensu stricto* area (Oliveira and Gibbs 2000), seasonally dry forest in Costa Rica (Bawa 1974), secondary forest in Venezuela (Ruiz and Arroyo 1978, woody species only), cloud forest in Venezuela (Sobrevilla and Arroyo 1982), dry forest in Mexico (Bullock 1985), and rainforest in Costa Rica (Bawa et al. 1985a). Estimates for self-incompatible (SI) and inbreeding species are based on limited samples of the total surveyed flora (tested species). Inbreeding included self-compatible, autogamous, and apomictic species.

(1980) for tropical moist forests, dioecy in cerrado *sensu lato* is correlated with small, structurally simple unisexual flowers which utilize a broad spectrum of small insects capable only of unspecialized pollination interactions. Seasonal drought, high temperatures, and distance between conspecific trees may limit the efficiency of small insect pollination and occurrence of dioecious species in open cerrado areas (Oliveira 1996).

Where studies of breeding systems of cerrado species have combined controlled pollinations with observations of pollen tube growth, a notable feature has emerged: the scarcity of taxa with conventional homomorphic or heteromorphic self-incompatibility (SI). Homomorphic SI, with inhibition of self pollen at the stigma or within the stylar transmitting tract, has been reported only for the genera *Vochysia* (Oliveira and Gibbs 1994) and *Miconia* (Goldenberg and Shepherd 1998), while heteromorphic SI in the cerrado occurs only in the genera *Erythroxylum* (Erythroxylaceae) and *Palicourea* (Rubiaceae), two families in which this breeding system predominates (Barros 1998; Oliveira and Gibbs 2000).

Most of the cerrado species studied for breeding system and postpollination events show "late-acting self-incompatibility" (*sensu* Seavey and Bawa 1986) or "ovarian" sterility (cf. Sage et al. 1994). Despite failure to set fruits following selfing, self-pollen tubes grow apparently successfully to the ovary where ovule penetration usually occurs, as in *Tabebuia caraiba* and *T. ochracea* (Gibbs and Bianchi 1993), *Vellozia squamata*

Table 17.3 Pollination and Breeding Systems of Woody Species in Cerrado Plant Formations

	Campo cerrado CPAC	Cerrado sensu stricto CPAC	Cerrado sensu stricto FAL	Cerrado sensu stricto BBG	Cerradão FAL	Cerradão CPAC
Pollination						
Small insects	35 (9/26)	38 (22/58)	38 (23/61)	44 (26/59)	45 (27/60)	40 (25/63)
Large bees	35 (9/26)	33 (19/58)	39 (24/61)	32 (19/59)	37 (22/60)	36 (23/63)
Bats	8 (2/26)	5 (3/58)	5 (3/61)	3 (2/59)	3 (2/60)	5 (3/63)
Breeding systems						
Dioecy	12 (3/26)	7 (4/58)	11 (10/89)	15 (9/59)	16 (16/99)	27 (18/67)
Monoecy	0 (0/26)	5 (3/58)	4 (3/89)	5 (3/59)	3 (3/99)	4 (3/67)
Hermaphrodite	88 (23/26)	88 (51/58)	85 (76/89)	80 (47/59)	81 (80/99)	69 (46/67)
SI	77 (14/16)	63 (20/28)	68 (29/36)	66 (25/30)	65 (33/41)	56 (18/22)
Inbreeding	11 (2/16)	25 (8/28)	17 (7/36)	14 (5/30)	16 (8/41)	12 (4/22)
Outbreeding	89	70	79	81	81	83

Note: There is a gradient of woody species density from *campo cerrado*, to cerrado *sensu stricto* to *cerradão* that is characteristic of cerrado landscape near Brasília, central Brazil. Data express the percentage of species. Number of species showing each feature and the total number of species are given in parentheses. Breeding system frequency was estimated from limited samples. Outbreeding included dioecious plus self-incompatible (SI) species. Lists of identified species were used for each site. CPAC is a cerrado *sensu lato* reserve in the Centro de Pesquisa Agropecuária de Cerrado (Ribeiro et al. 1985). FAL is the University of Brasília experimental reserve (Ratter 1985). BBG is the Brasília Botanic Garden area (Oliveira and Gibbs 2000).

(Oliveira et al. 1991), *Eriotheca gracilipes* (Oliveira et al. 1992), *Dalbergia miscolobium* (Gibbs and Sassaki 1998), *Gomidesia lindeniana* (Nic-Lughadha 1998), *Hymenaea stigonocarpa* (Gibbs et al. 1999), *Qualea multiflora, Q. parviflora* (Oliveira 1998b), and *Callisthene fasciculata* (Oliveira 1998b). "Late-acting self-incompatibility" (LSI) is a poorly understood phenomenon that almost certainly encompasses diverse mechanisms, such as early acting lethal recessives (Nic-Lughadha 1998) as well as possibly novel recognition systems (see Lipow and Wyatt 2000). The seemingly widespread occurrence of LSI in woody cerrado taxa once again underlines the similarity between the reproductive biology of these woodlands and that of moist forest communities, where LSI is also common (Bawa, Perry, and Beach 1985). But both communities may reflect the widespread but hitherto underestimated occurrence of this outbreeding "mechanism" (Gibbs and Bianchi 1999).

Brazilian cerrados also parallel the La Selva situation (Kress and Beach 1994), since the understory stratum of herbaceous and hemixyle species seem to be dominated by self-compatible species (Barbosa 1997).

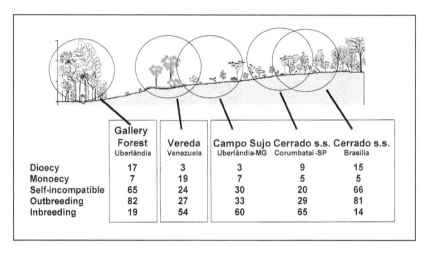

	Gallery Forest Uberlândia	Vereda Venezuela	Campo Sujo Uberlândia-MG	Cerrado s.s. Corumbataí-SP	Cerrado s.s. Brasília
Dioecy	17	3	3	9	15
Monoecy	7	19	7	5	5
Self-incompatible	65	24	30	20	66
Outbreeding	82	27	33	29	81
Inbreeding	19	54	60	65	14

Figure 17.2 Frequency of breeding systems in different cerrado formations of central Brazil (%). Outbreeding species are the sum of dioecious and self-incompatible ones. Inbreeding are self-compatible, autogamous or apomictic species. Data for *Mauritia flexuosa* palm swamp areas (*vereda, buritizal,* or *morichal*) which occurs all over the cerrado region and also in Venezuela, are presented for comparison. Sources: P. E. Oliveira, unpublished data (gallery forest); Ramirez and Brito 1990 (*vereda*); Barbosa 1997 (*campo sujo*); Saraiva et al. 1996 (cerrado *sensu stricto* at Corumbataí-SP); Oliveira and Gibbs 2000 (cerrado *sensu stricto* at Brasília). See chapter 6 for description of vegetation physiognomies.

Although self-compatibility does not necessarily imply a predominantly selfing mating system (e.g., Pascarella 1997), self-compatibility has been related to situations such as scarcity of pollinators, small population sizes, and dominance of herbaceous species (Sobrevilla and Arroyo 1982, Bullock 1995). However, differences in floristic composition between tree *versus* understory strata may also introduce a family bias that increases the incidence of selfing taxa in the latter (Bianchi et al. 2000). Apomixis is rare among cerrado *sensu lato* species (Barbosa 1997; Oliveira and Gibbs 2000): the few apomictic taxa include possible apospory in *Miconia* spp. and other Melastomataceae (Goldenberg and Shepherd 1998) and adventitious embryony in *Eriotheca pubescens* (Oliveira et al. 1992).

Various inconclusive explanations have been offered for the predominantly outbreeding behavior of tree species of moist forest and cerrado woodlands. In general, long-lived woody taxa in all communities exhibit outbreeding (Stebbins 1958), allegedly to maintain genetic heterozygosity in species with long reproductive life cycles and high seedling mortality. Predominant outbreeding has consequences for population structure. Genetic variability in the cerrado *sensu lato* should be similar to that of other tropical communities in which intrapopulation variability is greater than interpopulation variability (Kageyama 1990). Recent data using genetic markers in *Caryocar brasiliense*, a common bat-pollinated cerrado tree, showed high outcrossing rates in this species but indicated that habitat fragmentation and disturbance may limit gene flow and create small, relatively homogeneous population subunits (Collevatti et al. 2000).

VICARIANCE AND CERRADO-FOREST BOUNDARIES

Climatic changes, particularly during the Pleistocene, have caused repeated expansions and contractions of the cerrado *sensu lato* and moist forest areas, which have been important for the evolution of the Neotropical flora (Prance 1992; Ratter 1992; Oliveira-Filho and Ratter 1995; see chapter 3). Given such historical interactions, it is not surprising that many pairs of vicariant species between cerrado and forest can be identified (Rizzini 1963; Sarmiento 1983). They comprise species pairs with morphological similarities but which may show marked differences in growth habit.

Gottsberger (1986) studied the differences between forest and cerrado *sensu lato* species in some taxa. He found that the reproductive biology was similar in species of the Bignoniaceae and Malpighiaceae of cerrado and moist forest. But in the Annonaceae and other beetle-pollinated groups, there seemed to be a more or less distinct fauna of pollinators in cerrado and forest plants. Moreover, for Malvaceae, he proposed an evolutionary trend from allogamous ornithophilous woody groups in the forest to predominantly endogamous and mellitophilous herbaceous genera and species in the cerrado.

In contrast, no such differences could be observed in other more recently studied taxa. *Vochysia*, the largest genus in the neotropical family Vochysiaceae, is mostly distributed in rainforest but has some 20% of the species occurring in the cerrado region. Six such species were studied near Brasília, in central Brazil (Oliveira and Gibbs 1994), where both shrub and tree species occur in cerrado *sensu lato*, while other species are large gallery forest trees. Floral morphology, pollination, and xenogamous breeding systems were found to be fairly uniform in these *Vochysia* species despite the variation in life form and habitat.

A similar situation occurs in *Hymenaea* (Caesalpiniaceae), another forest genus with savanna species. In central Brazil *Hymenaea courbaril* var. *stilbocarpa* (sometimes treated as *H. stilbocarpa*) is a forest tree up to 25 m with bat-pollinated flowers and a xenogamous breeding system (Bawa, Perry, and Beach 1985; Bawa, Bullock, Perry et al. 1985; for Central America trees of *Hymenaea courbaril*). The smaller cerrado vicariad, *Hymenaea stigonocarpa* (up to 8 m), is also bat-pollinated and self-incompatible (Gibbs et al. 1999).

These studies indicate that the adaptive shift between forest and cerrado habitats, accompanied in some taxa by dramatic changes in life form from large forest trees to small cerrado shrubs, does not necessarily involve changes in floral biology and breeding systems. The same large-bee and bat species, the main pollinators of forest trees, may be found visiting and pollinating congeneric species in neighboring cerrado areas.

BREEDING BIOLOGY AND CONSERVATION IN CERRADO

The data emerging for the reproductive biology of cerrado plants have important consequences for conservation and understanding of the organization of cerrado communities. Open cerrado formations have been viewed as communities maintained by disturbance and limited in reproductive output, in which vegetative reproduction was considered more important than sexual reproduction (Rizzini 1965). Other studies, however, have indicated a more or less stable mosaic of communities, whose primary production is limited by nutritional levels, soil depth, seasonality, and even the occurrence of fire (Sarmiento et al. 1985). Cerrado plant communities are adapted to these conditions in terms of physiology, phenology, and reproductive output (Sarmiento and Monasterio 1983; Sarmiento et al. 1985; Oliveira and Silva 1993). Sexual reproduction and regeneration via seed is as important in this region as any other tropical woodland.

The reproductive biology data discussed here support this latter point of view. The similarities in pollination and breeding systems indicate that the cerrado is an environment where the genetic variability provided by outcrossing is as important as it is in forest communities. Cerrado species seem to present potentially large mating populations and mostly relatively generalist pollination systems. These systems rely on a diversity of visitors and should be relatively resilient to environmental disturbances. However, such theoretical considerations need to be investigated, since to our knowledge there have been no studies on the reproductive success of cerrado species in disturbed and fragmented areas.

Moreover, despite this potential resilience, anthropomorphic changes in the intensity and periodicity of fire and disturbance (chapters 4, 5) may represent strong selective factors in a community basically dependent on pollinators to reproduce. Dioecious species and other taxa pollinated by less mobile vectors and/or dependent on animal dispersal might be particularly sensitive. Another consequence of such disturbances may be increased vegetative regeneration (Hoffmann 1998). Vegetative regeneration by resprouting may also result in islands of clonal plants with consequences for sexual reproduction of allogamous species.

The combined effects of fire and disturbance are likely to cause an expansion of open physiognomies and consequently a possible reduction of suitable sites for some ground nesting bees, as observed in Costa Rica dry forests (Frankie et al. 1990). Such changes may also affect carpenter bees, *Xylocopa* spp., which nest in dry dead timber (Camillo and Garofalo 1982). Drastic disturbance affecting important groups of pollinators could reduce the efficiency of the pollination systems and affect reproduction of many species in the community despite the possible resilience due to generalist systems. These woody species both depend on and sustain a rich fauna of pollinators dependent on them for survival. Herbaceous taxa, on the other hand, are mostly capable of selfing, and some are wind pollinated (Silberbauer-Gottsberger and Gottsberger 1988; Barbosa 1997; chapter 7). Many also produce dry autochorous or epizoochorous fruits (Silberbauer-Gottsberger 1984), so that open grasslands, which result from increasing disturbance, would possibly maintain an impoverished sample of the original fauna of the area.

Diversity of flowering phenology seems to be a characteristic of cerrado vegetation (Oliveira 1998a), and so it is possible to find some species in flower throughout the year. Such diversity of flowering phenology has important consequences for resource availability for pollinators. Biomass gradients, which may or may not be accompanied by floristic differentiation, are an important feature of cerrado communities (Goodland 1971; Ribeiro et al. 1985). Differences in species composition and importance along these gradients provoke dramatic changes in resource availability in both temporal and spatial terms (Oliveira 1998a). Simple changes in species density from place to place may shift the peak of flowering in cerrado and *cerradão* areas (see fig. 17.3) and may result in patchy resource availability for pollinators along vegetation gradients.

The emerging data for the reproductive biology of the woody stratum of the cerrado is essentially similar to that established for the moist and seasonally dry forests of Central America (Kress and Beach 1994; Bullock 1995). This applies equally to the spectrum of pollinators, with special

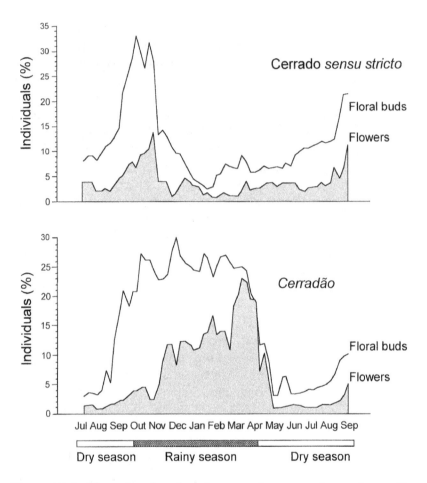

Figure 17.3 Flowering phenology in a cerrado *sensu stricto* and *cerradão* denser woodland in Brasília, central Brazil. Intensity of flowering as a percentage of individuals was obtained from species phenology and relative density in each area (redrawn from Oliveira 1998a). See chapter 6 for description of vegetation physiognomies.

importance of medium- to large-bee species, and to the prevalence of outbreeding taxa, commonly promoted by "late-acting" type mechanisms. The reproductive biology of the gallery forests and moist forest areas inside the Cerrado Biome are as yet poorly studied. However, it is likely that the pollinators of cerrado *sensu lato* and those of the adjacent forests have multiple, mutual interactions. It follows that conservation of biodiversity in the cerrados is intimately linked with that of moist forest areas (see chapter 18). Each may facilitate the other by providing resources throughout the year to maintain the local flux of pollinators. Excessive perturbation of either component could lead to loss of reproductive efficiency and diversity.

REFERENCES

Barbosa, A. A. 1997. "Biologia Reprodutiva de uma Comunidade de Campo Sujo, Uberlândia-MG." Ph.D. thesis, Universidade Estadual de Campinas, Brazil.

Barros, M. A. G. 1998. Sistemas reprodutivos e polinização em espécies simpátricas de *Erythroxylum* P.Br. (Erythroxylaceae) do Brasil. *Rev. Bras. Bot.* 21:159–166.

Bawa, K. S. 1974. Breeding systems of tree species of a lowland tropical community. *Evolution* 28:85–92.

Bawa, K. S. 1980. Evolution of dioecy in flowering plants. *Ann. Rev. Ecol. Syst.* 11:15–39.

Bawa, K. S. 1990. Plant-pollinator interactions in tropical rain forests. *Ann. Rev. Ecol. Syst.* 21:399–422.

Bawa, K. S., S. H. Bullock, D. R. Perry, R. E. Coville, and M. H. Grayum. 1985. Reproductive biology of tropical lowland rain forest trees: II. Pollination systems. *Amer. J. Bot.* 72:346–356.

Bawa, K. S., D. R. Perry, and J. H. Beach. 1985. Reproductive biology of tropical lowland rain forest trees: I. Sexual systems and incompatibility mechanisms. *Amer. J. Bot.* 72:331–345.

Bianchi, M. B., P. E. Gibbs, D. E. Prado, and J. L. Vesprini. 2000. Studies on the breeding systems of understorey species of a Chaco woodland in NE Argentina. *Flora* 195:339–348.

Bullock, S. H. 1985. Breeding systems in the flora of a tropical deciduous forest. *Biotropica* 17:287–301.

Bullock, S. H. 1994. Wind pollination of Neotropical dioecious trees. *Biotropica* 26:172–179.

Bullock, S. H. 1995. Plant reproduction in neotropical dry forests. In S. H. Bullock, H. A. Mooney, and E. Medina, eds., *Seasonally Dry Tropical Forests*, pp. 277–303. Cambridge: Cambridge University Press.

Camillo, E. and C. A. Garófalo. 1982. On the bionomics of *Xylocopa frontalis* (Oliver) and *Xylocopa grisescens* (Lepeletier) in Southern Brazil: I. Nest construction and biological cycle. *Rev. Bras. Biol.* 42:571–582.

Carvalho, A. M. C. and L. R. Bego. 1996. Studies on Apoidea fauna of cerrado vegetation at the Panga Ecological Reserve, Uberlândia, MG. Brazil. *Rev. Bras. Ent.* 40:147–156.

Collevatti, R. G., D. Grattapaglia, and J. D. Hay. 2000. Microsatellite markers provide evidence for the effect of fragmentation on the genetic structure of *Caryocar brasiliense*, an endangered tropical tree species of the Brazilian Cerrado. *Mol. Ecol.* 10:349–356.

Frankie, G. W., W. A. Haber, P. A. Opler, and K. S. Bawa. 1983. Characteristics and organization of the large bee pollination system in the Costa Rican dry forest. In C. E. Jones and R. J. Little, eds., *Handbook of Experimental Pollination Biology*, pp. 411–447. New York: Van Nostrand Reinhold Co.

Frankie, G. W., S. B. Vinson, L. E. Newstrom, J. F. Barthell, W. A. Harber, and J. K. Frankie. 1990. Plant phenology, pollination ecology, pollinator behaviour and conservation of pollinators in neotropical dry forest. In K. S. Bawa and M. Hadley, eds., *Reproductive Ecology of Tropical Forest Plants*, pp. 37–47. Carnforth: Parthenon.

Furley, P. A. and J. A. Ratter. 1988. Soil resources and plant communities of the central Brazilian cerrado and their development. *J. Biogeogr.* 15:97–108.

Gibbs, P. E. and M. Bianchi. 1993. Post-pollination events in species of *Chorisia* (Bombacaceae) and *Tabebuia* (Bignoniaceae) with late-acting self-incompatibility. *Bot. Acta* 106:64–71.

Gibbs, P. E. and M. B. Bianchi. 1999. Does late-acting self-incompatibility (LSI) show family Clustering? Two more species of Bignoniaceae with LSI: *Dolichandra cynanchoides* and *Tabebuia nodosa*. *Ann. Bot.* 84:449–457.

Gibbs, P. E., P. E. Oliveira, and M. Bianchi. 1999. Post-zygotic control of selfing in *Hymenaea stigonocarpa* (Leguminosae-Caesalpinoideae), a bat-pollinated tree of the Brazilian cerrados. *Int. J. Plant Sci.* 160:72–78.

Gibbs, P. E. and R. Sassaki. 1998. Reproductive biology of *Dalbergia miscolobium* Benth. (Leguminosae-Papilionoideae) in SE Brazil: The effects of pistilate sorting on fruit-set. *Ann. Bot.* 81:735–740.

Goldenberg, R. S. and G. J. Shepherd. 1998. Studies on the reproductive biology of Melastomataceae in "cerrado" vegetation. *Pl. Syst. Evol.* 211: 13–29.

Goodland, R. 1971. A physiognomic analysis of the cerrado vegetation of Central Brazil. *J. Ecol.* 59:411–419.

Gottsberger, G. 1986. Some pollination strategies in neotropical savannas and forests. *Pl. Syst. Evol.* 152:29–45.

Gottsberger, G. 1989. Beetle pollination and flowering rhythm of *Annona* spp. (Annonaceae) in Brazil. *Pl. Syst. Evol.* 167:165–187.

Gribel, R., P. E. Gibbs, and A. L. Queiróz. 1999. Flowering phenology and pollination biology of *Ceiba pentandra* (Bombacaceae) in Central Amazonia. *J. Trop. Ecol.* 15:247–263.

Gribel, R. and J. D. Hay. 1993. Pollination ecology of *Caryocar brasiliense* (Caryocaraceae) in Central Brazil cerrado vegetation. *J. Trop. Ecol.* 9:199–211.

Heithaus, E. R. 1979. Community structure of neotropical flower visiting bees and wasps: diversity and phenology. *Ecology* 60:190–202.

Hoffmann, W.A. 1998. Post-burn reproduction of in a Neotropical savanna: The relative importance of sexual and vegetative reproduction. *J. Appl. Ecol.* 35:422–433

Kageyama, P. Y. 1990. Genetic structure of tropical tree species of Brazil. In K. S, Bawa and M. Hadley, eds., *Reproductive Ecology of Tropical Forest Plants*, pp. 375–387. Paris: UNESCO.

Kress, W. J., and J. H. Beach. 1994. Flowering plant reproductive systems. In L. A. McDade, K. S. Bawa, H. A. Hespenheide, and G. S. Hartshorn, eds., *La Selva: Ecology and Natural History of a Neotropical Rainforest*, pp. 161–182. Chicago: University of Chicago Press.

Lipow, S. R. and R. Wyatt, 2000. Single gene control of postzygotic self-incompatibility in poke milkweed, *Asclepias exaltata* L. *Genetics* 154:893–907.

Mantovani, W. and F. R. Martins. 1988. Variações fenológicas das espécies do cerrado da Reserva biológica de Mogi-Guaçú, Estado de São Paulo. *Rev. Bras. Bot.* 11:101–112.

Mendonça, R. C., J. M. Felfili, B. M. T. Walter, S.-J. M.C., A. V. Rezende, T. S. Filgueiras, and P. E. Nogueira. 1998. Flora vascular do cerrado. In S. M. Sano and S. P. Almeida, eds., *Cerrado: Ambiente e Flora*, pp. 289–556. Brasília: Empresa Brasileira de Pesquisa Agropecuária.

Nic-Lughadha, E. 1998. Preferential outcrossing in *Gomidesia* (Myrtaceae) is maintained by a post-zygotic mechanism. In S. Owens and P. Rudall, eds., *Reproductive Biology: In Systematics, Conservation and Economic Botany*, pp. 363–379. Richmond: Royal Botanic Gardens at Kew.

Oliveira, P. E. 1996. Dioecy in the cerrado vegetation of Central Brazil. *Flora* 191:235–243.

Oliveira, P. E. 1998a. Fenologia e biologia reprodutiva das espécies de cerrado. In S. M. Sano and S. P. Almeida, eds., *Cerrado: Ambiente e Flora*, pp. 169–192. Brasília: Empresa Brasileira de Pesquisa Agropecuária.

Oliveira, P. E. 1998b. Reproductive biology, evolution and taxonomy of the vochysiaceae in Central Brazil. In S. Owens and P. Rudall, eds., *Reproductive Biology: In Systematics, Conservation and Economic Botany*, pp. 381–393. Richmond: Royal Botanic Gardens at Kew.

Oliveira, P. E. and P. E. Gibbs. 1994. Pollination and breeding systems of six *Vochysia* species (Vochysiaceae) in Central Brazil. *J. Trop. Ecol.* 10:509–522.

Oliveira, P. E. and P. E. Gibbs. 2000. Reproductive biology of woody plants in a cerrado community of Central Brazil. *Flora* 195:311–329.

Oliveira, P. E. and J. C. S. Silva. 1993. Reproductive biology of two species of *Kielmeyera* (Guttiferae) in the cerrados of Central Brazil. *J. Trop. Ecol.* 9:67–79.

Oliveira, P. E., P. E. Gibbs, A. A. Barbosa, and S. Talavera. 1992. Contrasting breeding systems in two *Eriotheca* (Bombacaceae) species of the Brazilian cerrados. *Pl. Syst. Evol.* 179:207–219.

Oliveira, P. E., P. E. Gibbs, and M. B. Bianchi. 1991. Pollination and breeding biology of *Vellozia squamata* (Liliales-Velloziaceae): A species of the Brazilian cerrados. *Bot. Acta* 104:392–398.

Oliveira-Filho, A. T. and J. A. Ratter. 1995. A study of the origin of Central Brazilian forests by the analysis of plant species distributions. *Edinb. J. Bot.* 52:141–194.

Pascarella, J. B. 1997. The mating system of the tropical understorey shrub *Ardisia escallonoides* (Myrsinaceae). *Amer. J. Bot.* 84:456–460.

Prance, G. T. 1992. The phytogeography of savanna species of neotropical Chrysobalanaceae. In P. A. Furley, J. Proctor, and J. A. Ratter, eds., *Nature and Dynamics of Forest-Savanna Boundaries*, pp. 295–330. London: Chapman and Hall.

Ramirez, N. and Y. Brito. 1990. Reproductive biology of a tropical palm swamp community in the Venezuelan llanos. *Amer. J. Bot.* 77:1260–1271.

Ratter, J. A. 1985. *Notas Sobre a Vegetação da Fazenda Água Limpa*. Edinburgh: Royal Botanic Garden.

Ratter, J. A. 1992. Transitions between cerrado and forest vegetation in Brazil. In P. A. Furley, J. Proctor, and J. A. Ratter, eds., *Nature and Dynamics of Forest-Savanna Boundaries*, pp. 417–430. London: Chapman and Hall.

Ratter, J. A., S. Bridgewater, R. Atkinson, and J. F. Ribeiro. 1996. Analysis of the floristic composition of the Brazilian cerrado vegetation: II. Comparison of the woody vegetation of 98 areas. *Edinb. J. Bot.* 53:153–180.

Ratter, J. A., J. F. Ribeiro, and S. Bridgewater. 1997. The Brazilian cerrado vegetation and threats to its biodiversity. *Ann. Bot.* 80:223–230.

Ribeiro, J. F., J. C. S. Silva, and G. J. Batmanian. 1985. Fitossociologia de tipos fisionômicos de cerrado em Planaltina–DF. *Rev. Bras. Bot* 8:131–142.

Rizzini, C. T. 1963. A flora do cerrado: Análise florística das savanas centrais. In M. G. Ferri, ed., *Simpósio sobre o Cerrado*, pp. 127–177. São Paulo: Editora da Universidade de São Paulo.

Rizzini, C. T. 1965. Experimental studies on seedling development of cerrado woody plants. *Ann. Miss. Bot. Gard.* 52:410–426.

Roubik, D. W. 1989. *Ecology and Natural History of Tropical Bees*. Cambridge: Cambridge University Press.

Ruiz, T. Z. and M. T. K. Arroyo. 1978. Plant reproductive ecology of a secondary deciduous tropical forest. *Biotropica* 10:221–230.

Sage, T. L., R. I. Bertin, and E. G. Williams. 1994. Ovarian and other late-acting self-incompatibility systems. In E. G. Williams, A. E. Clarke, and R. B. Knox, eds., *Genetic Control of Self-Incompatibility and Reproductive Development in Flowering Plants*, pp. 116–140. Netherlands: Kluwer Academic Publishers.

Saraiva, L., O. Cesar, and R. Monteiro. 1996. Breeding systems of shrubs and trees of a Brazilian savanna. *Arq. Biol. Tecnol. (Brazil)* 39:751–763.

Sarmiento, G. 1983. The savannas of tropical America. In F. Bouliere, ed.,

Ecosystems of the World: Tropical Savannas, pp. 245–288. Amsterdam: Elsevier.

Sarmiento, G., G. Goldstein, and F. Meinzer. 1985. Adaptive strategies of woody species in neotropical savannas. *Biol. Rev.* 60:315–355.

Sarmiento, G. and M. Monasterio. 1983. Life forms and phenology. In F. Bouliere, ed., *Ecosystems of the World: Tropical Savannas*, pp. 79–108. Amsterdam: Elsevier.

Sazima, I., S. Buzato, and M. Sazima. 1996. An assemblage of hummingbird-pollinated flowers in a montane forest in Southeastern Brazil. *Bot. Acta* 109:149–160.

Sazima, M., S. Buzato, and I. Sazima. 1999. Bat-pollinated flower assemblages and bat visitors at two Atlantic forest sites in Brazil. *Ann. Bot.* 83:705–712.

Sazima, M. and I. Sazima. 1975. Quiropterofilia em *Lafoensia pacari* St. Hil (Lythraceae), na Serra do Cipó, Minas Gerais. *Ciênc. Cult.* 27:406–416.

Seavey, S. R. and K. S. Bawa. 1986. Late-acting self-incompatibility in Angiosperms. *Bot. Rev.* 52:195–219.

Silberbauer-Gottsberger, I. 1984. Fruit dispersal and trypanocarpy in Brazilian cerrado grasses. *Pl. Syst. Evol.* 147:1–27.

Silberbauer-Gottsberger, I. and G. Gottsberger. 1975. Uber sphingophile Angiospermen Brasiliens. *Pl. Syst. Evol.* 123:157–184.

Silberbauer-Gottsberger, I., and G. Gottsberger. 1988. A polinização das plantas do Cerrado. *Rev. Bras. Biol.* 48:651–663.

Sobrevilla, C. and M. T. K. Arroyo. 1982. Breeding systems in a montane tropical cloud forest in Venezuela. *Pl. Syst. Evol.* 140:19–37.

Stebbins, G. L. 1958. Longevity, habitats and release of genetic variability in higher plants. *Cold Spring Harbor Symp. Quant. Biol.* 23:365–378.

Warming, E. 1908. *Lagoa Santa—Contribuição para a Geografia Phytobiologica.* (Portuguese translation of *Lagoa Santa—et Bidrad til den Biologiske Plantegeographi*, Copenhagen, 1882). Bello Horizonte: Imprensa Official do Estado de Minas Gerais.

Part V
The Conservation
of the Cerrados

18

Biodiversity and Conservation Priorities in the Cerrado Region

Roberto B. Cavalcanti and Carlos A. Joly

THE CONSERVATION OF THE CERRADO BIOME AND ITS ECOSYS-tems has been neglected until very recently, for two related reasons. First, central Brazil was very sparsely occupied until the mid-twentieth century, and therefore perceived threats to the environment were low (chapter 5). Second, the native cerrado had little apparent economic value and was often unattractive even in the eyes of specialists, due to the scrubby nature of the vegetation, low faunal densities, and a pronounced dry season with frequent fires. The region was colonized by Europeans systematically since the 18th century, towns were started at strategic points by prospectors seeking gold and diamonds, and the countryside was occupied by large farms focused primarily on extensive cattle ranching using native pasture. The low nutrient content and high acidity of most soils and the lack of railroad access restricted agriculture. For an excellent description of much of the cerrado in the early 1800s by one of the world's leading botanists of that time, see St. Hilaire (1847–1848).

The mechanization of Brazilian agriculture after 1950 and construction of major highways through central Brazil increased human impact dramatically (chapter 5). The use of lime and new fertilization techniques, together with development of high yield/drought-resistant varieties of soybean, rice and corn, helped open the cerrados as Brazil's new agricultural frontier. The low cost of land, abundant rainfall during the growing season, and deep soils on gently rolling terrain suitable for mechanization were key factors in the development of large-scale agribusiness operations in the region. At least 67% of the cerrado region had been converted to intensive human use by the early 1990s, with current estimates placing conversion at 80% (Myers et al. 2000). In the state of São Paulo in south-eastern Brazil the cerrado vegetation has been reduced from 14% to nearly 1.2% of the state's area (Secretaria do Meio Ambiente de São Paulo 1997). Further impacts on the cerrado may be an undesirable side effect of the pressure by national and international organizations to halt defor-estation in Amazonia: in the search for alternative sites for agricultural development, and especially for planting forests for wood pulp or to act as carbon sinks, the Cerrado Biome has often been a target (Ab'Saber et al. 1990). A detailed account, including a historical overview, of the human-induced transformation of the cerrados is given in chapter 5.

In this chapter we discuss current methods and results for identifying priorities for biodiversity conservation in the cerrado region and provide perspectives on strategies for the future of the region.

THE CERRADO BIOME IS A RECOGNIZED GLOBAL BIODIVERSITY "HOTSPOT"

Brazil harbors outstanding biodiversity, sharing with Indonesia the top two slots of the world's richest "Megadiversity" countries (Mittermeier et al. 1997). Within Brazil, several biomes have merited individual global recognition. The Atlantic forest and the Brazilian cerrados are included in the world's 25 principal "hotspots," areas with great endemism and less than 30% remaining natural vegetation (Myers et al. 2000). Worldwide, these 25 "hotspots" cover 1.4% of the planet's land surface and harbor 44% of the world's vascular plant species and 35% of all species in four vertebrate groups (Myers et al. 2000).

Cerrado endemism is high for plants, on the order of 44%, but fairly low for vertebrates, estimated at 9.2% (Myers et al. 2000). A recent study for saturniid lepidopterans also indicated low endemism, 12.6% (Camargo and Becker 1999; see also chapter 11). Species richness is high, however. The cerrados harbor 13% of the butterfly species from the neotropics, 35% of the bees, and 23% of the termites (Brown 1996, Raw 1998 and references therein). Nearly half of the Brazilian birds occur in the cerrado (Silva 1995a; chapter 13). For plants, the cerrado is consid-ered the most diverse neotropical savanna (Lenthall et al. 1999).

The habitat diversity of the cerrado region is exceptionally high. Gallery forests, *Mauritia* palm groves, calcareous caves, mesophytic forests, dry forests in the "Mato Grosso de Goiás," and high-altitude rocky *campos rupestres*, among others, have many endemic species and provide a rich mixture of habitats for the region (see chapter 6). The gallery forests in particular provide physical connection with the Amazonian and Atlantic Forests, in addition to contributing to the maintenance of the open cerrado biota in dry periods (Prance 1982, Fonseca and Redford 1984; Redford and Fonseca 1986; Cavalcanti 1992). Clearing the cerrado is a nonrandom process which focuses on the areas most suitable for agriculture. This selective removal destroys the integrity of the original habitat matrix and precipitates extinctions due to a variety of related effects (Pimm and Raven 2000).

UNDERSTANDING BIODIVERSITY
IN THE CERRADO REGION

The Cerrado Priority-Setting Workshop in 1997 was developed and organized by a consortium consisting of the Fundação Pró-Natureza, Conservation International, the Universidade de Brasília, and the Fundação Biodiversitas. The *pantanal* wetlands were included in this Workshop, for several reasons: (1) there are extensive similarities in flora and fauna; (2) the *pantanal* headwaters are in the Cerrado; and (3) many scientists have expertise in both regions, giving economy of scale and organization (Cochrane et al. 1985). Major funding was provided by the Brazilian Ministry of the Environment, as part of the National Conservation and Sustainable Use of Biodiversity Project supported by the Global Environment Facility (Ministério do Meio Ambiente 1999). The design incorporated the experience of earlier priority-setting exercises in Amazonia and the Northeastern Atlantic Forest (Conservation International et al. 1991).

Preworkshop data gathering began in 1996 and 1997 with state-of-the-art reviews of knowledge by expert consultants for mammals, birds, reptiles and amphibians, invertebrates, fishes and aquatic biota, botany, soils and climate, conservation units, coverage of natural vegetation and socio-economic factors (Marinho Filho 1998; Silva 1998; Colli 1998; Araújo et al. 1998; Raw 1998; Diniz and Castanheira 1998; Britski 1998; M. Ribeiro 1998; J. F. Ribeiro 1998; Assad 1998; Pádua and Dias 1998; Mantovani and Pereira 1998; Sawyer 1998). Reports, databases, and digital maps were combined to generate products for use during the Workshop, including distribution maps of endangered and endemic species, maps of remaining and present vegetation, collection localities, protected areas, human population pressure and land use practices, and classification of Cerrado regions according to climate and soils.

PRIORITY AREAS FOR CONSERVATION
IN THE CERRADO REGION

During the workshop, priorities were set in three stages. First, thematic groups reviewed the information on all the Cerrado for a single theme, defined priority sites for conservation, and documented the relevant data for each site. Next, subregional groups took the recommendations from all thematic groups and overlaid, merged, enlarged, reduced, or added them to consolidate one map of priority areas per subregion. Finally, the subregional maps were combined into the overall cerrado biodiversity priority map with 87 areas (see fig. 18.1). Priority areas were selected on the

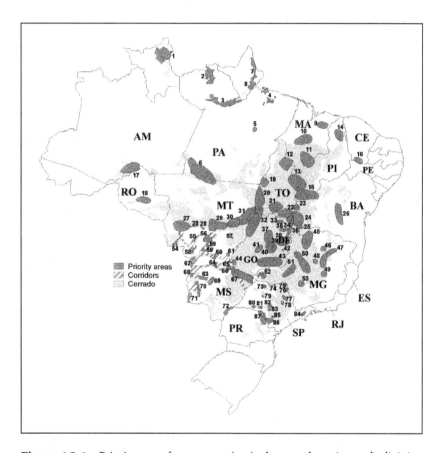

Figure 18.1   Priority areas for conservation in the cerrado region and adjoining *pantanal* (based on the results of the Priority-setting Workshop; see Brasil 1999). **List of localities:** (1) Roraima savannas; (2) Paru savannas; (3) Monte Alegre campos; (4) Ilha de Marajó savannas; (5) Serra dos Carajás; (6) Serra do Cachimbo; (7) North Amapá savannas; (8) South Central Amapá savannas;

basis of biological importance using the distribution of endemic, rare, threatened, or migratory species; species of economic or cultural value; species richness and composition of biological communities; and abiotic and landscape features crucial to conserving biodiversity (climate, rivers, geomorphology, soils). Priority actions for each site were determined by

Figure 18.1 (continued) (9) Northeastern Maranhão; (10) Maranhão semideciduous forests; (11) Mirador-Uruçuí; (12) Southeastern Maranhão; (13) Southeastern Piauí, Maranhão, and Tocantins tablelands; (14) Three biomes; (15) Rio Negro, Águas Emendadas, and Rio do Sono basins; (16) Chapada do Araripe, (17) Humaitá campos; (18) Pacaás-Guaporé-Ricardo Franco Corridor; (19) Middle Araguaia and Rio dos Cocos Basin; (20) Bananal Island; (21) Middle Tocantins; (22) Southern Tocantins-Conceição/Manuel Alves region; (23) Southeastern Tocantins semideciduous forests; (24) Goiás-Bahia hinterlands and São Domingos caves; (25) Posse-Correntina-São Domingos region; (26) Chapada Diamantina National Park; (27) Rio Papagaio; (28) Paraguay/Sepotuba headwaters; (29) Xingu headwaters; (30) Cascalheira and Querência streams; (31) Alto Boa Vista; (32) Araguaia valley and Rio das Mortes wetland; (33) Serra da Mesa/Niquelândia/Minaçu; (34) Pouso Alto; (35) Chapada dos Veadeiros; (36) Paranã valley and range; (37) Goiás-Rio das Almas/upper Tocantins; (38) Federal District and environs; (39) Pirenópolis; (40) Goiânia, Silvânia, Aparecida de Goiás, and Serra Dourada; (41) Serra Dourada dry forests; (42) Cristalina; (43) Upper Paraná; (44) Emas National Park and Araguaia headwaters; (45) Upper São Francisco basin (Peruaçu valley); (46) Grão Mogol; (47) Diamantina and Jequitinhonha valley; (48) Serra do Cabral; (49) Serra do Cipó; (50) Paracatu/Três Marias; (51) Upper Paranaíba/Patrocínio; (52) Minas Gerais Triangle; (53) Serra da Canastra National Park; (54) Serra de Santa Bárbara; (55) Serra das Araras; (56) Chapada dos Guimarães National Park and Cuiabá lowlands; (57) Nova Xavantina monodominant forest; (58) Paraguaizinho; (59) Chapada dos Guimarães and Barão de Melgaço; (60) Sucuriú; (61) Piquiri-Itiquira headwaters; (62) Pantanal west border A; (63) Rio Negro and Nhecolândia; (64) Taquari; (65) Emas/Taquari; (66) Emas-Jauru headwaters; (67) Jauru; (68) Pantanal west border B; (69) Taboco; (70) Bodoquena; (71) Chaco (Porto Murtinho Pantanal); (72) Mouth of the Ivinhema; (73) São José do Rio Preto; (74) Barretos; (75) Northeastern São Paulo; (76) Patrocínio Paulista; (77) Araraquara; (78) Campinas; (79) Araçatuba; (80) Presidente Prudente; (81) Marília; (82) Bauru; (83) Botucatu; (84) Paraíba valley; (85) Itapeva; (86) Itararé; (87) Jaguariaíva and Sengés (Paraná).
Key to state codes: Amazonas (AM), Bahia (BA), Ceará (CE), Distrito Federal (DF), Espírito Santo (ES), Goiás (GO), Maranhão (MA), Minas Gerais (MG), Mato Grosso (MT), Mato Grosso do Sul (MS), Pará (PA), Paraná (PR), Pernambuco (PE), Piauí (PI), Rio de Janeiro (RJ), Rondônia (RO), São Paulo (SP), Tocantins (TO).

analyzing the human pressures at the local level, which included demography, vulnerability of natural ecosystems to agriculture, cattle, industries, urban expansion, and various other types of economic use (BRASIL 1999).

Conservation urgency and opportunity were derived from data on human occupation and from the estimates of land cover produced by the first study to survey the entire cerrado and dry areas in the *pantanal* using satellite imagery (Mantovani and Pereira 1998; see fig. 18.2). Using images from 1992 and 1993, with occasional gaps filled with images from 1987 to 1991, Mantovani and Pereira (1998) estimated that one-third of the cerrado remained little or not disturbed. Well-preserved areas were found in 3 distinct regions: (a) the border between the states of Piauí, Maranhão and Bahia; (b) the border between the states of Tocantins, Mato Grosso, and Goiás; and (c) the border between Tocantins, Goiás,

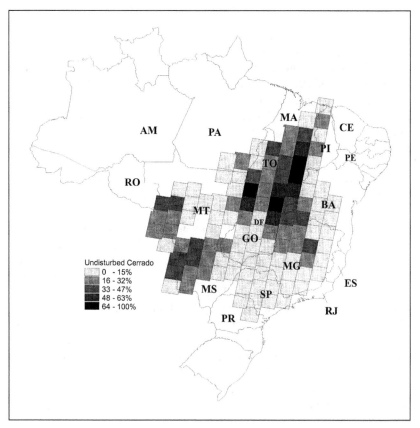

Figure 18.2   Human disturbance to the cerrado landscape, as assessed by satellite imagery (from Brasil 1990; Mantovani and Pereira 1998). State codes as in figure 18.1.

and Bahia. The regions of heaviest human impact were in the states of Mato Grosso do Sul, Goiás, and São Paulo; the border of São Paulo and Paraná; and the "Triangle" of Minas Gerais (figs. 18.1, 18.2). In these areas, 50% to 92% of the cerrado surface was under heavy human pressure (Mantovani and Pereira 1998).

Presently it is estimated that under 20% of the cerrado remains undisturbed (Myers et al. 2000). The agricultural frontier continues to expand. Major soybean cultivation projects are underway in regions classified as little disturbed by Mantovani and Pereira (1998), particularly at the borders of the states of Tocantins, Maranhão and Piauí. Roadbuilding and waterway development, and hydroelectric dams on the Tocantins river, are opening the last inaccessible regions of the cerrado.

The cerrado is poorly known for many taxonomic groups. Large areas of the states of Bahia and Tocantins are unsurveyed. For birds, approximately 70% of the region never has been adequately inventoried (Silva 1995a, b, 1998). For reptiles, 97% of the localities recorded were insufficiently surveyed (Colli 1998). Plants were arguably the group best inventoried in geographical completeness, with an active program to fill gaps in the entire region (Felfili et al. 1994; Ratter et al. 1996). For insects, coverage varied among groups, with collections generally concentrated where researchers are most active, such as Brasília (central Brazil) and the state of São Paulo (Diniz and Castanheira 1998). Data for aquatic organisms were even more limited. Many species of copepods are known only from one locality (Reid 1994). Sampling effort for fishes varied 20-fold between major river basins such as the São Francisco, the Tocantins, and the Paranaíba (Ribeiro 1998).

The existing federal and state conservation system in the cerrado region is insufficient both in extent and representation, covering approximately 1.6% of the surface area (Pádua and Dias 1997). Major subdivisions of the biome are highly endangered and poorly represented, such as the dry forests of the states of Goiás, Minas Gerais, Maranhão and Mato Grosso do Sul (fig. 18.1; areas 24, 25, 41, 45, 70). Of the approximately 100 different environmental units mapped by the Brazilian Institute for Agricultural Research (EMBRAPA), 73% have no protected parks or reserves (Pádua and Dias 1997). Considering that about 20% of the cerrado region is still undisturbed, an interim goal would be to protect half of that, or 10% of the biome, which would require expanding sixfold the existing units. Workshop participants gave high priority to 17 locations for creating new major conservation units. Since the Workshop, two have been decreed: the Peruaçu valley in the state of Minas Gerais and the Serra da Bodoquena in the state of Mato Grosso do Sul (fig. 18.1, areas 45, 70).

PRIORITY AREAS AND CONSERVATION ACTION IN THE CERRADO/PANTANAL INTERFACE

The cerrado highlands border the *pantanal* floodplain on the east and north sides and include headwaters of the pantanal's major rivers. The conservation of natural vegetation on these border crests is essential for the viability of the world's largest tropical freshwater wetland. Workshop participants recommended creating a biodiversity corridor system that would join the priority areas for conservation in the *pantanal* basin with those in the cerrado, following the main river drainages and climbing the floodplain escarpments (hatched area of fig. 18.1). Major cerrado habitats in this corridor occur in Emas National Park (area 44), Chapada dos Guimarães National Park (56), Bodoquena National Park (70), and the headwaters of the Paraguay river (28). Despite most of these areas harboring parks, the landscape is being rapidly fragmented by clearing for pasture and agriculture, threatening the integrity and potential expansion of the protected areas system. Since 1999, Conservation International, Fundação Emas, IBAMA, the state secretariats for the environment of Mato Grosso do Sul and Goiás have been implementing the Emas-Rio Negro Biodiversity Corridor (fig. 18.1, areas 63–66) with the support of international agencies including USAID. This five year program aims to work with private landowners and public agencies to manage the landscape to enhance and restore connections between natural habitats throughout the corridor (Cavalcanti et al. 1999).

PRIORITY AREAS AND CONSERVATION ACTION IN THE CENTRAL AND NORTHEASTERN CERRADOS

The central region of the cerrados has some of the highest species diversities and the best studied sites for various taxa. The Federal District has 503 of the 820 bee species of the Cerrado, 63 of the 129 social wasps, and 80% of the lepidopterans (Raw 1998 and data cited therein). The bird list of the Federal District and environs includes at least 439 species (Negret et al. 1984, updated by Bagno 1996), with 837 listed for all the biome (Silva 1995a).

Priority areas in this region include the Federal District and a group of contiguous areas ranging north in the state of Goiás near the border of Bahia and into the state of Tocantins (fig. 18.1, areas 22, 24, 25, 34, 35, 36, 38). In addition to the main cerrado vegetation types, these areas include the highly endemic plant communities of the Chapada dos Veadeiros (35), the dry forests of the Paraná valley (36), the extensive calcareous caves of São Domingos (24), the parkland and sandy savannas of the Bahia-Tocantins interface (21–25), and several areas on the Tocantins drainage (21–23). On the whole the areas are largely unprotected, with the exception of the Federal District, a national park in Chapada dos Veadeiros, and a state park in São Domingos. The denser dry forests and *cerradão* (chapter 6) are almost entirely destroyed, together with the habitat of several endemic bird subspecies (Silva 1989).

The Rio Araguaia drainage with its unique wetland habitats, sandy beaches with nesting turtles, pink dolphins, and mix of Amazonian and cerrado influences is highly threatened by agricultural erosion and pesticide runoff, former gold mining tailings, erosion and sedimented mercury in its tributaries, and lack of protection of the headwaters. A string of areas of high biological importance occurs in the basin, with the only notable protected areas in the Ilha do Bananal and a recently established state park in Tocantins (fig. 18.1, areas 19, 20, 32). The Xingu headwaters and Rio das Mortes (areas 29–31) areas are also highly important, studied in the 1950s and 1960s by the Roncador-Xingu and Royal Society expeditions (Sick 1955; Fry 1970).

The extreme northeastern cerrados offer one of the last opportunities to set up very large protected areas on the order of 200,000+ hectares. The states of Maranhão, Piauí and the Bahia/Tocantins border have six high-priority areas, including the Rio do Sono/Formoso do Rio Preto/Jalapão complex (fig. 18.1, areas 13, 15), the Mirador region (11), southwest and central Maranhão (10, 12), and the "three biomes" region in Piauí (14). This hinterland still has few roads and a sparse population. Botanical analyses suggest that the northeastern cerrados are a center of diversity (Castro and Martins 1998).

PRIORITY AREAS AND CONSERVATION ACTION
IN THE EASTERN AND SOUTHEASTERN CERRADOS

This region is the most highly fragmented and occupied for agriculture, cattle ranching, and urban expansion. In Minas Gerais Triangle (fig. 18.1, area 52) and in the state of São Paulo (SP) the native cerrado is reduced to small remnants rarely exceeding 100 hectares. Of the eight high biodiversity priority areas in the region, two contain significant national parks, Serra da Canastra and Serra do Cipó (fig.18.1, areas 53, 49). There are still opportunities for conservation in central and northern state of Minas Gerais, in the areas around Diamantina, Grão Mogol, in the valley of the

Jequitinhonha, and in the valley of the São Francisco. The Espinhaço range of Minas Gerais and Bahia (fig.18.1, areas 26, 46–48), rich in plant and animal endemics, was included in several priority areas of the workshop and is recognized as of global importance for bird conservation (International Council for Bird Preservation 1992).

In the state of São Paulo (SP), where cerrados originally covered 14% of the area, about 1% still has this vegetation. These 238,400 ha of remnant cerrado are fragmented into 8,353 islands, over half of which (4,372) are smaller than 10 ha and probably will disappear in the next few years. Only 47 fragments exceed 400 ha, many of them (17) in the administrative region of Ribeirão Preto (Kronka et al. 1998). The State of São Paulo Secretary of the Environment organized a workshop in 1995 to establish a policy for the conservation of the cerrado in the state (Secretaria do Meio Ambiente de São Paulo 1997). As a result, all permits to deforest cerrado remnants are now reviewed on a case by case basis, by a state commission coordinated by the São Paulo Biodiversity Conservation Program–PROBIO/SP. Several research initiatives are underway with the Biota Program of the Fundação de Amparo à Pesquisa do Estado de São Paulo, which deals with biodiversity conservation and sustainable use (FAPESP 1999). The conservation of cerrado remnants in São Paulo is of importance since they occur near the southern limit of the biome and have large areas of contact with the Atlantic forest region (see fig. 18.1). Ironically, much of the pioneering work on cerrado ecology was done in the state of São Paulo, in sites that are now relictual testimonials of the original landscape (Labouriau 1966; Ferri 1971).

The priority areas for conservation extend to the state of Paraná (fig. 18.1, area 87), where cerrado apparently occurred in five places; the largest protected area is a state park with 430 hectares (Straube 1998).

CONSERVATION POLICY

The priority setting workshop helped cerrado conservation policy at the federal, state, and local levels. The Minister of the Environment instituted a working group to develop a cerrado conservation strategy based on the workshop recommendations, including setting a target of expanding the conservation units to cover 10% of the biome by the year 2002 (Ministério do Meio Ambiente 1999). The state of Minas Gerais did its own conservation assessment using priority setting methodology and incorporating data from the biome level work (Costa et al. 1998). A long-term effort to amend the Constitution to include the Cerrado Biome as a national heritage has been recently revived by a parliamentary front. Several new or proposed parks are being implemented in priority areas.

Many policy recommendations emerged from the workshop, especially for incorporating conservation of biodiversity into the land use planning instruments, such as river basin management plans, road and electricity infrastructure plans, and urban and agricultural expansion initiatives. The focus of environmental action must move from environmental mitigation to a proactive conservation approach. Specific recommendations (BRASIL 1999) include the following:

- *Develop ecological corridors.* Federal, state, and local governments should develop programs to stimulate restoration and connection of natural habitat fragments, combining public and private protected areas in biodiversity corridors.
- *Coordinate government agencies and policies.* More effective integration between Ministries can help develop common environmental policies for land tenure and use, energy, waters, and health, including agricultural financing and subsidies; management of water resources to avoid depletion during the dry season; incentives to use the cleared areas more efficiently; restrictions on further clearing of native habitat; encouragement of economic use of native species; and economic incentives to landowners undertaking ecological corridor restoration.
- *Legislation.* The current laws, if well applied, could go a long way to protect the cerrados. The government should strengthen the cerrado technical subcommittee of the National Environmental Council and enhance protection of critical habitats within the cerrado region such as the dry forests, the *Mauritia* palm groves (*veredas*), the rocky high altitude *campos rupestres*, karstic zones, Amazonian savannas, and the floodplains of major rivers (chapter 6). The Forestry Code, currently under revision, should increase protection of cerrado vegetation to exceed current requirements of 20% in each property.
- *Consolidation of conservation units.* The existing public protected areas must be secured, through resolving ousting land claims, implementing management plans, and staffing parks and reserves. New large parks of 300,000 hectares or more should be created in the remaining blocks of natural cerrado. The protected area system should be extended to include all landscapes and subregions of the biome. The tax abatement Private Reserve Program (RPPN) should be expanded at the state and municipal level and provide additional

financial incentives and technical support for landowners to create and maintain reserves.

- *Research and inventories.* A scientific network for cerrado research is needed to support biodiversity inventory and monitoring. A rapid assessment program and a detailed inventory project should be deployed, focusing on high biodiversity areas and habitats and regions underrepresented in existing datasets, using standardized protocols for collection and documentation. A fund to finance this work should be drawn from the monies paid for environmental mitigation of large infrastructure projects.
- *Support for scientific collections.* A network of reference collections at local institutions should be set up, complemented by holdings at major museums. A Cerrado Museum in Brasília (core area of cerrados) is highly recommended. Existing collections need electronic cataloging to facilitate research and publication of updated lists of cerrado fauna and flora. Training of taxonomists and collecting in poorly sampled areas is urgently needed.
- *Monitoring.* A set of indicator groups should be chosen for continuous monitoring, drawn from the endemic biota, from rare and/or endangered species, and from species of economic importance. The whole biome should be surveyed with satellite data at regular intervals to measure fragmentation and expansion of human occupation, and to assess effectiveness of corridor and protected areas programs.

ACKNOWLEDGMENTS

We appreciate the help and comments of Gustavo Fonseca, Russel Mittermeier, Bráulio Dias, Maria T. J. Pádua, Lauro Morhy, and the coordinators and participants of the Cerrado/Pantanal Priority setting Workshop. Keith Brown made invaluable suggestions and greatly improved the manuscript. We are grateful to Paulo Oliveira for his invitation to write this chapter, editorial comments, and patience with deadlines. We would also like to recognize the Ministry of the Environment/ Secretariat for Biodiversity and Forests, Conservation International, Funatura, Fundação Biodiversitas, Universidade de Brasília, the Global Environment Facility (GEF), the Brazilian Research Council (CNPq), the World Bank, Unibanco Ecologia, and Fundação o Boticário de Proteção à Natureza, among others, in developing and supporting biodiversity conservation strategies in the cerrado. C. A. Joly was supported by a CNPq grant (520334/99-0).

REFERENCES

Ab'Saber, A., J. Goldenberg, L. Rodés, and W. Zulauf. 1990. Identificação de áreas para o florestamento no espaço total do Brasil. *Estudos Avançados* 4:63–119.

Araújo, A., C. S. Verano, and R. Brandão. 1998. Biodiversidade do cerrado: Herpetofauna. *Workshop Ações Prioritárias para Conservação da Biodiversidade do Cerrado e Pantanal*. Brasília: Ministério do Meio Ambiente, FUNATURA, Conservation International, Fundação Biodiversitas, Universidade de Brasília. Web: www.bdt.org.br/workshop/cerrado/br.

Assad, E. (ed). 1998. Relatório final do grupo temático fatores abióticos. *Workshop Ações Prioritárias para Conservação da Biodiversidade do Cerrado e Pantanal*. Brasília: Ministério do Meio Ambiente, FUNATURA, Conservation International, Fundação Biodiversitas, Universidade de Brasília. Web: www.bdt.org.br/workshop/cerrado/br.

Bagno, M. 1996. *Atualização da Lista de Aves do Distrito Federal*. Relatório e Banco de Dados. Departamento de Zoologia, Universidade de Brasília, Brasília, Brazil. Web: www.bdt.org.br/zoologia/aves/avesdf/texto.

BRASIL. 1999. *Ações Prioritárias para Conservação da Biodiversidade do Cerrado e Pantanal*. Brasília: Ministério do Meio Ambiente, FUNATURA, Conservation International, Fundação Biodiversitas, Universidade de Brasília.

Brown Jr., K. S. 1996. Diversity of Brazilian Lepidoptera: History of study, methods for measurement, and use as indicator for genetic, specific, and system richness. In C. A. Bicudo and N. A. Menezes, (eds.), *Biodiversity in Brazil: A First Approach*, pp. 121–154. São Paulo: CNPq and Instituto de Botânica.

Britski, H. 1998. Peixes do cerrado e pantanal: Informações prévias. *Workshop Ações Prioritárias para Conservação da Biodiversidade do Cerrado e Pantanal*. Brasília: Ministério do Meio Ambiente, FUNATURA, Conservation International, Fundação Biodiversitas, Universidade de Brasília. Web: www.bdt.org.br/workshop/cerrado/br.

Camargo, A. A. and V. O. Becker. 1999. Saturniidae (Lepidoptera) from the Brazilian Cerrado: Composition and biogeographic relationships. *Biotropica* 31:696–705.

Castro, A. A. J. F. and F. R. Martins. 1998. Cerrados do Brasil e do nordeste: Caracterização, área de ocupação e considerações sobre sua fitodiversidade. *Workshop Ações Prioritárias para Conservação da Biodiversidade do Cerrado e Pantanal*. Brasília: Ministério do Meio Ambiente, FUNATURA, Conservation International, Fundação Biodiversitas, Universidade de Brasília. Web: www.bdt.org.br/workshop/cerrado/br.

Cavalcanti, R. B. 1992. The importance of forest edges in the ecology of open country cerrado birds. In P. A Furley, J. Proctor, and J. A. Ratter, (eds.), *The Nature and Dynamics of Forest-Savanna Boundaries*, pp. 513–517. London: Chapman and Hall.

Cavalcanti, R. B., L. P. Pinto, J. M. C. da Silva. 1999. Criteria for Establishing Protected Areas. 14 pp. *International Experts Meeting on Protected Forest Areas*. San Juan, Puerto Rico. Sponsored by the Governments of Brazil and the United States of America.

Cochrane, T. T., L. G. Sánchez, J. A. Porras, L. G. de Azevedo, and C. L. Garver. 1985. *Land in Tropical America = La Tierra en América Tropical = A Terra na América Tropical*, Vol. 1. Cali and Planaltina: Centro Internacional de Agricultura Tropical and Empresa Brasileira de Pesquisa Agropecuária.

Colli, G. R. (ed.) 1998. Biogeografia e conservação da herpetofauna no cerrado, pantanal e savanas amazônicas: Relatório final do grupo temático Herpetofauna. *Workshop Ações Prioritárias para Conservação da Biodiversidade do Cerrado e Pantanal*. Brasília: Ministério do Meio Ambiente, FUNATURA, Conservation International, Fundação Biodiversitas, Universidade de Brasília. Web: www.bdt.org.br/workshop/cerrado/br.

Conservation International. 1991. Workshop 90. *Biological Priorities for Conservation in Amazonia*. Map. In association with IBAMA and INPA. Washington, DC., U.S.

Conservation International. 1993. *Biodiversity Priorities for Papua New Guinea*. Map. In association with Government of Papua New Guinea Dept. of Environment and Conservation, and Biodiversity Support Program. Washington, DC., U.S.

Costa, C. M. R., G. Hermann, C. S. Martins, L. V. Lins, and I. R. Lamas (eds.). 1998. *Biodiversidade em Minas Gerais: Um Atlas para sua Conservação*. Belo Horizonte: Fundação Biodiversitas.

Diniz, I. R. and Castanheira, H. C. (eds.). 1998. Relatório final do grupo temático Invertebrados. *Workshop Ações Prioritárias para Conservação da Biodiversidade do Cerrado e Pantanal*. Brasília: Ministério do Meio Ambiente, FUNATURA, Conservation International, Fundação Biodiversitas, Universidade de Brasília. Web: www.bdt.org.br/workshop/cerrado/br.

Felfili, J. M., T. S. Filgueiras, M. Haridasan, M. C. da Silva Jr., Mendonça, R. C., and A. V. Rezende. 1994. Projeto biogeografia do bioma Cerrado: Vegetação e solos. *Cad. Geoc.* 12:75–166.

Ferri, M. G. (ed.) 1971. *Simpósio sobre o Cerrado*, pp. 127–177. São Paulo: Editora da Universidade de São Paulo.

FAPESP. 1999. BIOTA/FAPESP—*The Biodiversity Virtual Institute*. Web: www.biota.org.br.

Fonseca, G. A. B. and K. H. Redford. 1984. The mammals of IBGE's Ecological Reserve, and an analysis of the role of gallery forests in increasing diversity. *Rev. Bras. Biol.* 44:517–523.

Fry, C. H. 1970. Ecological distribution of birds in north-eastern Mato Grosso State, Brazil. *An. Acad. Bras. Ciênc.* 42:275–318.

International Council for Bird Preservation.. 1992. *Putting Biodiversity on the Map*. Cambridge, UK: International Council for Bird Preservation.

Kronka, J. N. F., M. A. Nalon, C. K. Matsukuma, M. Pavão, J.R. Guillau-mon, A. C. Cavalli, E. Giannotti, M. S. S. Ywane, L. M. P. R. Lima, J. Montes, I. H. D. Cali, and P. G. Haack. 1998. *Áreas de domínio do Cerrado no Estado de São Paulo*. São Paulo: Secretaria do Meio Ambi-ente.Web: http://www.bdt.org.br/images/bdt/sma/frag.gif.

Labouriau, L. G. (ed.) 1966. *II Simpósio Sobre o Cerrado*. Rio de Janeiro: Academia Brasileira de Ciências.

Lenthall, J. C., S. Bridgewater, and P. Furley. 1999. A phytogeographic analy-sis of the woody elements of new world savannas. *Edinb. J. Bot.* 56:293–305.

Mantovani, J. E. and A. Pereira. 1998. Estimativa da integridade da cober-tura vegetal do Cerrado/Pantanal através de dados TM/LANDSAT. *Workshop Ações Prioritárias para Conservação da Biodiversidade do Cerrado e Pantanal*. Brasília: Ministério do Meio Ambiente, FUNATURA, Conservation International, Fundação Biodiversitas, Uni-versidade de Brasília. Web: www.bdt.org.br/workshop/cerrado/br.

Marinho-Filho, J. (ed.) 1998. Relatório final do grupo temático Mastofauna. *Workshop Ações Prioritárias para Conservação da Biodiversidade do Cer-rado e Pantanal*. Brasília: Ministério do Meio Ambiente, FUNATURA, Conservation International, Fundação Biodiversitas, Universidade de Brasília. Web: www.bdt.org.br/workshop/cerrado/br.

Ministério do Meio Ambiente. 1999. *Termo de Compromisso Visando a For-mulação e a Implementação do Plano de Ação Integrado para os Biomas Cerrado e Pantanal*. Brasília: Ministério do Meio Ambiente, SECEX, Sec-retaria de Biodiversidade e Florestas.

Mittermeier, R., P. R. Gil, and C. G. Mittermeier. 1997. *Megadiversity: Earth's Biologically Wealthiest Nations*. Mexico: CEMEX.

Myers, N., R. A. Mittermeier, C. G. Mittermeier, G. A. B. da Fonseca, and J. Kent. 2000. Biodiversity hotspots for conservation priorities. *Nature* 403:853–858.

Negret, A., J. Taylor, R. C. Soares e R. B. Cavalcanti. 1984. *Aves da Região Geopolítica do Distrito Federal*. Brasília: Ministério do Interior, Secre-taria do Meio Ambiente.

Pádua, M. T. J. and B. F. S. Dias. 1997. Representatividade de unidades de conservação do cerrado e do Pantanal Matogrossense. *Workshop Ações Prioritárias para Conservação da Biodiversidade do Cerrado e Pantanal*. Brasília: Ministério do Meio Ambiente, FUNATURA, Conservation International, Fundação Biodiversitas, Universidade de Brasília. Web: www.bdt.org.br/workshop/cerrado/br.

Pimm, S. and P. Raven. 2000. Extinction by numbers. *Nature* 403:843–845.

Prance, G. T. 1982. A review of the phytogeographic evidence for Pleistocene climate changes in the neotropics. *Ann. Miss. Bot. Gard.* 69:594–624.

Ratter, J.A., S. Bridgewater, R. Atkinson, and J. F. Ribeiro. 1996. Analysis of the floristic composition of the Brazilian Cerrado vegetation: II. Com-parison of the woody vegetation of 98 areas. *Edinb. J. Bot.* 53:153–180.

Ratter, J.A. and T. C. D. Dargie. 1992. An analysis of the floristic composition of 26 Cerrado areas in Brazil. *Edinb. J. Bot.* 49:235–250.

Raw, A. 1998. Relatório sobre número de insetos, a riqueza de espécies e aspectos zoogeográficos nos cerrados. *Workshop Ações Prioritárias para Conservação da Biodiversidade do Cerrado e Pantanal.* Brasília: Ministério do Meio Ambiente, FUNATURA, Conservation International, Fundação Biodiversitas, Universidade de Brasília. Web: www.bdt.org.br/workshop/cerrado/br.

Redford, K. H., and G. A. B. Fonseca. 1986. The role of gallery forests in the zoogeography of the Cerrado's non-volant mammalian fauna. *Biotropica* 18:126–135.

Reid, J. W. 1994. The harpacticoid and cyclopoid copepod fauna in the cerrado region of Central Brazil: 1. Species composition, habitats, and zoogeography. *Acta Limnol. Bras.* 6:56–68.

Ribeiro, J. F. (ed.) 1998. Relatório final do grupo temático Botânica. *Workshop Ações Prioritárias para Conservação da Biodiversidade do Cerrado e Pantanal.* Brasília: Ministério do Meio Ambiente, FUNATURA, Conservation International, Fundação Biodiversitas, Universidade de Brasília. Web: www.bdt.org.br/workshop/cerrado/br.

Ribeiro, M. 1998. Conservação e uso sustentável da biota aquática do cerrado e pantanal: Relatório para o grupo temático Biota Aquática. *Workshop Ações Prioritárias para Conservação da Biodiversidade do Cerrado e Pantanal.* Brasília: Ministério do Meio Ambiente, FUNATURA, Conservation International, Fundação Biodiversitas, Universidade de Brasília. Web: www.bdt.org.br/workshop/cerrado/br.

Saint-Hilaire, A. 1847–1848. *Voyages Dans l'interieur du Brésil: III part. Voyage aux Sources du Rio S. Francisco et Dans la Province de Goyaz.* Paris: Arthus Bertrand.

Sawyer, D. (ed.). 1998. Diagnósticos sobre temas econômicos e sociais na região do Cerrado/Pantanal: Relatório final, grupo temático Sócio-Economia. *Workshop Ações Prioritárias para Conservação da Biodiversidade do Cerrado e Pantanal.* Brasília: Ministério do Meio Ambiente, FUNATURA, Conservation International, Fundação Biodiversitas, Universidade de Brasília. Web: www.bdt.org.br/workshop/cerrado/br.

Secretaria do Meio Ambiente de São Paulo. 1997. *Cerrado: Bases para Conservação e Uso Sustentável das Áreas de Cerrado do Estado de São Paulo.* São Paulo: Secretaria de Estado do Meio Ambiente, SP, Brazil.

Sick, H. 1955. O aspecto fisionômico da paisagem do médio rio das Mortes, Mato Grosso, e a avifauna da região. *Arq. Mus. Nac.* 42:541–576.

Silva, J. M. C. 1989. "Análise Biogeográfica da Avifauna de Florestas do Interflúvio Araguaia-São Francisco." Master's thesis, Universidade de Brasília, Brasília, Brazil.

Silva, J. M. C. 1995a. Birds of the Cerrado Region, South America. *Steenstrupia* 21:69–92.

Silva, J. M. C. 1995b. Avian inventory of the Cerrado Region, South America: Implications for biological conservation. *Bird. Conserv. Int.* 5:291–304.

Silva, J. M. C. (ed.). 1998. Relatório final do grupo temático Aves. *Workshop Ações Prioritárias para Conservação da Biodiversidade do Cerrado e Pantanal.* Brasília: Ministério do Meio Ambiente, FUNATURA, Conservation International, Fundação Biodiversitas, Universidade de Brasília. Web: www.bdt.org.br/workshop/cerrado/br.

Straube, F. C. 1998. O cerrado no Paraná: ocorrência original e atual e subsídios para sua conservação. *Cadernos de Biodiversidade, Instituto Ambiental do Paraná* 1(2):12–24.

Contributors

Alexandre F. B. Araujo
Departamento de Zoologia, C. P. 04631
Universidade de Brasília
70919-900 Brasília DF
Brasil
araujo@unb.br

Rogério P. Bastos
Departamento de Biologia Geral
Universidade Federal de Goiás, C. P. 131
74001-970 Goiania GO
Brasil
bastos@icb1.ufg.br

Keith S. Brown Jr.
Departamento de Zoologia, C.P. 6109
Universidade Estadual de Campinas
13083-970 Campinas SP
Brasil
ksbrown@unicamp.br

Mercedes M. C. Bustamante
Departamento de Ecologia, C. P. 04631
Universidade de Brasília,
70919-900 Brasília DF
Brasil
mercedes@unb.br

Roberto B. Cavalcanti
Departamento de Zoologia
Instituto de Ciências Biológicas, C. P. 04631
Universidade de Brasília
70919-900 Brasília DF
Brasil
rbcav@unb.br
and
Conservation International do Brasil
Av. Antonio Abrahão Caram 820, Conjunto 302
31275-000 Belo Horizonte MG
Brazil

Guarino R. Colli
Departamento de Zoologia
Instituto de Ciências Biológicas, C. P. 04631
Universidade de Brasília
70919-900 Brasília DF
Brasil
grcolli@unb.br

Nilton Curi
Departamento de Ciência do Solo
Universidade Federal de Lavras, Caixa Postal 37
37200-000 Lavras MG
Brasil
niltcuri@ufla.br

Kleber Del-Claro
Departamento de Biociências, C. P. 593
Universidade Federal de Uberlândia
38400-902 Uberlândia MG
Brasil
delclaro@ufu.br

Ivone R. Diniz
Departamento de Ecologia
Instituto de Ciências Biológicas, C. P.
04631
Universidade de Brasília
70919-900 Brasília DF
Brasil
ivone@rudah.com.br

Tarciso. S. Filgueiras
UPIS-Faculdades Integradas
Campus II, Caixa Postal 10743
73350-980 Planaltina DF
Brasil
tfilg@uol.com.br

Augusto C. Franco
Departamento de Botânica, C. P. 04631
Universidade de Brasília
70919-900 Brasília DF
Brasil
acfranco@unb.br

Donald P. Franzmeier
Agronomy Department
Purdue University
Lilly Hall of Life Sciences
West Lafayette, IN 47907-1150
USA
dfranzmeier@purdue.edu

André V. L. Freitas
Departamento de Zoologia, C.P. 6109
Universidade Estadual de Campinas
13083-970 Campinas SP
Brasil
baku@atribuna.com.br

Peter E. Gibbs
School of Environmental and
    Evolutionary Biology
University of St. Andrews
St. Andrews, KY16 9AL
Scotland, UK
peg@st-andrews.ac.uk

David R. Gifford (deceased)
Programa de Ecologia
Universidade de Brasília
Brasília DF
Brasil

John D. Hay
Departamento de Ecologia, C. P.
04631
Universidade de Brasília
70919-900 Brasília DF
Brasil
jhay@unb.br

Raimundo P. B. Henriques
Departamento de Ecologia, C. P.
04631
Universidade de Brasília
70919-900 Brasília DF
Brasil
henriq@unb.br

William A. Hoffmann
Departamento de Engenharia Flores-
    tal, C. P. 04631
Universidade de Brasília
70919-900 Brasília DF
Brasil
hoffmann@unb.br

Carlos A. Joly
Departamento de Botânica
Universidade Estadual de Campinas
13083-970 Campinas SP
Brasil
cjoly@unicamp.br

Keila M. Juarez
Fundação Polo Ecológico de
    Brasília/Zoo Brasília
Av. das Nações s/n
70910-900 Brasília DF
Brasil
macfadem@zaz.com.br

Carlos A. Klink
Departamento de Ecologia, C. P.
    04631
Universidade de Brasília
70919-900 Brasília DF
Brasil
klink@unb.br

Marie-Pierre Ledru
Departamento de Geologia Sedimen-
    tar e Ambiental
Universidade de São Paulo, USP/IRD
Instituto de Geociências
Rua do Lago 562
05508-900 São Paulo SP
Brasil
ledru@usp.br

Regina H. F. Macedo
Departamento de Zoologia, C. P.
    04631
Universidade de Brasília
70919-900 Brasília DF
Brasil
rhmacedo@unb.br

Jader Marinho-Filho
Departamento de Zoologia, C. P.
    04631
Universidade de Brasília
70919-900 Brasília DF
Brasil
jmarinho@unb.br

Robert J. Marquis
Department of Biology
University of Missouri-St. Louis
8001 Natural Bridge Road
St. Louis, MO 63121-4499
USA
robert_marquis@umsl.edu

Antonio C. Miranda
Departamento de Ecologia, C. P.
    04631
Universidade de Brasília
70919-900 Brasília DF
Brasil

Heloisa S. Miranda
Departamento de Ecologia, C. P.
    04631
Universidade de Brasília
70919-900 Brasília DF
Brasil
hmiranda@unb.br

Helena C. Morais
Departamento de Ecologia, C. P.
    04631
Universidade de Brasília
70919-900 Brasília DF
Brasil
morais@unb.br

Adriana G. Moreira
Environmentally and Socially Sustain-
    able Development (ESSD)
The World Bank
SCN Quadra 02, lote A
Ed. Corporate Financial Center, conj.
    303/304 70712-900 Brasília DF
Brasil
amoreira@worldbank.org

Paulo E. F. Motta
EMBRAPA-CNPS
Rua Jardim Botânico 1024
Jardim Botânico
22460-000 Rio de Janeiro RJ
Brasil

Paulo E. Oliveira
Departamento de Biociências, C. P.
    593
Universidade Federal de Uberlândia
38400-902 Uberlândia MG
Brasil
poliveira@ufu.br

Paulo S. Oliveira
Departamento de Zoologia, C.P. 6109
Universidade Estadual de Campinas
13083-970 Campinas SP
Brasil
pso@unicamp.br

Ary T. Oliveira-Filho
Departamento de Ciências Florestais
Universidade Federal de Lavras
37200-000 Lavras MG
Brasil
aryfilho@ufla.br

James A. Ratter
Royal Botanical Garden Edinburgh
Inverleith Row
Edinburgh, EH3 5LR
Scotland
UK

Flávio H. G. Rodrigues
Pós-Graduação em Ecologia
Instituto de Biologia, C.P. 6109
Universidade Estadual de Campinas
13083-970 Campinas SP
Brasil
fhgr@mymail.com.br

Index

Printed in the USA
CPSIA information can be obtained
at www.ICGtesting.com
JSHW011520221024
72172JS00014B/114